The Physics–Astronomy Frontier

The Physics–Astronomy Frontier

Fred Hoyle

CALIFORNIA INSTITUTE OF TECHNOLOGY

Jayant Narlikar

TATA INSTITUTE OF FUNDAMENTAL RESEARCH

John Faulkner

EDITORIAL CONSULTANT

W. H. Freeman and Company
San Francisco

Cover photo courtesy of High Altitude
Observatory, Boulder, Colorado.

Sponsoring Editor: Arthur C. Bartlett
Project Editor: Larry Olsen
Copyeditor: Ruth Cottrell
Designer: Robert Ishi
Production Coordinator: William Murdock
Illustration Coordinator: Audre Loverde
Artists: Evan Gillespie and John and Jean Foster
Compositor: York Graphic Services
Printer and Binder: Arcata Book Group

Library of Congress Cataloging in Publication Data

Hoyle, Fred, Sir.
    The physics-astronomy frontier.

    Includes index.
    1. Astronomy. 2. Physics. I. Narlikar, Jayant
Vishnu, 1938–   joint author. II. Title.
QB43.2.H7   523   80-11708
ISBN 0-7167-1160-5

33,855

Printed in the United States of America

9 8 7 6 5 4 3

Contents

Preface

The universe is necessarily wider in its range of phenomena than anything we can hope to experience here on Earth. Humans have ever looked to the heavens for clues to the nature and order of the world, and this is still true today. Physicists in search of the laws that order our world have turned to astronomy and cosmology for evidence to support their theories, and astronomers draw upon the experiments of physicists in earthly laboratories to understand the phenomena of the heavens. This fundamental interplay between physics and astronomy has inspired us to write *The Physics-Astronomy Frontier.*

This book is concerned with astronomy from the point of view of the physicist, which is why we discuss such topics as the nature of atoms, quantum mechanics, and ideas about radiation right at the beginning. Unfortunately, human beings are not equipped with an instinctive understanding of these matters because everyday life is not much concerned with them. Indeed, one must go to a great deal of trouble and expense to become concerned with the structure of atoms, and the study of quantum mechanics was not widespread until the modern era of microelectronics. Yet, it is scarcely possible to proceed a step in our understanding of astronomy without coming to grips with these deeper parts of physics.

Order is maintained in the world through what is often referred to as "cause and effect." Cause and effect arise through interactions between particles, of which there are four kinds—electrical, strong, weak, and gravitational. These four interactions are discussed in the opening chapter, and each part of the book is devoted to the phenomena that fall within the scope of each kind of interaction. Part I, on the electrical interaction, describes astronomical knowledge gained from the radiation spectrum—optical, radio, millimeter-wave, infrared, and x-ray astronomy. Part II, on the strong and weak interactions, focuses on the properties of the interiors of stars. Part III, on the gravitational interaction, discusses the theories of Newton and Einstein as a basis for understanding black holes and the problems of cosmology. In each part, the fundamental physical discoveries are presented as the basis for understanding astronomical phenomena. We hope this method of organization will provide a fresh view of astronomy not provided by most astronomy books, which focus on the order of discoveries or the objects encountered at increasing distances from Earth. By focusing on the fundamental forces at work in the universe, we provide a basis for understanding astronomical phenomena rather than simply describing them.

In this book we have limited ourselves to an almost entirely nonmathematical treatment. The most difficult mathematics is some simple algebra in Chapter 11. Without the algebra, we would have found it difficult to discuss the physics of black holes in a meaningful way. This book should easily be within the scope of anyone who has a strong high school background in general science.

We are very greatly indebted to Dr. John Faulkner for his careful and constructive evaluation of our manuscript and for the numerous suggestions he made for its improvement.

May 1980

Fred Hoyle
Jayant Narlikar

The Physics–Astronomy Frontier

Chapter 1
Spacetime Diagrams and the Structure of Matter

Some years ago, one of the authors made a journey to Chicago to give a paper at a meeting of the American Physical Society. The trip was memorable because during the flight the Sun rose in the west. The plane flew out over the Atlantic Ocean, heading to the northwest. The short January afternoon drew to its close in the usual way, with the Sun sinking below the southwestern horizon. The sky grew dark. People in the cabin slept, talked, drank cocktails, and listened to music. But then, in a slight and subtle way, the quality of the light in the cabin began to change. The glow in the western sky grew brighter, not fainter as it was supposed to. And the sky just went on growing brighter and brighter, until, miraculously it seemed, the fierce golden disc of the Sun appeared again on the western horizon.

The Sun rose in the west that day because the route had gone so far north over Greenland that the speed of the plane was overcompensating the spin of the Earth. It was like being on a planet that rotated east to west instead of our west to east. With the coming of supersonic commercial flights, this phenomenon will become well known. At the time of this flight, however, it seemed distinctly peculiar, as if the Earth and Sun had reversed their motions in space and time.

Few of us in our daily lives have occasion to question our common-sense concepts of space and time. To most of us, time is sequential clocktime, one

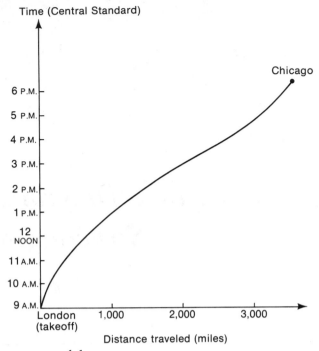

FIGURE 1-1
The time–distance plot for an airplane going from London to Chicago. Note that the plot is not a straight line, because the plane goes more slowly at the beginning and at the end than it does in midflight.

second following another, presumably to infinity, and space is unbounded and presumably likewise of infinite extent. But to the physicist and astronomer attempting to understand the laws that govern the physical universe, space and time are concepts with a precisely connected meaning. Since the time of Albert Einstein, space and time have been thought to be parts of a more fundamental reality called *spacetime*. To illustrate this concept, let us attempt to plot on a graph the motion of the airplane flying from London to Chicago. Assuming the path of the plane to be smooth, without any jinks, we can display its motion in the manner of Figure 1-1, with time plotted one way and distance traveled the other way. This graphic method of displaying the journey is known in physics as a *spacetime diagram*.

Actual flight paths are not quite smooth, of course, but this additional detail causes no difficulty, since we can generalize Figure 1-1 to a more realistic form in which two space dimensions are used as well as the time dimension. The two space dimensions can be used to show the variation of latitude and longitude along the path of the aircraft. And if we wish to show the variations in height of the plane aboveground, we can add a third dimension of space; our complete spacetime diagram would then have three dimensions of space and one dimension of time. So we pass from the simple concept of Figure 1-1 (one time dimension, one space dimension) to the more complex idea of a four-dimensional diagram (one time dimension, three space dimensions). According to the physicist, the whole drama of the universe is enacted in such four-dimensional spacetime.

It is important to notice the difference between a spacetime diagram and a purely spatial diagram, a difference illustrated by Figures 1-2 and 1-3. In Figure 1-2, we have a purely spatial diagram showing the Earth's orbit around the Sun. The dot representing the Earth moves around the Sun as time goes along. But this way of thinking is vague, because it does not show how the spatial motion is related to the passage of time. If we wish to display the passage of time explicitly, as in Figure 1-3, we must show the orbit of the Earth in the resulting spacetime diagram as a spiral having the Sun as its axis.

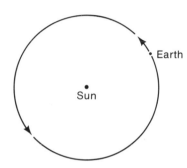

FIGURE 1-2
A purely spatial diagram of the orbit of the Earth around the Sun. At any explicit moment of time, the Earth is at a particular place in the orbit. As time goes on, the Earth moves in the manner indicated.

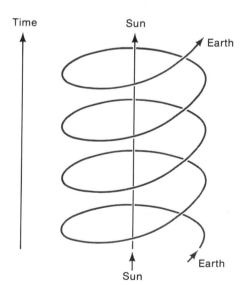

FIGURE 1-3
When the time dimension is added to Figure 1-2, the path of the Earth becomes a spiral. There is no representation of the present moment in this diagram; all moments exist together.

Here we run into a curious problem. We tend to prefer Figure 1-2 to Figure 1-3, because we attach subjective importance to explicit moments of time. We think of Figure 1-2 as an explicit moment, with other orbital configurations of the Earth and Sun occurring at other explicit moments. In Figure 1-3, on the other hand, we have the *whole history* of the Earth's movement, without any one configuration being singled out as having special significance. There is nothing in Figure 1-3 related to the concept of the *present moment of time*. In physics as it is normally understood, there is no explicit moment denoting the present. All moments of time exist together, with the whole world occupying four-dimensional spacetime.

It can be argued that physicists gain clarity at the cost of interesting detail. Figure 1-1 gives nothing of the makeup of the airplane that went from London to Chicago. The headphones, the cocktails, the crew, and the passengers have all been lost. To this objection, physicists respond by saying that conventional descriptions achieve little because the details they give simply *do not go far enough*. What is a cocktail? How does the human brain function? Physicists look for answers to these further questions, and if we press them for details, they are only too likely to reply with an avalanche of information.

MACROSCOPIC BODIES ARE COMPOSED OF ATOMS

To the physicist, a body like an airplane or the Earth is composite—that is, it can be divided into smaller pieces. How small? Without giving a clear-cut answer to this question, the ancient Greeks believed there had to be an ultimate limit to the smallness of the pieces into which any body could be divided, and they referred to these ultimate pieces as atoms. Scientists of the nineteenth century held the same belief, and they set about trying to find out how many different kinds of atoms there are in the world. In this inquiry, they were reasonably successful, as will be seen from Table 1-1, which lists all the atoms now known, together with their cosmic abundances and the year in which each was discovered.

Separating bodies into their constituent atoms was an important problem in the nineteenth century. Scientists learned techniques for preparing standard samples of the different kinds of atoms—standard in the sense that each sample contained the same number of atoms irrespective of the kind of atom. For example, a sample of hydrogen contained the same number of hydrogen atoms as there were carbon atoms in a sample of carbon. With this done, they could compare the weights of the different samples and so calculate the relative weights of the different atoms. Except for the neighboring pairs cobalt and nickel and tellurium and iodine, Table 1-1 is arranged according to increasing weight. Hydrogen is the lightest atom, helium the next lightest, and so on. Taking the weight of a hydrogen atom to be 1, they found the weight of a helium atom to be about 4, that of a carbon atom to be about 12, oxygen 16, aluminum 27, iron 56, and so on. They then referred to such relative weights as *atomic masses*.

TABLE 1-1
The elements

Z	Name	Chemical symbol	Date of discovery	Abundance in cosmic material[a]
1	Hydrogen	H	1766	3.18×10^{10}
2	Helium	He	1895	2.21×10^9
3	Lithium	Li	1817	49.5
4	Beryllium	Be	1798	0.81
5	Boron	B	1808	350
6	Carbon	C	Old	1.18×10^7
7	Nitrogen	N	1772	3.64×10^6
8	Oxygen	O	1774	2.14×10^7
9	Fluorine	F	1771	2,450
10	Neon	Ne	1898	3.44×10^6
11	Sodium	Na	1807	6.0×10^4
12	Magnesium	Mg	1755	1.06×10^6
13	Aluminum	Al	1827	8.5×10^5
14	Silicon	Si	1823	10^6
15	Phosphorus	P	1669	9,600
16	Sulfur	S	Old	5.0×10^5
17	Chlorine	Cl	1774	5,700
18	Argon	A	1894	1.17×10^5
19	Potassium	K	1807	4,205
20	Calcium	Ca	1808	7.2×10^4
21	Scandium	Sc	1879	35
22	Titanium	Ti	1791	2,770
23	Vanadium	V	1830	262
24	Chromium	Cr	1797	1.27×10^4
25	Manganese	Mn	1774	9,300
26	Iron	Fe	Old	8.3×10^5
27	Cobalt	Co	1735	2,210
28	Nickel	Ni	1751	4.8×10^4
29	Copper	Cu	Old	540
30	Zinc	Zn	1746	1,245
31	Gallium	Ga	1875	48
32	Germanium	Ge	1886	115
33	Arsenic	As	Old	6.6
34	Selenium	Se	1817	67
35	Bromine	Br	1826	13.5
36	Krypton	Kr	1898	47
37	Rubidium	Rb	1861	5.88
38	Strontium	Sr	1790	26.8
39	Yttrium	Y	1794	4.8
40	Zirconium	Zr	1789	28

TABLE 1-1 (*continued*)

Z	Name	Chemical symbol	Date of discovery	Abundance in cosmic material[a]
41	Niobium	Nb	1801	1.4
42	Molybdenum	Mo	1778	4
43	Technetium	Tc	1937	unstable
44	Ruthenium	Ru	1844	1.9
45	Rhodium	Rh	1803	0.4
46	Palladium	Pd	1803	1.3
47	Silver	Ag	Old	0.45
48	Cadmium	Cd	1817	1.42
49	Indium	In	1863	0.189
50	Tin	Sn	Old	3.59
51	Antimony	Sb	Old	0.316
52	Tellurium	Te	1782	6.41
53	Iodine	I	1811	1.09
54	Xenon	Xe	1898	5.39
55	Cesium	Cs	1860	0.387
56	Barium	Ba	1808	4.80
57	Lanthanum	La	1839	0.445
58	Cerium	Ce	1803	1.18
59	Praseodymium	Pr	1879	0.149
60	Neodymium	Nd	1885	0.779
61	Promethium	Pm	1947	unstable
62	Samarium	Sm	1879	0.227
63	Europium	Eu	1896	0.085
64	Gadolinium	Gd	1880	0.297
65	Terbium	Tb	1843	0.055
66	Dysprosium	Dy	1886	0.351
67	Holmium	Ho	1879	0.079
68	Erbium	Er	1843	0.225
69	Thulium	Tm	1879	0.034
70	Ytterbium	Yb	1878	0.216
71	Lutetium	Lu	1907	0.0362
72	Hafnium	Hf	1923	0.210
73	Tantalum	Ta	1802	0.0210
74	Tungsten	W	1781	0.160
75	Rhenium	Re	1925	0.0526
76	Osmium	Os	1803	0.745
77	Iridium	Ir	1803	0.717
78	Platinum	Pt	1735	1.40
79	Gold	Au	Old	0.202
80	Mercury	Hg	Old	0.40
81	Thallium	Tl	1861	0.192

TABLE 1-1 (*continued*)

7

Z	Name	Chemical symbol	Date of discovery	Abundance in cosmic material[a]
82	Lead	Pb	Old	4.0
83	Bismuth	Bi	1753	0.143
84	Polonium	Po	1898	unstable
85	Astatine	At	1940	unstable
86	Radon	Rn	1900	unstable
87	Francium	Fr	1939	unstable
88	Radium	Ra	1898	unstable
89	Actinium	Ac	1899	unstable
90	Thorium	Th	1828	0.058
91	Protoactinium	Pa	1917	unstable
92	Uranium	U	1789	0.0262
93	Neptunium	Np	1940	unstable
94	Plutonium	Pu	1940	unstable
95	Americium	Am	1945	unstable
96	Curium	Cm	1944	unstable
97	Berkelium	Bk	1950	unstable
98	Californium	Cf	1950	unstable
99	Einsteinium	Es	1955	unstable
100	Fermium	Fm	1955	unstable
101	Mendelevium	Md	1955	unstable
102	Nobelium	No	1958	unstable
103	Lawrencium	Lw	1961	unstable

[a] Abundances from a compilation by A. G. W. Cameron (*Space Science Reviews,* 15 (1970), 121-146). Notice that the abundances are *relative* to each other, with 10^6 for Si taken as the standard of reference.

The mass of a large body is simply the mass of its constituent atoms. Imagine the body divided into atoms of various kinds. Count the number of atoms of each kind, allowing for the fact that different atoms have different masses. Work always in terms of hydrogen, the atom of least mass. Thus for each hydrogen atom count 1, but for each carbon atom count 12, because each carbon atom has twelve times the mass of a hydrogen atom. For each oxygen atom, count 16, and so on. At the end of this counting process, you have a measure of the mass of the body in question, reckoned in terms of the hydrogen atom as the standard unit. We can imagine this process being carried out for any body, for any planet, or for any star.

Coming back for a moment to our spacetime diagram for a body like an airplane, we can achieve greater precision by thinking of the airplane not as a single path in the diagram but as a bundle of paths with a separate path for each atom. We can think of such a bundle as a cable formed from a large number of fine threads.

There is no need to draw a separate diagram for each different body. Each body can be represented in the same diagram by its own cable. Sometimes a thread will emerge out of a cable, as when an atom escapes from the Sun into the wind of atoms the Sun emits all the time. Sometimes a thread will emerge from one cable and, after wandering by itself for a while, enter another cable, as when an atom in the solar wind enters the atmosphere of the Earth. Although the concepts have not been changed very much, the apparently simple picture of Figure 1-1 has suddenly become quite complex. The cable representing the airplane contains on the order of 10^{31} threads, so small are the atoms compared to the much larger bodies of our everyday world.

ATOMS ARE COMPOSED OF STILL MORE BASIC PARTICLES

We are still very far from having answered our original question: How small are the pieces into which matter can ultimately be divided? One achievement of physics in the first half of this century was to prove that the atoms of Table 1-1 do not constitute the answer to this question, for atoms themselves are composite structures built from more basic particles. Most of the weight of an atom lies in a small central region, the *nucleus,* which contains two kinds of particles, *protons* and *neutrons.* Outside this nucleus are much lighter particles called *electrons.* In a normal atom, the number of electrons and protons is equal. The number of neutrons is not always fixed, however, even for atoms of the same kind. When there are alternatives for the number of neutrons present in a specific kind of atom, the alternatives are called *isotopes.* Thus, the atom of chlorine has two isotopes, one with 17 protons and 18 neutrons and the other with 17 protons and 20 neutrons. Each kind of atom has a definite number of protons, however, usually denoted by Z. In Table 1-1, we classified the atoms in terms of the increase in Z, with the proton number increasing by one at each step.

Looking at atoms in this way, we need only three particles to describe our physics instead of the many kinds of atoms that were thought to be basic in the nineteenth century. However, although we thereby gain a better understanding of the nature of atoms, our spacetime cables of threads become still more complex. An atom, instead of being represented by a single thread, has itself become a bundle of threads, with a separate thread for each electron, proton, and neutron. The threads for the protons and neutrons are gathered together to form the nucleus of the atom, but the threads for the electrons weave patterns that extend far outside the nucleus. To understand the difference in scale, think of the nucleus of an atom (say, an atom of oxygen) as being the size of a golf ball; then the patterns woven by the electron threads would occupy a baseball park. The study of these electron patterns forms the science of *atomic physics,* whereas the study of the compact protons and neutrons forms the science of *nuclear physics.*

It is possible to build artificial atoms, atoms whose nuclei contain numbers of protons and neutrons that are different from the numbers that occur naturally. Artificial atoms have the curious property that either some of their protons change into neutrons or some of their neutrons change into protons, thereby altering the atom itself. Such unstable atoms are said to be radioactive.

Artificial atoms eventually change through proton/neutron interchanges into atoms that occur in nature. Naturally occurring atoms are usually stable, no interchange taking place within their nuclei. There are specified cases, however, in which the proton/neutron interchanges require an enormous span of time, comparable to the age of the solar system. Such specific cases are found in the natural state together with their products. The proton/neutron interchanges are said to generate *natural* radioactivity in these cases.

The fact that neutrons and protons are interchangeable according to the formulas

$$\text{neutron} \longrightarrow \text{proton} + \text{other particles}$$

or

$$\text{proton} \longrightarrow \text{neutron} + \text{other particles}$$

suggests that protons and neutrons are not really different particles but different manifestations of some more fundamental entity. This same suggestion came from the discovery of other new particles. From about 1940 onward, more and more of these new particles were found. Six of them were given the rather odd designations Λ, Σ^+, Σ^0, Σ^-, Ξ^-, Ξ^0.* These six, together with the proton (p) and the neutron (n), which the six resemble in some ways, form a family of eight particles called *baryons*.

The six members of this eightfold family are not found in the atoms of our everyday world, because each of the six changes into either a neutron or a proton, according to the formula

$$\Lambda, \Sigma^+, \Sigma^0, \Sigma^-, \Xi^-, \Xi^0 \longrightarrow n \text{ or } p + \text{other particles}$$

in an exceedingly small interval of time, on the general order of 10^{-10} second. The other particles involved in these transitions are not like p, n, Λ, Σ^+, Σ^0, Σ^-, Ξ^-, Ξ^0. They are either electrons or particles like electrons. These other particles are called *leptons,* and they are distinct from the eightfold family. (For one thing, every member of the eightfold family is of greater mass than the

---

*Pronounced respectively lambda, sigma, and xci (as in *excite*).

leptons.) Yet there are indeed other baryon families akin to the eightfold family. There is a family of ten, of which one member, the Ω^- (Omega minus), is rather famous, because its properties were predicted before its existence was confirmed by experiments in the laboratory.

The idea that permitted the existence of Ω^- to be predicted came in the early 1960s. It was suggested independently by Murray Gell-Mann and George Zweig that all these heavy particles are composite. The eightfold and the tenfold families can be built out of three more basic particles, which have become known as *quarks*. The three different quarks are sometimes referred to as the up quark, the down quark, and the strange quark. The members of the eightfold and tenfold families are all obtained by combining quarks three at a time. Thus, the proton is a combination of 2 up quarks and 1 down quark; the neutron is 1 up quark and 2 down quarks; the Λ is a combination of 1 up quark, 1 down quark, and 1 strange quark. The Ω^- is a combination of 3 strange quarks.

The use of peculiar words to describe these new particles is deliberate. All words have associated meanings, and if physicists wish to avoid everyday associations, they choose a new and hitherto unfamiliar word like *quark*. On the other hand, the word *strange* does have well-known associations, and this again is deliberate. The strange quark is not present in either the proton or the neutron, the two members of the eightfold family that are found in our everyday world. Particles containing the strange quark are all evanescent, belonging to the "strange" new world revealed by modern experiments.

During the past two or three years, additional particles, different from the members of the previously known families, have been discovered. To explain them, physicists are currently suggesting that there may be more than three quarks. A fourth kind, called a charmed quark, when associated with two of the others (up, down, and strange), gives rise to a new array of particles, which were actually found subsequent to their prediction. At the time of writing, a fifth quark has been detected. Ultimately, the total number of quarks may turn out to be six or eight—these being the numbers that have been predicted in various theories. The marked tendency for predicted particles to be found in actual experiments shows that, however esoteric the ideas of physicists may have become, a satisfactory measure of truth lies within them.

THE WORLD PICTURE IS SUBJECT TO RESTRICTIONS

Let us look once again at our spacetime picture of cables and threads. We formerly had a thread for each neutron and each proton in the nucleus of every atom. Now each neutron and each proton must be represented by a bundle of three quarks. We have thus arrived at an exceedingly intricate tapestry of cables and threads—the avalanche of detail that we mentioned before.

Complex as the profusion of cables and threads in a spacetime diagram may be, it is by no means a freewheeling picture. The cables, representing large bodies like the Earth, and the threads, representing basic particles, do not go just anywhere. They are subject to important restrictions. Only certain pictures can be woven in the tapestry of the world. It is by means of these restrictions that cause and effect become related, that order and structure become established.

Earlier we saw that an airplane and its contents—the cocktails, the food, the crew, and the passengers—are made from an exceedingly large number of atoms, on the order of 10^{31}. The astonishing thing is that so enormous a quantity, 10,000,000,000,000,000,000,000,000,000,000 atoms, can organize themselves into the complex components of our everyday lives, the airplane and its structure, the seat you sit in, the book or newspaper you read, the conversation of a neighbor. All this is done by the restrictions that we shall now discuss.

The nature of the restriction on a typical particle, *a*, is illustrated in Figure 1-4. At a point *A* on the path of particle *a*, an influence—or interaction, as it is usually called—is received from other particles. In Figure 1-4, an interaction is received, for example, from point *B* on the path of particle *b*. Such an interaction affects the form of the path of particle *a*. Thus, the path that any particle follows is influenced by its interaction with other particles. The particle paths are thus not chosen arbitrarily. The paths are all interlinked with one another, and it is the business of the mathematician (or of the physicist wearing a mathematical hat) to calculate how the linkages go and how the restrictions operate on the paths of the particles.

Four distinct forms of interaction have been discovered. These are the four known forces of the physical world. The four forces are the gravitational force, the electrical force, the weak force, and the strong force.

The *gravitational force* is important for the large cables in our tapestry but not so important for the fine threads. The gravitational interaction holds the

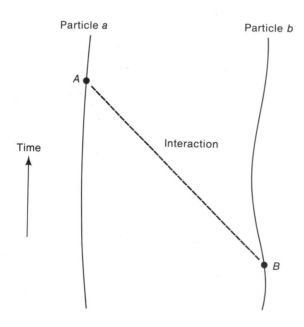

FIGURE 1-4
A typical point *A* on the path of particle *a* experiences an interaction from a typical point *B* on the path of particle *b*.

FIGURE 1-5
Apollo 17: Rover vehicle on the desolate lunar surface.
(Courtesy of NASA.)

Earth in its orbit around the Sun. It holds us on the surface of the Earth and on the surface of the Moon, as in Figure 1-5.

The *electrical force* is perhaps the most widely studied of the four. The electrical interaction of the protons in the nucleus of an atom, acting on the surrounding electrons, prevents the electrons from flying off and leaving the nucleus bare. The electrical interaction also causes atoms to bond together to form *molecules;* for example, two atoms of hydrogen and one of oxygen form H_2O, the molecule of water. The electrical interactions between molecules of water are highly important and unusual, since they give bulk liquid water its remarkable properties as a solvent—properties that appear to be essential for the development of life. The detailed study of the properties of atoms and molecules forms the science of chemistry.

The *weak force* permits a quark to change from one kind to another, whereas the *strong force* binds together the three quarks that form a proton, neutron, or one of the other particles of the eightfold and tenfold families that we have already discussed. The strong interaction also binds together the protons and neutrons in the nucleus of an atom. Without the strong interaction, the protons and neutrons in a nucleus would simply fly apart.

The four forces operate on very different scales. The weak and the strong forces are important over distances of 10^{-13} to 10^{-12} cm, this being the scale of the nuclei of atoms. The electrical force is especially important on the scale of the electrons that surround the nuclei of atoms and on the scale of molecules, 10^{-8} to 10^{-7} cm. The gravitational interaction is important on the scale of large bodies like the Earth, the Sun, and the distant universe.

THE RESTRICTIONS ON THE WORLD PICTURE ARE MATHEMATICAL

All four forces are expressed mathematically, that is, by mathematical equations. All mathematical equations are of the form $x = y$, where x and y are the same number. Why is such an equation considered important? If x and y are the same number, then obviously $x = y$. However, x may be a number calculated or measured by one procedure, and y a number calculated or measured by another procedure. Initially, we may have no reason to suppose that numbers obtained in two apparently different ways will be the same. When we find they are indeed the same, we express our surprise and delight by triumphantly writing $x = y$! This is the essence of mathematics. For example, x might be a number calculated from the geometrical form of the path of particle a in Figure 1-4, and y might be a number calculated from the interaction coming from particle b. Then the statement $x = y$ relates the form of the path of particle a to the interaction on it. This was precisely the logic by which Isaac Newton arrived at his theory of gravitation.

Knowing the form of the path of particle a by this technique, we can predict where particle a is going to be. Thinking of a as the Moon in its orbit around the Earth enables us to predict the next eclipse of the Sun. The reader may inquire as to exactly what prediction has actually been made and check the prediction when the time arrives. It will assuredly be found that the prediction works. *No other way of successfully predicting the future behavior of the world has ever been found.*

The practical success in predicting the next eclipse of the Sun by the Moon occurs because the mathematical form of the gravitational interaction is known. The problem is simple to calculate, since it involves only a few bodies; if very slight effects due to the gravitational force of planets are omitted, only the Sun, the Moon, and the Earth are involved. Such practical calculations cannot be carried through when many bodies need to be considered, however. The trouble does not come from mathematical ignorance; we know all the equations of the form $x = y$. The trouble is that it takes far too long to do all the elementary multiplications, divisions, additions, and subtractions that are needed to work out all the numbers x and y. Even with the help of the most powerful digital computers, we still fail to cope with more than a few hundred bodies, which is nothing like the full measure of complexity of our tapestry of cables and threads. We think we know the mathematical structure of the whole tapestry, but, when challenged to work through the mathematics, we have to admit that we can manage only little bits of the picture.

ENERGY AND MOMENTUM REPRESENT OVERALL TRUTHS ABOUT THE WORLD PICTURE

Even though we can manage only bits of the picture, there are important things we can say about the *whole picture,* or about any disconnected part of it, no matter how complex that part may be. Although we must remain ignorant of most of the practical details of how the cables and threads interact with each other, mathematics tells us that certain things must be true, *irrespective of the practical details.* For any system of cables and threads, interacting in the most complex way—that is, with all forms of interaction being involved—there are four quantities associated with the system that *do not change* with time. One of these quantities is called the *energy* of the system, and the other three quantities are called the components of its *momentum.*

Energy takes many forms. A body has energy because of its motion (*kinetic energy*) and because of its position (*potential energy*). In addition, a body contains energy because of its atoms and the arrangements of its atoms. The *nature* of atoms is changed in nuclear reactions, and the *arrangements* of atoms are changed in chemical reactions. The former yields *nuclear energy,* and the latter yields *chemical energy.* All matter inherently possesses these types of energy. Energy is also involved in the interactions of matter. Light and heat, for example, are forms of energy from the electrical interaction, as we shall see in Chapter 2.

When *all* the forms of energy in a system are added together, the total stays always the same. Of course, over an interval of time, energy may change from one form to another—we may derive energy of motion from the chemical energy of atoms, as we do when gasoline is burned to power an automobile. But when we compare the total energy in its various forms at the beginning and end of the time interval, we find the amounts to be the same. This is true provided that one important condition is satisfied, that the part we elect to consider is self-contained. Energy must not be exchanged with other systems. The mechanical energy to be derived from the gasoline in the tank of a car matches exactly the chemical energy of the gasoline, but this match will obviously be destroyed if we stop to fill the tank or provide the car with an electrical battery.

The three components of momentum are quantities of a kind similar to energy. Because momentum has three components that correspond to the three dimensions of space, it possesses a quality that is lacking from energy alone. If we are hit by some object, the impact of the blow is determined by the object's kinetic energy. But if we happen to be standing on the edge of a cliff, the energy of the blow may not be the most crucial aspect. An important consideration is then the direction of the blow, whether it is over the cliff or back from the edge of the cliff. Direction is thus an important component of momentum that energy lacks.

As with energy, momentum has many forms, and all the forms can be calculated mathematically. When this is done, we find again that each of the three components of momentum will be the same at the beginning and at the end of any time interval for any isolated system.

ALL FORMS OF ENERGY CAN BE MEASURED WITH RESPECT TO THE SAME UNIT

We are probably all familiar with the commercial unit of electricity known as the kilowatt-hour (kWh). When an electric motor with a 1-kW rating is used for 1 hour, a conversion of energy from electricity to mechanical forms and to heat occurs, the amount of the energy used being 1 kWh. Exactly the same amount of electricity is used when a device with a 10-kW rating is used for 0.1 hour (energy usage equals power rating multiplied by time used).

How much energy (in all its forms) do you need to maintain your personal way of life? In the United States, this amount is about 250 kWh per person per day. This amounts to a conversion to heat of nearly 100,000 kWh for each year of life. Only a small fraction of this 100,000 kWh is contained in the food we eat. Most of the energy we use (that we cause to change from one form to another) is for driving automobiles, heating houses and other buildings, smelting metals, and manufacturing goods. Our society today differs from earlier human societies because the food we eat forms only a small fraction of the total energy we use (a few percent). In early times, and even in the days of Greece and Rome, food energy made up a much higher fraction of total energy consumption than it does now. Some people think that we differ from the past because of such things as law, the United Nations, the Congress. But these social and political forms are only the surface details of modern society, whereas we differ fundamentally in energy consumption.

Suppose we seek to raise the standard of living of everybody in the world to the level in the United States. This would require annually about 100,000 kWh for each of the 4,000 million people now living in the world. The total annual energy requirement would therefore be 400 million million (4×10^{14}) kWh. It is interesting to compare this requirement with the annual energy conversion in the Sun, which results in the emission of light and heat into space. The annual energy from the Sun is about 3×10^{27} kWh, a vast quantity compared to the needs of the human species. The hugeness of this solar energy flow suggests to many people that solar energy will prove to be the power source of future societies. However, there is a difficulty with this idea. We can collect sunlight over small areas, but we cannot collect it over big areas. Nor do we yet know whether collection over big areas will ever be possible in a useful, practical way.

Although the kilowatt-hour is the simplest everyday unit of energy, it is more convenient for scientific purposes to use the kilowatt-second (kWs) as our energy unit, and we shall do so in this book. Just as 1 kWh is the energy converted by a 1-kW device for 1 hour, so 1 kWs is the energy converted by a 1-kW device for 1 second. Since there are 3,600 seconds in an hour, it is clear that 1 kWh = 3,600 kWs.

An important reason for preferring the kilowatt-second to the kilowatt-hour is that the kilowatt-second is related in a simple way to a unit used frequently in scientific literature—the *erg*. The relationship of the erg to the kilowatt-second is

$$1 \text{ kWs} = 10^{10} \text{ ergs}.$$

Thus, any statement of energy use given in terms of kilowatt-seconds can be converted into ergs simply by adding ten zeros to the number. Or, we can convert energy given in ergs to kilowatt-seconds by taking ten zeros off the number, or multiplying by 10^{-10}. Thus, 3 ergs is the same as 3×10^{-10} kWs.

Of the energy units discussed so far, the kilowatt-hour and the kilowatt-second are well chosen for everyday practical purposes, and the erg is well chosen for the cables of our spacetime diagrams. But none of these are useful for the fine threads of the diagrams. For the threads representing electrons, the *electron volt* (eV) is often used, with the equivalence

$$1 \text{ eV} \cong 1.6 \times 10^{-12} \text{ erg.*}$$

For protons, neutrons, and the other baryons, millions or even thousands of millions of electron volts arise in experimental work. The MeV or the GeV is then used, with the equivalences

$$1 \text{ MeV} = 10^6 \text{ eV} \cong 1.6 \times 10^{-6} \text{ erg}$$

and

$$1 \text{ GeV} = 10^9 \text{ eV} \cong 1.6 \times 10^{-3} \text{ erg}.$$

THE LAYOUT OF THIS BOOK FOLLOWS BASIC PHYSICS RATHER THAN TRADITIONAL ASTRONOMY

The following chapters deal with the many ways the four interactions of matter lead to the diverse phenomena observed by astronomers. Instead of sequencing our discussions according to the chance order in which astronomical discoveries have been made, we have chosen an order that depends on the basic properties

---

*The symbol $\cong$ means that, for convenience in writing, the number 1.6 has been abridged from 1.6021.

of matter and the four basic forces. Part I presents more details about the electrical interaction and optical astronomy, radio astronomy, millimeter-wave astronomy, infrared astronomy, and x-ray astronomy (Chapters 2-7).

Part II focuses on the strong and weak interactions. In Chapters 8 and 9, we show, first, how the internal properties of stars are affected by the strong and weak interactions, and, second, that the properties of stars can be used to determine distance scales over the whole universe.

Part III is concerned with the gravitational interaction. Chapter 10 leads from the simple ideas of Newtonian gravitation up to Einstein's general theory of relativity. This material forms the groundwork for the discussion of black holes in Chapter 11. The remaining chapters are concerned with the subject of cosmology.

Part I:
The Electrical Interaction

Chapter 2
Radiation, Quantum Mechanics, and Spectrum Lines

§2-1. Radiation from Macroscopic Particles

In Chapter 1, we saw how the physicist is apt to think of a macroscopic object (an object containing very many atoms) as a single "particle," and we considered an airplane flying from London to Chicago as a line between two points on a graph. In Figure 2-1, P and Q are "particles" in this sense. Radiation travels from one particle to another in the manner shown schematically in Figure 2-1. From a general point P on the world line of an electrically charged particle a, radiation travels to particle b, reaching it at some point Q. Provided particle b also has an electric charge, the motion of b will be affected by the radiation from a. In this way, the motion of one particle can influence the motion of another. We have an *interaction* between particles.

Although radiation is often thought of as having an existence independent of particles, a little thought shows we are never aware of radiation except through its effect on particles. All problems involving radiation can be dealt with in terms of the interaction picture of Figure 2-1; they can be worked out in a clear-cut way. Let us consider an important example.

*Radiation, Quantum
Mechanics, and
Spectrum Lines*

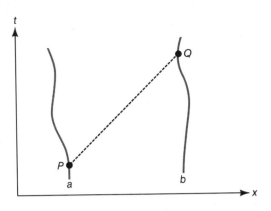

FIGURE 2-1
The particles *a* and *b* are both considered to have electric charge. The radiative influence from a point *P* on the path of *a* reaches particle *b* at some point *Q*, the time associated with *Q* being later than that associated with *P*.

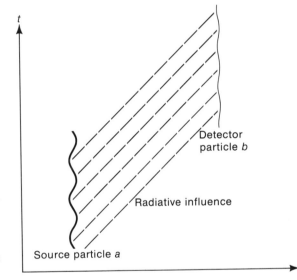

FIGURE 2-2
Particle *a* is set in regular oscillation with frequency ν. Provided the speed of motion of particle *a* is small compared to the speed of light, the radiative effect of the oscillation of *a* is to cause *b* also to oscillate with the same frequency ν. Because the oscillations of *b* occur later than those of *a*, we refer to *a* as the source particle and to *b* as the detector particle.

Suppose particle *a* moves only in one spatial dimension, say, *x*, and suppose the motion is a simple oscillation, as in Figure 2-2. We can count the number of oscillations that occur in some specified time interval. Then, we take the ratio, which we denote by ν, to be

$$\nu = \frac{\text{number of oscillations}}{\text{time interval}}$$

This ratio gives the number of oscillations per unit time. The quantity ν is known as the *frequency* of the oscillation.

Next, we insert the important condition that the speed of motion of the particle a is to be significantly less than the speed of light. (The case in which the speed of motion approaches that of light will be considered later.)

With both particles a and b having electric charge, the radiative interaction from a causes b also to oscillate with frequency ν, as indicated in Figure 2-2, where for simplicity it is supposed that particle b has no motion other than that which arises from the influence of a. Depending on the value of ν, we say that the radiation falls into one or another of the categories set out in Table 2-1. These categories are man-made and have arisen from the differing experimental procedures used to examine the various ranges of ν.

TABLE 2-1
Forms of radiation according to conventional designations

Name	Frequency (oscillations per second)	Method of detection
Radio	less than 3×10^9	electronic
Short waves (Microwaves)	3×10^9 to 3×10^{11}	electronic
Infrared	3×10^{11} to 3.75×10^{14}	effects on crystals
Visible	3.75×10^{14} to 7.5×10^{14}	eye, photography, electronic
Ultraviolet	7.5×10^{14} to 3×10^{16}	photography, electronic
Soft x rays	3×10^{16} to 2×10^{17}	electronic, photography
Harder x rays	2×10^{17} to 3×10^{19}	ionization of gases
γ rays	above 3×10^{19}	ionization of gases

DETECTOR PARTICLES WITH AN APPROPRIATE SPATIAL ARRANGEMENT CAN PRODUCE A FOCUSING EFFECT

We will refer from here on to particle a as the source particle and to particle b as the detector particle. In practice, many particles, not just one, are involved at the detector. This is illustrated in Figure 2-3, in which several detector particles have the same frequency as the source particle.

In Figure 2-4(a), the detector particles b are considered to be activated by a distant source particle not shown in the figure. The resulting motions of particles b activate a further particle, c, and they do so much more strongly than the distant source particle. How is this secondary dominance of particles b achieved? Through appropriate positioning. For this, we must take into account the three dimensions of space, as in Figure 2-4(b), where the detector particles b lie at the surface of a mirror. Figure 2-4(b) is not a spacetime diagram but a

24

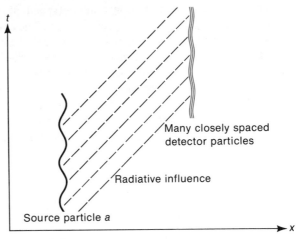

FIGURE 2-3
A source particle sets many detector particles in oscillation, the amplitude of the latter being in general very small compared to the amplitude of the source particle.

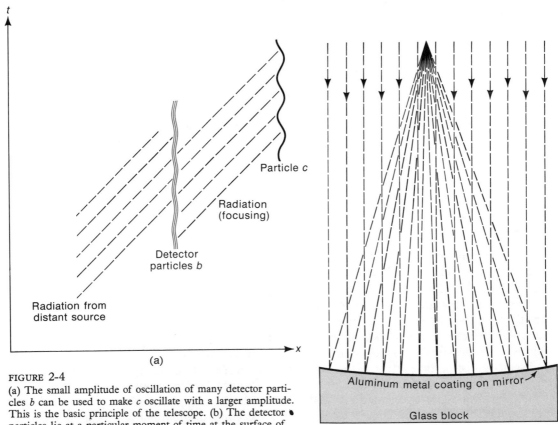

(a)

FIGURE 2-4
(a) The small amplitude of oscillation of many detector particles b can be used to make c oscillate with a larger amplitude. This is the basic principle of the telescope. (b) The detector particles lie at a particular moment of time at the surface of the mirror, and particle c lies instantaneously at the focus.

(b)

representation of the detector particles at one particular moment of time. The particle c at the same particular moment lies at the focus of the mirror. The two parts of Figure 2-4 illustrate the principle of the *telescope*.

25

§2-1. Radiation from Macroscopic Particles

DETECTOR PARTICLES WITH AN APPROPRIATE SPATIAL ARRANGEMENT CAN ALSO PRODUCE AN ANTIFOCUSING EFFECT

It is possible to arrange detector particles in such a way that they produce an effect very different from that of the telescope. They can be so positioned that they do *not* produce an oscillation of particle c in Figure 2-5, *unless v happens to be close to some known value, say, ω.* The value of ω in this kind of device is sometimes determined by the positioning of the detector particles, sometimes by the position of c, and sometimes in an electronic way. The important aspect of such a device is that ω should be known and should be variable at the observer's command. *Spectroscopes, diffraction gratings,* and *electronic receivers* are all devices of this kind. The position you select for the tuner of a radio receiver decides ω. If you select a tuning position that is "off" the local radio station, nothing happens, and you have the situation of Figure 2-5. Then by varying the tuner you can find the frequency v of the local station.

This type of device has the important property of determining v. We simply vary ω until particle c is observed to oscillate. In this way, the frequency of radiation from a distant source particle can be determined—for example, from a particle in a distant star.

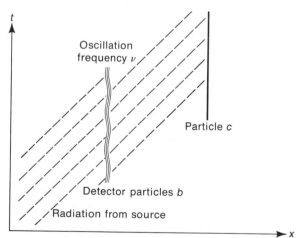

FIGURE 2-5

When particle c of Figure 2-4(a) is *prevented* from moving, unless the oscillation frequency v of detector particles b is close to some assigned frequency ω, which can be determined by the observer, we have an arrangement which forms the principle of the spectroscope. In the situation depicted here, v is not the same as the assigned value of ω, and so particle c does not oscillate.

MANY SOURCE PARTICLES CAN ACT EITHER INDEPENDENTLY (INCOHERENTLY) OR IN CONCERT (COHERENTLY) WITH EACH OTHER

Just as we must consider more than one detector particle, we must also consider many source particles. Remarkable effects can occur when many source particles are spatially positioned to act in concert with one another. Such positioning is deliberate and man-made in a radio transmitter, but in astronomy these deliberate adjustments do not occur. Nevertheless, there are cases, both in astronomy and in the laboratory, in which the source particles arrange themselves automatically in a coherent way. The *laser* is probably the best-known example of a process of this kind. Similar cases occur occasionally in astronomy, in the emission from gas clouds in our own galaxy. The frequency v of the radiation in such cases is low. Unlike the laser, where v falls in the range of visible light, v for coherent radiation from the galactic gas clouds falls in the radio and microwave ranges. The process for these wavelength ranges, physically similar to the laser, is referred to as a *maser*.

FAST-MOVING PARTICLES PRODUCE A MORE COMPLEX FORM OF ELECTRICAL INTERACTION

We have been concerned so far only with particles that move slowly compared to light. When a source particle oscillates at a speed close to that of light, the detector particles b no longer oscillate with the frequency v of the source particle. The situation, illustrated in Figure 2-6, can be understood by considering the variable-frequency device of Figure 2-5, which had the property that

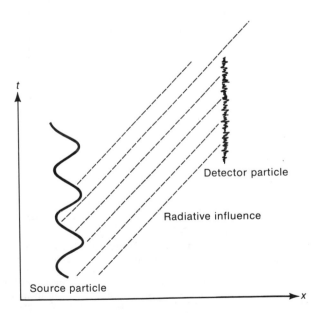

FIGURE 2-6
The speed of the source particle is here considered to be close to the speed of light. The oscillations of a detector particle are now much faster and more complex than those of the source particle.

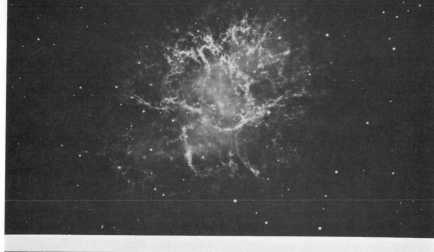

FIGURE 2-7
The Crab nebula, about 5 light-years in diameter. The light from the inner part of the Crab, shown separately in the lower half of the figure, comes from electrons within the Crab that are moving at speeds close to that of light. The outer, irregular part of the nebula shows up because of light emitted by hydrogen atoms and is different in nature from the inner part. (Courtesy of Hale Observatories.)

oscillations were *not* set up in the particle c unless the oscillation frequency of the detector particles b was close to an assigned frequency ω. For source particles of high speed, it is found that oscillations occur in the particle c over a wide range of assigned values of ω, even if ω is much larger than ν.

Figure 2-7 shows an object, the Crab nebula, containing electrons moving at speeds close to that of light (see also Plate I). The electrons follow helical paths in a magnetic field (see Figure 2-8). In typical cases, the oscillations implied by this helical motion have source frequencies in the range from 10 cycles per

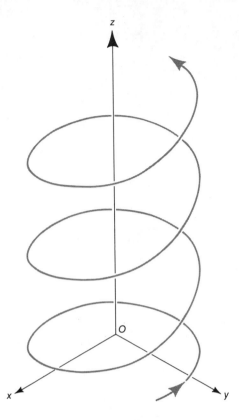

FIGURE 2-8
The helical path of an electron moving in a uniform
magnetic field that points in the direction *OZ*.

second up to perhaps 10^3 cycles per second. Yet, oscillations of particle c in a detecting device like that in Figure 2-5 are observed for assigned values of ω from about 10^7 cycles per second in the radio band up to frequencies in the infrared and optical bands, and even up to frequencies for x rays and γ rays, that is, up to more than 10^{18} cycles per second. The blue light seen in the inner part of Plate I arises from just this process.

How such particle speeds arise is a problem to be considered in a later section. Here, we simply note that fast-moving particles occur not only in our galaxy but also in other galaxies, especially in galaxies that are observed to be strong sources of radio waves (radio galaxies). Fast-moving particles also occur in a strange kind of object discovered about 15 years ago, the quasi-stellar objects, or *quasars*. These we shall study in Chapter 4.

§2-2. Time-Sense and Causality

We come now to a deeper question than any we have considered so far. What determines which is the source particle and which the detector particle? The answer to this question is that radiative interactions always go forward in time.

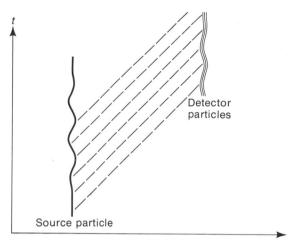

FIGURE 2-9
The separation in time of the motions of source particles and detector particles establishes causality in the
world; the detectors oscillate *later* than the source.

In Figure 2-1, the motion of particle a at point P influences the motion of
particle b at Q, because Q is later than P. The motion of b at Q does not
influence the motion of a at P, because this would imply a radiative interaction
that traveled backward in time. The answer to our question, therefore, is that we
distinguish the source particle from the detector particle through the *time-sense*
of the interaction, as in Figure 2-9.

SOME PROBLEMS CAN BE SOLVED IN EXACT MATHEMATICAL TERMS

Two problems concerning the electrical interaction can be solved exactly. The
usual way of stating the first problem is the following: Given a point P on the
path of a particle, given the direction of the path at P (the velocity of the particle
at P), and given all the interactions coming with a forward time-sense from
other particles, we can determine the rest of the path completely. We can
calculate every point of the path, and we know when the particle is at every such
point.

A somewhat less familiar but equivalent statement of this same problem,
which will be useful later, is: Given two points, say, P_1 and P_2, on the path of
a particle, and given all the interactions coming with a forward time-sense from
other particles, we can determine the form of the path between P_1 and P_2
completely. We can calculate every point on the path, and we know the time
when the particle is at every point.

The second problem that can be solved exactly is: Given the path of a particle,
we can calculate the radiative interaction excited by the particle on other
particles completely.

Notice the relation of these problems to time-sense. The interactions we need to know in order to solve the first problem come from the past. The interactions that the particle exerts on other particles, calculated in the second problem, go to the future, as with the source particle in Figure 2-9.

We can ask, in relation to the first problem, how can we ever know all the interactions that come from the past? We could seek to answer this question by considering the particles giving rise to the required interactions, by attempting to use the second problem for them. But then we would have to know the paths of these other particles, and so would have to know the interactions to which they themselves were subjected. So the question is simply pushed a stage farther back. Indeed, we have the chicken-and-egg situation of Figure 2-10, in which we have one set of particles being influenced by another set of particles, which in turn were influenced by a further set of particles, and so on. How far back in such a chain must we go before we can reasonably be certain that what happened

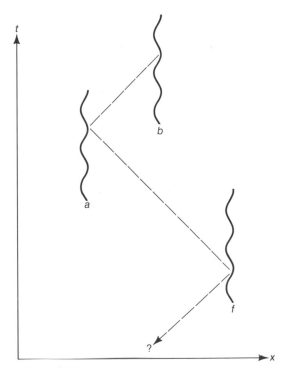

FIGURE 2-10
The causal chain, in which the motion of *b* depends on *a*, the motion of *a* depends on *f*, and the motion of *f* depends on still other particles at still earlier moments of time. In principle, this causal chain cannot be broken until we have pursued time backward to the origin of the universe—if there was an origin to the universe.

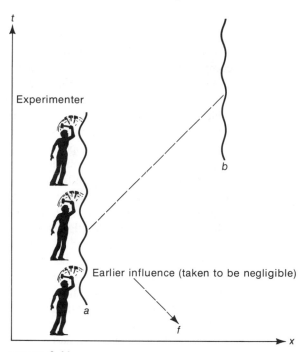

Experimenter

Earlier influence (taken to be negligible)

FIGURE 2-11
The assumption of the experimenter is that the causal chain of
Figure 2-10 can be broken by imposing a large and known
influence on particle a—for example, by hitting it with a hammer. This is taken to be so large and so drastic that the effect
of particle f can be ignored in comparison.

before was irrelevant to the specific physical system we wish to investigate at the
present moment of time?

In principle, there is no limit to how far back we must go, unless it is argued
that the universe had a definite beginning, in which case there is nothing before
a certain time in the past. In practice, however, it is often possible to study a
local aggregation of particles under conditions where the effects of preceding
interactions are small. This is the kind of situation physicists seek to set up.
They design experiments to be essentially independent of preceding effects and
so succeed in breaking the causal dilemma of needing to go ever farther and
farther backward in the sense of Figure 2-10. The experimental procedure,
illustrated in Figure 2-11, is made possible by the local restrictions imposed by
physicists themselves. Such a breaking of the causal chain is never strictly
accurate. Yet, physicists are satisfied as long as any preceding interactions have
only a negligible effect on the results. The situation is less satisfactory, however,
when we are concerned not just with a local setup but with the whole universe.
The causal chain then cannot be broken, and the logical problem of a "beginning" manifests itself. We will encounter this same problem repeatedly as we go
along.

§2-3. Quantum Mechanics

The considerations of the previous section hold good for macroscopic "particles"—that is, for bodies composed of many elementary particles like electrons and protons. For example, the source "particle" a in Figure 2-2 is required to contain many such elementary particles. The considerations also hold good for an elementary source particle, such as the electron, provided many electrons have similar motions. For instance, the many electrons in the Crab nebula have the helical motion shown in Figure 2-8. But what happens when there are only a few elementary source particles or perhaps only a single electron? Now the situation is different, unexpectedly and drastically different.

Until the end of the nineteenth century, the situation was thought to be still the same, however, and well understood. In the year 1899, the British physicist Sir Oliver Lodge made the injudicious remark: "Everything seems to be working itself out splendidly." This was just before the storm broke that was to lead to a profound crisis in physics, a crisis that was at last resolved in the year 1925 by the emergence of what became known as *quantum mechanics*. Quantum mechanics seemed so much at variance with all the earlier concepts of physics that the whole scientific world was deeply shocked by it. Einstein, in particular, never became reconciled to it, while Schrödinger, who played a big part in the development of quantum theory, said, "I don't like it, and I'm sorry I ever had anything to do with it."

IN QUANTUM MECHANICS, PARTICLES DO NOT FOLLOW UNIQUE PATHS IN A SPACETIME DIAGRAM

We can bring out the essential conceptual difference between quantum and classical physics immediately. In the older physics, given two points P_1 and P_2 on the path of a particle (macroscopic *or* elementary) and given all the interactions on the particle coming from earlier times, we can determine the form of the path between P_1 and P_2 completely. We can calculate every point on the path, and we know when the particle is at every such point. This can be done by solving equations of the type $x = y$ discussed in Chapter 1. The technique of the calculation is quite similar to that used to determine the next eclipse of the Sun by the Moon.

When the particle is elementary, such as an electron, the problem takes a quite different form in quantum theory. There is then no unique path connecting P_1 and P_2. It is possible in principle for the particle to go from P_1 to P_2 *by any path*. We can think that the particle must follow some one specific path, but we have no means of knowing which one. The situation is illustrated in Figure 2-12. The system of rules describing quantum mechanics permits a quantity, known as the *amplitude,* to be calculated for each such path, from which the probability that the particle will follow the path is easily obtained; the amplitude is a kind of square root of the probability. This new procedure, startlingly radical in conception, contains a thankful simplification. Amplitudes can be added. By adding

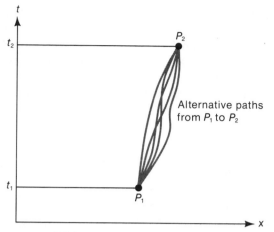

FIGURE 2-12
In quantum physics, there are alternative paths by
which a particle may go from point P_1 to point P_2.

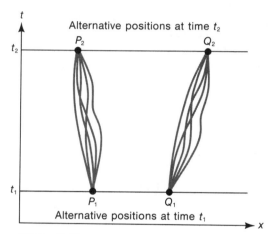

FIGURE 2-13
In going from time t_1 to time t_2, not only may the
particle go from P_1 to P_2 by any path, but P_1 may
be any point at time t_1, and P_2 may be any point at
t_2. The situation is in principle completely general.

all the amplitudes for all the paths going from P_1 to P_2, we obtain a sum that
represents the total amplitude for paths between P_1 and P_2, from which the
probability that the particle will go from P_1 and P_2 is immediately obtained.
Notice that we do not add probabilities. It is the "square roots" of the probabilities that we add.

A typical situation in quantum mechanics is illustrated in Figure 2-13. We
have two values of the time, denoted by t_1 and t_2, with t_1 earlier than t_2. In the

situation of Figure 2-13, we do not even know where the particle is located to begin with. It may be in any spatial position, P_1 and Q_1 being examples. It may go to any other spatial position, for example, P_1 going to P_2, or Q_1 going to Q_2, at time t_2, and it may do so by any path. There is certainly a sweeping grandeur in the very generality of this concept. The problem now lies in getting anything precise out of it. In the world of experience, at any rate in the everyday macroscopic world, particles do seem to be precisely located, and they do seem to be able to proceed along definite paths from one place to another. How are we to cope with this?

We begin to answer this question by emphasizing that, although all paths are in principle possible, all paths are not equally probable. To illustrate this point, we can imagine two classes of path, one of comparatively high probability, the other of low probability. If we redraw Figure 2-13, showing only the paths of high probability, the situation can be much more manageable, as in Figure 2-14, where the high-probability paths are confined to a single bundle. And if we imagine the bundle to become more compressed, as in Figure 2-15, we approximate a situation in which the particle can be thought of as going from one point to another by a unique path. In such a situation, it is found that quantum mechanics leads to exactly the same path as did the older theory of Maxwell, Lorentz, and Einstein. This at least is a port of refuge in the storm. The situation of Figure 2-15 indeed turns out to apply to all large bodies. The Earth's motion around the Sun or the motion of any macroscopic lump of material is the same as it was before.

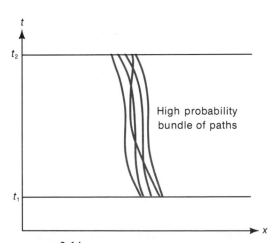

FIGURE 2-14
The completely general situation of Figure 2-13 is controlled by the circumstance that the particle has a high probability of following a member of only a restricted bundle of paths. The particle *may* follow some other path, but it has only a low probability of doing so.

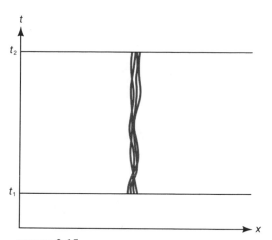

FIGURE 2-15
As the bundle of paths of high probability becomes more and more constricted, the situation approximates the classical situation, in which the particle is considered to follow a uniquely specified path.

The situation becomes very different from the earlier theory, however, when only a comparatively few particles are involved, such as individual atoms. Let us consider the simplest atom, hydrogen, consisting of a single electron and a single proton. Because the proton has a much larger mass than the electron, the bundle of paths for the proton is less spread out than the bundle for the electron. For simplicity, we can think of the proton as following a definite path, like a classical particle. But the electron has no unique path; it cannot be thought of as pursuing a definite orbit around the proton. Elementary discussions of atomic physics sometimes state that the electron in the hydrogen atom moves around the proton rather like a planet moving around the Sun, except that, unlike a planet, which can move in any elliptic orbit, the electron can move only in certain specific orbits. This statement is quite wrong, however. The electron can move around the proton *in any orbit*. The orbit does not even have to be an ellipse, as that of a planet does. The most we can say is that some orbits are more probable than others.

ATOMS ARE USUALLY FOUND IN ONE OF A NUMBER OF WELL-DEFINED STATES

This still seems a grotesquely vague picture. Yet, the problem turns out to assume a wonderful regularity. If, instead of worrying about specific paths, we worry about bundles of paths, the electron in the hydrogen atom will usually have one or another of a certain standard set of bundles, some of which are

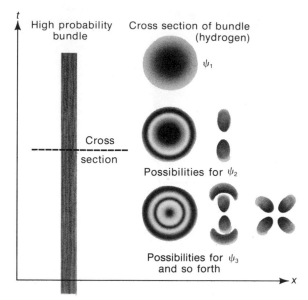

FIGURE 2-16
The paths of electrons in atoms can be considered in terms of alternative bundles ψ_1, ψ_2, ψ_3, and so forth, in the sense that the electrons will *usually* be following paths in one or another of these bundles. Sometimes, however, there is a swap from one bundle to another, in which case the atom is said to undergo a *transition*. The cross sections of ψ_1, ψ_2, ψ_3, and so forth are taken for the single electron in the hydrogen atom. As noted in the text, there are several possibilities for ψ_2, ψ_3, and so forth. In the right-hand diagrams, the paths run through the shaded regions.

shown in Figure 2-16. These bundles can be referred to notationally as ψ_1, ψ_2, and so forth, each symbol in this sequence denoting one of the standard bundles. Expressed in this way, the standard bundles are called *states*.

THE EMISSION AND ABSORPTION OF RADIATION ACCORDING TO QUANTUM MECHANICS ARISES WHEN PARTICLES CHANGE THEIR STATES

The electron of the hydrogen atom will usually, but not always, be in one or another of the states ψ_1, ψ_2, and so forth. Not always, because it is possible for an electron to jump from one state to another, $\psi_2 \longrightarrow \psi_1$, for example. The sequence ψ_1, ψ_2, and so forth, can be arranged so that we always have a jump leftward if radiation is emitted by the atom and a jump rightward if radiation is absorbed. The states are then said to be arranged in sequence with respect to *energy*. So we think of the electron as being mostly in one or another of the states $\psi_1, \psi_2, \ldots$, but occasionally changing its state by jumping to the left or to the right.

A change of state can be caused not only by the emission or absorption of radiation, but also by collisions between particles, and such changes can also go both ways, to the left or to the right in the sequence $\psi_1, \psi_2, \ldots$. In a hot gas, there are, at any moment of time, atoms in each of these standard states, which are called *stationary states* because the atom halts for a while in any one of them before jumping to another. In a hot gas, there is a shuffling of atoms among the states, some jumping one way in the sequence $\psi_1, \psi_2, \ldots$, and others jumping the opposite way, like a vast circus of fleas. If all changes to the right are in exact balance with corresponding changes to the left, $\psi_{37} \longrightarrow \psi_{53}$ being in balance with $\psi_{53} \longrightarrow \psi_{37}$, for example, the gas is said to be in equilibrium, and a temperature can be assigned to it by the procedure to be described in Chapter 3.

For a gas that contains a large number of atoms, suppose that many of the atoms undergo a certain change of state in a given time interval, say the transition $\psi_2 \longrightarrow \psi_1$. The radiation emitted by a large number of atoms in such a leftward step in our sequence of states has properties similar to the radiation from the simple particle oscillation of Figure 2-2. However, we must take care to remember that, whereas the frequency ν of oscillation of the classical particle in Figure 2-2 could be assigned to suit ourselves, *we now have a situation in which ν is determined by the transition in question.* A hydrogen atom does not emit radiation at any and all frequencies. It only emits frequencies corresponding to the changes of state to the left in the sequence $\psi_1, \psi_2, \ldots$.

In Figure 2-17, we have the *energy-level diagram* of the hydrogen atom, each level corresponding to one of the states $\psi_1, \psi_2, \ldots$, with ψ_1 the bottom level, ψ_2 the next level, and so on. The diagram is constructed so that the spacing between adjacent levels is proportional to ν for the transition between the adjacent states in question. For example, the spacing between the bottom level and the level immediately above it is proportional to ν for the transition $\psi_2 \longrightarrow \psi_1$, that between the second and third levels is proportional to ν for the transition $\psi_3 \longrightarrow \psi_2$, and so on.

The levels crowd together as we go to states far to the right in the sequence $\psi_1, \psi_2, \ldots$. This means that a transition like $\psi_{105} \longrightarrow \psi_{104}$ would correspond to

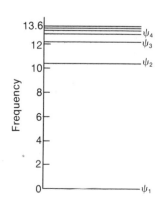

FIGURE 2-17
The energy diagram of hydrogen, with the states ψ_1, ψ_2, ψ_3, and so forth given levels that are determined by the frequencies of the radiation emitted in the transitions $\psi_2 \longrightarrow \psi_1$, $\psi_3 \longrightarrow \psi_1$, and so forth. The unit of frequency is 2.418×10^{14} oscillations per second. The discrete levels do not go above 13.6.

a quite low frequency, one that actually falls into the radio band and has been observationally detected in the radiation from hot gas clouds in our galaxy.

The bundle of paths corresponding to a member of the sequence $\psi_1, \psi_2, \ldots,$ becomes larger and larger in its spatial dimensions as the state in question lies more and more to the right in the sequence. This means that the electron has an appreciable probability of following paths that take it farther and farther away from the proton. Ultimately, there are paths that take it completely away from the proton. The electron is then said to be *free*. The level corresponding to such a free state lies above the regular sequence of "bound" states in Figure 2-17, as in Figure 2-18. Transitions can occur between one free state and another and between a free state and a bound state, involving emissions of radiation when the transition is downward in Figure 2-18, and involving absorption when the transition is upward.

We next cover a point of detail. There are always two or more states at each of the levels of Figure 2-17. Thus, we should think of ψ_1 as representing two states, ψ_2 as representing several states, and so on, as shown already in Figure 2-16. To be very precise, the bottom level of Figure 2-17 is actually split slightly, as in Figure 2-19. Transitions involving a frequency $\nu = 1.42 \times 10^9$ cycles per second occur between the two states of the split bottom level. Radiation from these transitions is widely observed by radio-astronomers and shows up in the spiral pattern of the distribution of hydrogen atoms in our galaxy, as we shall discuss in Chapter 6.

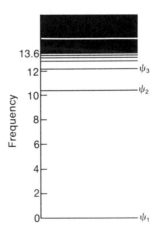

FIGURE 2-18
There are states such that the electron paths go far from the proton of the hydrogen atom; these lie above the 13.6 level. Electrons with paths in the bundles corresponding to these states are said to be *free*. The levels to be given to these free states, using the same radiation criterion as in Figure 2-17, form a continuous band in the energy-level diagram. The unit of frequency is the same.

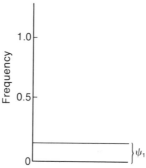

FIGURE 2-19

The ψ_1 state of the hydrogen atom is actually split into two very closely spaced levels. Because of this small spacing, the radiation emitted in transitions from the slightly higher member of this doublet to the slightly lower member has an abnormally low frequency of 1.42×10^9 oscillations per second. This low-frequency radiation is very important to studies in radio astronomy. The unit of frequency here is 10^{10} cycles per second.

THE FREQUENCIES IN THE ENERGY-LEVEL DIAGRAMS OF ATOMS ARE EXPLICITLY KNOWN

The energy-level diagram for hydrogen, shown in Figure 2-17, can be constructed in the following way. First, draw a level to represent the limiting situation in which the electron just becomes free. Next, mark the states ψ_1, ψ_2, ψ_3, . . . , below this initial level, at depths that are in the ratios $1 : \frac{1}{4} : \frac{1}{9} : \ldots$, the denominators in the fractions being just the squares of the natural numbers, $2^2 = 4$, $3^2 = 9$, ... The spacings between the bottom level, corresponding to ψ_1, and the levels corresponding to ψ_2, ψ_3, . . . , are then in the ratios $1 - \frac{1}{4} : 1 - \frac{1}{9} : \ldots$. These same ratios appear in the frequencies observed for the transitions $\psi_2 \longrightarrow \psi_1, \psi_3 \longrightarrow \psi_1, \ldots$. Neglecting small effects, these frequencies are

$$3.29 \times 10^{15} \left[1 - \tfrac{1}{4}, 1 - \tfrac{1}{9}, \ldots \right] \text{ cycles per second.}$$

Hence, the relative spacings of the levels in our constructed diagram correspond to the actual frequency ratios observed for the transitions $\psi_2 \longrightarrow \psi_1, \psi_3 \longrightarrow \psi_1, \ldots$. The radiation emitted in these transitions is known as the *Lyman series*. This series is illustrated in Figure 2-20. Absorption of radiation at the same frequencies occurs for the upward transitions $\psi_1 \longrightarrow \psi_2, \psi_1 \longrightarrow \psi_3, \ldots$, and this is illustrated in Figure 2-21.

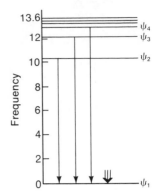

FIGURE 2-20
The radiation emitted by the transitions $\psi_2 \longrightarrow \psi_1$, $\psi_3 \longrightarrow \psi_1$, and so forth is called the *Lyman series*. The unit of frequency is the same as in Figure 2-17.

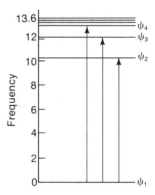

FIGURE 2-21
Radiation can be absorbed in transitions that are the reverse of those in Figure 2-20. (Same unit of frequency.)

Similarly, we can have downward transitions $\psi_3 \longrightarrow \psi_2, \psi_4 \longrightarrow \psi_2, \ldots$. The frequencies associated with these transitions are observed to be

$$3.29 \times 10^{15} \ [\tfrac{1}{4} - \tfrac{1}{9}, \tfrac{1}{4} - \tfrac{1}{16}, \ldots] \text{ cycles per second.}$$

The observed frequencies again have the same ratios as the spacings of the levels in our constructed diagram. Light emitted at these frequencies by hot hydrogen gas is shown in Figure 2-22, the transitions being illustrated diagramatically in Figure 2-23. The sequence of Figure 2-23 is known as the *Balmer series*. Upward transitions, involving absorption, can also take place at the same frequencies. These are illustrated in Figures 2-24 and 2-25.

It was one of the early triumphs of quantum theory actually to predict the frequency values just given, the number 3.29×10^{15} being obtained by precise calculation and being verifiable by observation.

We are in a position now to understand how hydrogen can emit radiation both at discrete frequencies and with continuous values. Continuous values come from transitions between states in at least one of which the electron is free,

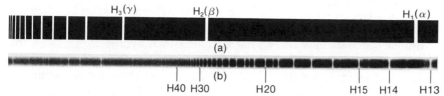

FIGURE 2-22

In an experimental arrangement that separates radiation according to its frequency, as in (a), with the frequency increasing toward the *left*, the radiation of the Balmer series falls at discrete places with certain regular spacings. Employing such an arrangement for the star HD 193182 gave many of the quite high members of the Balmer series, $\psi_{13} \longrightarrow \psi_2$, $\psi_{14} \longrightarrow \psi_2$, $\psi_{15} \longrightarrow \psi_2$, and so forth, as can be seen in (b). (Courtesy of Hale Observatories.)

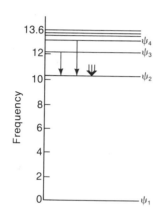

FIGURE 2-23

The radiation emitted by the transitions $\psi_3 \longrightarrow \psi_2$, $\psi_4 \longrightarrow \psi_2$, and so forth is called the *Balmer series*. (Same unit of frequency.)

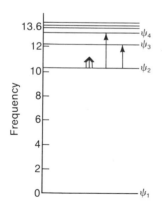

FIGURE 2-24

Radiation can also be absorbed at the characteristic frequencies of the Balmer series by $\psi_2 \longrightarrow \psi_3$, $\psi_2 \longrightarrow \psi_4$, and so forth. (Same unit of frequency as in Figure 2-17.)

FIGURE 2-25

In this photographic negative, we have an observational arrangement that separated the radiation from the quasar 3C 273 according to its frequency, with the frequency increasing toward the left. Which are the Balmer lines of hydrogen? (Courtesy of Dr. R. Lynds, Kitt Peak National Observatory.)

whereas discrete frequencies come from transitions between states in which the electron is bound to the proton.

Similar considerations apply to atoms other than hydrogen. Although the same concepts concerning bundles of paths continue to apply, leading to the same idea of a sequence of states $\psi_1, \psi_2, \ldots$, among which transitions occur, the problem becomes much more complex in detail, because more than one electron has to be considered. Even for the helium atom, which has only two electrons, the situation is already much more intricate, as can be seen from the helium energy-level diagram shown in Figure 2-26. Here, the levels have been classified into two branches. The symbols attached to the various levels are those used in the classification and need not concern us here.

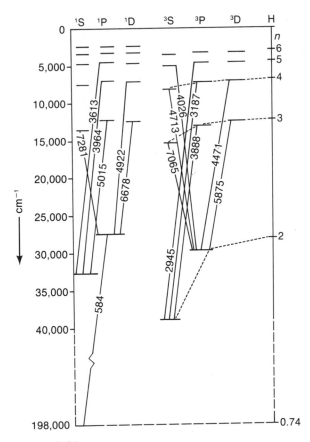

FIGURE 2-26

The helium energy-level diagram. Spectroscopists often express frequencies in cm^{-1}, instead of in oscillations per second. Here frequency increases *downwards*, with the unit step of 5000 cm^{-1} being an increase in frequency by 1.5×10^{14} oscillations per second; the dotted part of the scale is equal to 31.6 such units. Terms of hydrogen have been added for comparison and for assignment of n values. (From H. G. Kuhn, *Atomic Spectra*, New York, Academic Press, 1962.)

The study of complex diagrams like that of Figure 2-26 forms the subject of *atomic physics*. The radiation from each kind of atom has a characteristic pattern of discrete frequencies that distinguishes it from the patterns of other atoms, a circumstance of great importance in astronomy. When the characteristic radiation pattern of a certain kind of atom is found in the light from an astronomical object, then we know that this kind of atom is present in the object. From Figure 2-25, for example, we know that the quasar 3C 273 contains hydrogen. In this way, the kinds of atoms present in various objects can be discovered. If atoms emitted only radiation with continuous frequencies, as they would do according to the older, prequantum theory, we could never infer the compositions of distant stars, galaxies, or quasars. Without quantum mechanics, most of the power of modern astronomical techniques would be lost.

RADIATION EMITTED BY ONE ATOM IS OFTEN ABSORBED BY ANOTHER ATOM

The radiation emitted in a specific transition, for example, $\psi_2 \longrightarrow \psi_1$ of hydrogen, cannot be absorbed in any transition between the discrete levels of hydrogen except $\psi_1 \longrightarrow \psi_2$, as is illustrated in Figure 2-27, but it could be absorbed from ψ_2 into a state in which the electron was free. Occasionally, the discrete radiation emitted by one kind of atom may be absorbed in a transition between discrete levels of another atom, as in Figure 2-28, but such absorption is

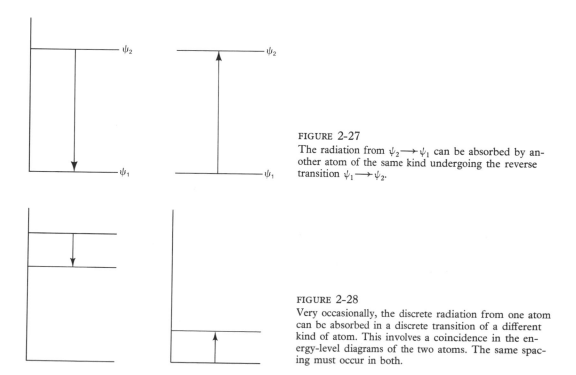

FIGURE 2-27
The radiation from $\psi_2 \longrightarrow \psi_1$ can be absorbed by another atom of the same kind undergoing the reverse transition $\psi_1 \longrightarrow \psi_2$.

FIGURE 2-28
Very occasionally, the discrete radiation from one atom can be absorbed in a discrete transition of a different kind of atom. This involves a coincidence in the energy-level diagrams of the two atoms. The same spacing must occur in both.

accidental and therefore rare. The discrete frequencies of one atom do not usually coincide with those of another atom. A more common situation is illustrated in Figure 2-29. Here, a downward transition of one atom produces radiation that frees an electron from a second atom. The second atom is then said to become *ionized*—it loses an electron. This process of *ionization* is widely employed in practice in all manner of electronic devices. It is known as the *photoelectric effect.*

Calcium atoms, after being ionized, have an energy-level diagram whose lower part is shown in Figure 2-30. The upward transitions marked H and K in

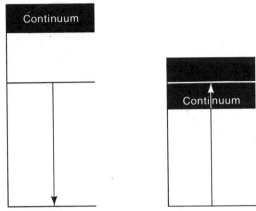

FIGURE 2-29
The radiation from one kind of atom is more usually absorbed by another kind of atom in a way that causes an electron to become free. This process is called *ionization.* The absorbing atom is said to become *ionized.*

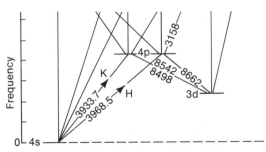

FIGURE 2-30
The lower part of the energy-level diagram for the calcium atom with one electron removed, showing the transitions that give the so-called H and K lines. (Same unit of frequency as in Figure 2-26.)

the figure involve the absorption of radiation at frequencies $\nu = 7.554 \times 10^{14}$ cycles per second and $\nu = 7.621 \times 10^{14}$ cycles per second. These two absorptions play an important role in astronomy.

§2-4. Names, Units, and Measurements

The preceding sections may seem a complex way to begin a nontechnical book. But quantum mechanics is admitted to be among the most difficult parts of physics, and we have not wanted to gloss over the difficulties. Indeed, by thinking through the matter in a reasonably proper way, we are now in a position to gain some profit in a quick understanding of quite a number of issues. When an atom changes its state, as it does, for instance, in the change of a hydrogen atom from ψ_3 to ψ_2 (Figure 2-23), radiation is emitted. But the radiation from an individual atom is not of the familiar classical kind, the kind that causes an electrical detector particle to oscillate in the manner of Figure 2-2. To obtain the situation of Figure 2-2, *many* hydrogen atoms must undergo transitions from ψ_3 to ψ_2. When this happens, the oscillations of the detector particle can be counted, and a frequency ν can be attributed to the radiation. The frequency in cycles per second is actually 3.29×10^{15} multiplied by $(\frac{1}{4} - \frac{1}{9})$. We say in such a situation that the hydrogen atoms have emitted a *spectrum line,* the line in this particular case being called Hα. In the inverse situation shown in Figure 2-24, in which many hydrogen atoms absorb radiation to undergo the transition ψ_2 to ψ_3, the spectrum line Hα appears *in absorption.*

AN INDIVIDUAL TRANSITION OF AN ATOM GIVES EMISSION OR ABSORPTION OF A QUANTUM OF RADIATION

If the radiation emitted or absorbed in the transition of an individual atom cannot be interpreted in the sense of Figure 2-2 as a wave of frequency ν, how are we to refer to it? Evidently, a new description is needed, and it was precisely to give a new description that the word *quantum* was coined in the early years of this century. A quantum is emitted in an individual transition $\psi_3 \longrightarrow \psi_2$, and a quantum is absorbed in the reverse transition $\psi_2 \longrightarrow \psi_3$. A quantum emitted by one atom can be absorbed by another atom, as we have seen in Figure 2-29. In the years following 1925, physicists learned how to calculate the transitions of atoms, and the new science was referred to as *quantum mechanics,* the mechanics of being able to calculate the emission and absorption of quanta. As it turned out, the basic ideas of quantum mechanics have much wider applications than the emission and absorption of radiation. Similar ideas apply within the nuclei of atoms and quite probably also to gravitation, although this has yet to be verified by experiment.

Figure 2-29 shows that, when a quantum is first emitted by one atom and then absorbed by another atom, the amount of the drop of the first atom in its energy-level diagram is equal to the amount of the upward jump of the second atom in its own energy-level diagram. What property of a quantum of radiation determines each jump? To answer this question, we turn back to Figure 2-2, to the frequency ν, which can be measured from the oscillations of a detector particle when many similar quanta are involved, as when many atoms undergo the downward transition of Figure 2-29. The frequency ν is also an attribute of each *individual* quantum, although for an individual quantum it shows not in the sense of Figure 2-2 but in the energy-level diagram of the emitting atom in Figure 2-29. Thus, "frequency" is really the same as "energy" (see the discussion that follows), and, by actually counting the oscillations of the detector particle in Figure 2-2 (for a situation in which many quanta are involved), we have a straightforward procedure for actually determining frequency. We therefore have a procedure to determine the levels in diagrams like Figure 2-29 or the hydrogen and helium diagrams of Figures 2-17 and 2-26.

THE RELATION BETWEEN THE FREQUENCY AND THE ENERGY OF A QUANTUM OF RADIATION DEPENDS ON THE CHOICES OF UNITS OF TIME AND ENERGY

The answer we would obtain for ν by actually counting the oscillations of a detector particle (Figure 2-2) is dependent on our unit of time. We shall obviously count many more oscillations in a unit of time if we use the hour as our unit instead of the second. In fact, we would then count 3,600 times more oscillations. It follows from this simple point that, if frequency and energy are to be regarded as equivalent physical quantities, we must take care to measure them in appropriately related units. Thus, if we decide to measure time in seconds, we are not then free to measure energy in any unit we please, such as the units we discussed in Chapter 1 (the kilowatt-hour, the kilowatt-second, or the erg). But this is exactly what physicists did in the early years of this century. They chose the second as the unit of time and the erg as the unit of energy, a discrepant situation that prevented many people from understanding that frequency and energy are the same physical quantity. Instead of setting them equal, $E = \nu$, energy and frequency were made proportional to each other, and the constant of proportionality was written as h, $E = h\nu$. This h is known as *Planck's constant*. Many people still make the mistake of thinking that the numerical value we attach to Planck's constant has some fundamental significance, whereas it really appears in the relation $E = h\nu$ as a correction factor that has arisen because the units of energy and time have been chosen without respect to each other. When time is measured in seconds and energy in ergs, h has the numerical value 6.6252×10^{-27}, but there is nothing sacrosanct about this number. If we had chosen to measure time or energy in some other unit, the numerical value of h would be changed, and, if we had chosen the units concordantly, we would have had $h = 1$. The sheer smallness of 6.6252×10^{-27} suggests mystery to some, but there is no mystery about it. The smallness comes from the circumstance that the erg and the second were chosen for convenience in the everyday macroscopic world, a world consisting of a very large number of atoms. The smallness would disappear if the units were chosen for convenience in a microscopic world, a world containing only a few atoms.

QUANTUM MECHANICS IS BASIC TO AN UNDERSTANDING OF PRACTICAL METHODS FOR THE DETECTION OF RADIATION

To conclude our discussion of quantum mechanics, let us turn back to Table 2-1. We are now in a position to discuss the meaning of the third column of this table. The general rules for the detection of radiation are as follows:

1. At the lowest frequencies (radio waves, microwaves), detection depends on there being a large number of quanta at each frequency. The method is essentially that of Figure 2-2.
2. At the highest frequencies (γ rays, x rays), individual quantum events, like the photoionization of Figure 2-29, can be detected, and they form the basis of the methods used.
3. In the optical and ultraviolet, there is a choice of methods. Individual events can be detected, as in rule 2, or processes involving a large number of quanta at each frequency can be used. In the latter case, the method employed is not that of Figure 2-2, however. Many photoionization events of the type of Figure 2-29 can be used to change the structure of small crystals. This is the principle of photography, in which light falls on a film of emulsion made up of very many tiny crystals of silver bromide or silver chloride. The effect of the light on the crystals is to cause a subtle change in the arrangements of their constituent atoms. This change can be made to show up chemically under the action of a developing solution that reduces the affected crystals to fine particles of metallic silver. Thereafter, all the unchanged crystals, in the places where the light did *not* fall, are removed by a fixing solution. After fixing, one thus has a reverse of the original light distribution, a so-called negative picture.
4. A rather similar idea is used in the infrared. The arrangements of atoms in a crystal can be changed by many quanta being incident upon it, except that in the infrared the electrical properties of the crystal are changed, not the chemical properties. A crystal of lead sulfide is used at the higher infrared frequencies, whereas the strange half-metal germanium, deliberately made to contain an impurity, is used at lower frequencies. We shall meet these ideas again in Chapter 6 when we consider infrared astronomy in more detail.

It may seem surprising that it should be possible to detect a single event of the kind illustrated in Figure 2-29. Again, we shall consider this matter in more detail in the chapter on x ray astronomy. For the present, a simple and rather crude example will suffice. Light incident on a surface, called a *photocathode,* causes free electrons to be emitted from the material of the surface. Each such electron is then accelerated in an electric field—a simple form of radiation field—so that it strikes a second surface with sufficient energy to knock out several more free electrons from this second surface. The process is repeated, with each one of these further electrons being accelerated toward a third surface,

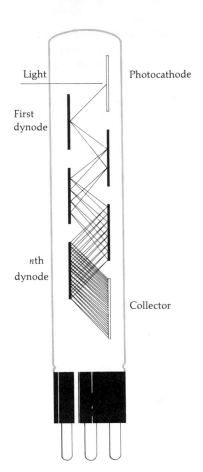

FIGURE 2-31
In the photomultiplier, light incident on a photocathode causes electrons to be released by the photoelectric effect. These free electrons are accelerated and focused to strike a metallic surface from which many more electrons are then knocked out. The process is repeated in a cascade until the original number of electrons from the photocathode has been increased a millionfold or more. (From J. C. Brandt and S. P. Maran, *New Horizons in Astronomy,* Second Edition. W. H. Freeman and Company. Copyright © 1979.)

where another multiplication of the number of free electrons takes place. After a suitable number of such stages, the resulting very large number of electrons is finally collected to yield an electric current, which can be measured without difficulty. Such a *photomultiplier* is illustrated in Figure 2-31.

Compared to the photographic plate, the single photomultiplier tube has major advantages and marked disadvantages. It is much more sensitive, and at low light intensities it responds to the intensity directly, instead of exhibiting the rather complex behavior of photographic emulsions. This makes the photomultiplier particularly suitable for accurate measurements of the magnitudes of stars, galaxies, and quasars. On the other hand, as long as the frequency v of the incident radiation is high enough to remove electrons from the photocathode (see Figure 2-29), the photomultiplier does not discriminate with respect to v. All frequencies above such a "threshold" contribute to the final current. Thus, to distinguish separate ranges of v, one must pass the incident light through suitable color filters or use some equivalent device. But this is a seriously wasteful procedure; while a filter is being employed, light is lost at all values of ν except in the range passed by the filter in question. The photographic plate, on the other hand, accepts all values of ν over a wide range simultaneously when used with a spectrograph, and this tends to compensate for the poorer sensitivity of photographic emulsions.

THE ELECTRICAL INTERACTION PLAYS A DOMINATING ROLE IN OUR SUBJECTIVE EXPERIENCE

You may well feel that the content of this chapter is difficult. Is the effort needed to understand it really worthwhile? Our answer is an emphatic yes, because our daily lives are entirely contingent on the electrical interaction of Figure 2-1 with which we began. The other interactions of physics play useful and essential practical roles. The strong interaction prevents atoms from flying apart. Without the strong interaction, atoms could not exist, and, without atoms, we could not exist. The gravitational interaction holds us down on the surface of the Earth (or on the surface of the Moon, as in Figure 1-5). Although we are not aware of these useful functions in our conscious thoughts, all aspects of our consciousness are affected by radiation and by the electrical interaction.

The behavior of the retina of the eye is quite similar in principle to the changes of arrangements of atoms that we have just discussed. Many quanta falling on the retinal molecules cause temporary changes of their shape, and these changes are translated into electrical terms recognized by the brain. Our thinking and our external senses are electrical in their operation. Even sounds and odors must likewise be translated into an electrical form before we become aware of them.

It is interesting to reflect on all the electrical processes involved in making a phonograph record of a piece of music and in listening to it. First, the composition of the music comes from electrical impulses in the brain of the composer. The composer's brain also uses electrical signals to direct the muscles of the hand in order to write the appropriate notes on paper. Once on paper, the music is fixed. It comes to life through interpretation by a performer. The performer scans the written notes with the eyes, which convert the notes back to an electrical form that the brain of the performer is able to recognize. The performer now directs corresponding electrical signals to the muscles to convert the music into sound, whether by manipulating the keys of a piano, by bowing strings, by blowing into a wind instrument, or by singing. The sound propagates through the air until it reaches a microphone, a device for converting the music back once more into electrical signals. For the production of a phonograph record, the electrical signals go after amplification to a crystal that changes its shape in correspondence with the signals. The crystal controls the cutting of grooves in the vinyl material of the record. Now the music is once more fixed. To come alive, the record must be played. As the phonograph needle follows the grooves of the record, a crystal within the pickup cartridge continually changes its shape by the pressure of the grooves on the needle. These shape changes are used still again to convert the music back to an electrical form. The electrical impulses are next amplified in their intensity and then sent along wires to loudspeakers, which convert the music for a second time into sound waves in the air. The sound propagates to the ears of the listener, and the ears reconvert (for the last time) the sound back to electrical signals. These signals now in the brain of the listener recreate, whether faithfully or not, the electrical signals as they first existed in the brain of the composer. To achieve the desired end, namely the transference of electrical brain impulses from one person to another, the sequence of events is astonishingly long and tortuous. The miracle is that, with care and attention and a sympathetic listener, it can be achieved with "high fidelity."

Leaving this digression on music, let us turn back to Figure 2-2 and notice how the simple conveying of an oscillation from source particle to detector particle forms the basis of most long-distance communication. Suppose the person with the hammer in Figure 2-11 causes the source particle to oscillate for a limited duration of time. The detector particle will then oscillate for a corresponding duration of time. By striking with the hammer, the person can send a series of such pulses of radiation interspersed with gaps. The pulses can be long or short, and a message can thus be conveyed by a system of dots and dashes. This indeed was the way in which long-distance messages were first transmitted all over the Earth, by what was known earlier in the twentieth century as *wireless telegraphy*. As early as 1835, Samuel Finley Breese Morse had invented an ordinary telegraph in which an electrical system of dots and dashes could be sent along wires, thus forming a kind of precursor to the telephone. The significance of the wireless lay in the use of radiation for conveying messages.

Nowadays, it is possible to make source particles oscillate with a combination of frequencies, one on top of another, in carefully adjusted proportions. This technical improvement beyond the crude method of Figure 2-11 permits much more complicated messages to be radiated, as in a television program, for example. Nevertheless, the root principle of all such transmissions is that of Figure 2-2. This also could be the principle whereby intelligent creatures communicate from one star system to another within our galaxy. Electrical radiation is thus very precious, since it is the mode of interaction of matter whereby intelligence is able to express itself.

General Problems and Questions

1. How can you describe the spatial path of a particle?

2. Discuss a method of representing the path of a particle in both space and time.

3. Show that the helical path of Figure 2-8 is followed by the particle in a right-handed sense.

4. Would the sense of this helical path be reversed by reversing the arrows of Figure 2-8?

5. Draw a figure corresponding to Figure 2-8 for a particle moving along a helical path in a left-handed sense.

6. Discuss the way in which the motion of one electrical particle influences the motion of another electrical particle.

7. How is the blue light from the inner part of the Crab nebula generated?

8. Assuming a distant source particle to be in oscillation with a speed small compared to the speed of light, how in principle can the frequency of the oscillation be determined?

9. What is the basic principle of the telescope?

10. Discuss the nature of causality.

11. Under a given external influence, in classical physics, a particle always moves from one point to another by a certain uniquely determined path. Is this true in quantum mechanics? Explain why or why not.

12. How does the electron in the hydrogen atom come to be described by a member of a certain set of states? What happens when many atoms undergo transitions from one such state to another?

13. What is the Lyman series?

14. What is the Balmer series?

15. Under what condition does the electron of the hydrogen atom become free?

16. How can an atom become ionized?

Chapter 3
Black Bodies, Stellar Spectra, and the H-R Diagram

§3-1. Temperatures and the Absolute Scale

When two bodies are placed in contact, energy passes from the hotter one to the cooler one, and energy continues to pass until the two bodies are the same temperature. A *thermometer* is a measuring device that changes its temperature appreciably for a comparatively small energy transfer of this kind. When placed in contact with a larger body whose temperature we wish to measure, a thermometer takes on the temperature of the larger body without disturbing it appreciably. The temperature of the larger body is inferred—for a mercury thermometer—from the length of the column of mercury. This column expands and contracts as the temperature rises and falls. Although practical temperature measurements with a mercury thermometer distinguish different temperatures from one another, the numerical value given to a particular temperature depends not only on the body in question but also on the expansion properties of mercury. The expansion properties of mercury are an extraneous factor that we would prefer to avoid.

The physical meaning of temperature can be better understood by considering a different kind of thermometer, one *consisting of a moderately diffuse* gas composed of suitably simple particles. (The gas must be adequately diffuse to prevent the need for a significant additional term on the right-hand side of the following equation. When an additional term is needed, the gas is said to be

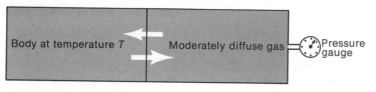

FIGURE 3-1

Measuring the temperature of a body with the aid of a gas thermometer. Heat can pass freely across the surface of separation of the body and the gas, but the other surfaces of the body are not considered to be losing or gaining heat.

imperfect.) We place a gas thermometer in contact with the body whose temperature T we wish to measure, as in Figure 3-1. Then we determine T from

$$T = (\text{constant}) \cdot (\text{pressure in gas thermometer}),$$

the pressure being measured with a gauge. The constant in this equation is fixed by the requirement that two determinations of T, one for melting ice and the other for boiling water, will differ by exactly 100.

Since the temperature of boiling water depends on atmospheric pressure— water boils at a lower temperature on a mountain than it does at sea level—a standard atmospheric pressure must be specified. The same standard pressure is also used when measuring T for melting ice; the temperature for melting also depends on atmospheric pressure, although not so markedly as it does for boiling water. A standard atmospheric condition corresponding to a barometric pressure of 76 cm of mercury is used.

Temperatures determined in this way are said to be *absolute* and are denoted by the symbol K. Each integral step of T is called a *degree*. Unlike the mercury thermometer, which would not give the same result if the mercury were replaced by another liquid, say, by water, the gas thermometer gives the same temperature determination for all gases composed of suitably simple particles.*

Hot surfaces emit radiation of different kinds according to their temperatures. Light emitted at 5,000 K is predominantly yellow. The light becomes redder at lower temperatures and bluer at higher temperatures. We judge an object that is 2,000 K to be very red and one that is 20,000 K to be very blue. Just from the observed color of a star, we can estimate the temperature at its surface. Very cool stars with surface temperatures below 2,000 K are known, as are hot blue stars with surface temperatures above 20,000 K.

TEMPERATURES APPLY TO CLOSED SYSTEMS IN WHICH THE ENERGY IS SUITABLY AVERAGED

We should be a little more cautious than we have been so far. We are so used to talking about temperatures in the everyday world that we are apt to think that a

---

*A gas thermometer cannot be used if changing the temperature changes the chemical nature or internal structure of the particles. This is the essential limitation on the nature of the particles forming the gas.

temperature can be assigned to every material system, but this thought is far from correct. The following discussion of the real physical meaning of temperature shows this fact very clearly.

Imagine a material system contained within a closed, perfectly insulated box that has no contact with the outside world, and suppose the particles making up the material can exchange energy among themselves, say, by colliding one with another. Then, *irrespective of how the system started,* the energy will eventually be shared equally among all the independent components of the system. (A particular component might, for example, be the speed of motion of a particular particle in a particular direction.) The masses of the particles make no difference; a typical particle of small mass has just as much energy as a typical particle of large mass, which means that the heavy particles eventually move more slowly than the light ones, irrespective of whether one kind of particle had more energy in the beginning.

It is this same energy, possessed by every component of the system, that determines its temperature. Only after an equal sharing of the energy has taken place can a temperature be assigned to the system. After defining temperature in this basic way, we have no reason to expect that there will be precisely 100 units between the temperatures of boiling water and melting ice. Consequently, to make a connection with our previous practical method of measuring temperature (Figure 3-1), we need to proceed as follows. We multiply the temperature by a constant, which we write as $\frac{1}{2}k$. Only after this multiplication, $\frac{1}{2}kT$, do we arrive at the average energy possessed by each independent component of the system,

$$\frac{1}{2}kT = \text{average energy.}$$

The right-hand side of this equation is a definite physical quantity, which means that the product of k and T is a fundamental physical quantity. Now, as long as we arrive at this required product, we are free to play around with our definition of T. In particular, we can define T so that there are 100 units of T difference between boiling water and melting ice. This, of course, decides the value we must assign to k so that the product kT will have the physically required value. With T decided in this way, and with energy measured with the erg as the unit (see Chapter 1), the value of k is 1.3805×10^{-16}. This constant has a special name—the *Boltzmann constant*.

§3-2. Black Bodies

ACTUAL PHYSICAL SYSTEMS ARE NEVER INSIDE RIGOROUSLY CLOSED BOXES

It is important to realize that the precise definition of temperature given in the preceding section is idealized, so idealized that it is doubtful whether there is a material system anywhere in the universe for which a temperature is defined with complete accuracy! Nowhere are material systems confined within closed, perfectly insulated boxes. In the laboratory, physicists use a device called a *calorimeter* to approximate as nearly as possible a closed box. At a lower level of precision, the conditions in buildings approximate a closed box. For example, on a cold winter day we can talk about the temperature inside a large building, and we can also talk about the temperature outside the building. It is not so useful to talk about the temperature near the entrances and exits since the situation is ill defined at these points. (Perhaps we should remind ourselves, when we listen to weather forecasters on television, that the temperatures they give are not really temperatures at all but crude approximations of temperatures because the Earth is not inside a closed, insulated box.)

The Earth constantly receives energy from the Sun, and it constantly radiates energy back into space—the latter in the form of invisible infrared radiation. The Earth is close enough to a confined box for the idea of temperature to be useful in describing climate because it does not matter too much that the visual radiation we receive from the Sun has much higher frequencies than the infrared radiation that the Earth returns to space. This difference in frequencies is, however, crucial for biology. The Earth is *never* close enough to a confined box for temperature to be a useful approximation for biology.

Perhaps the closest we can get to a strictly defined temperature is for an element buried deep inside a large body where the surrounding matter acts like the walls of a box. This is the case for matter inside the Earth and the Sun, yet even in these cases the difference from a strictly closed box is very important. Because the temperature of the material inside the Earth is not strictly defined, geothermal heat is able to escape from the interior of the Earth to the surface. Similarly, inside the Sun, energy escapes from the interior to the surface and supplies us with sunlight. If it were not for the outward flow of energy from inside the Sun, caused by a slight departure from a closed-box situation, the surface of the Sun would soon go dark, and life on Earth would quickly come to a dead stop.

MATERIAL SYSTEMS CONFINED IN SUFFICIENTLY CLOSED BOXES GENERATE BLACK-BODY RADIATION

Let us come back for a while to a strictly closed box. So far, we have said nothing about radiation inside the box, only about particles of matter. The situation we have considered to this point implies that the particles can interchange their energies without generating electrical radiation. In principle, this could happen through interactions other than the electrical interaction—through gravitation or through the strong interaction. In practice, however, the electrical interaction is overwhelmingly involved, and radiation is indeed generated.

For a closed box containing very many particles, the number of quantum transitions is large, and, from what we learned in Chapter 2, we can speak about the frequency v of radiation as shown in Figure 2-2. In a situation with very many transitions of atoms and many collisions, many values of the frequency v

are generated. Indeed, in a sufficiently chaotic situation, *all* frequencies up to quite high values of v are generated given time. The radiation is not only generated by source particles but also absorbed by detector particles. It is also absorbed by the walls of the box. Since the box is supposed to be completely insulated, the walls eventually generate just as much radiation as they absorb, and this radiation is reemitted back to the interior of the box, and not to the outside.

From such a totally chaotic situation, a marvellous simplicity emerges. If the particles can generate and absorb at all values of the frequency v, the radiation becomes independent of the details of its generation or absorption or on the nature of the particles themselves. Such a distribution is characteristic of a *black body*. The distribution still depends on the temperature T within the box, but even this dependence has a remarkable simplicity.

Suppose we consider all the radiation with frequencies lying in a narrow *bandwidth* between adjacent values v and $v + dv$, and suppose we write $E\,dv$ for the energy of the radiation in this bandwidth for each unit volume of the box. Then the intensity E depends on both v and T. Thus, if we plot graphs with E along one axis and v along the other axis, we should obtain a different graph for each temperature T. If, instead of plotting E and v in this way, we plot E divided by T^3 and v divided by T (that is, E/T^3 and v/T), *the resulting graph is the same for every temperature T.* We shall use this simplification when we plot the approximations to black-body radiation emitted by stars.

ONLY THE CONTINUOUS RADIATION FROM A STAR IS APPROXIMATELY LIKE THAT OF A BLACK BODY

Because the surface material of a star is not confined in a closed, insulated box, the radiation that a star emits into space is not precisely like a black-body distribution. The most important difference is that the star's radiation possesses spectrum lines, as we shall see in a moment.

Most of the emission from the Sun comes from material in a region called the *photosphere*. This material constantly receives energy from inside the Sun, and it constantly radiates energy into space. On the way out, the radiation passes through a thin atmosphere lying above the photosphere called the *chromosphere*. The chromosphere absorbs a little of the radiation emitted from the lower photosphere, and it is this absorption that adds spectrum lines to the nearly black-body emission of the photosphere.* For the present, we ignore such details and consider that a temperature can usually be specified for the material of the photosphere of a star, about 5,800 K in the case of the Sun. For this temperature, black-body radiation takes the form of the graph of Figure 3-2, and the continuous radiation from the Sun approximates this distribution. The

---

*Some authors refer to the part of the chromosphere immediately above the photosphere as the *reversing layer;* it is in the reversing layer that most of the spectrum lines arise.

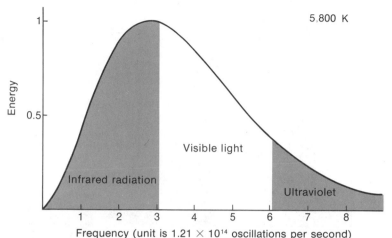

FIGURE 3-2

The frequency distribution of radiation from the Sun, with most energy as visible light, with somewhat less energy in the infrared (heat), and with a comparatively small contribution from the ultraviolet. To obtain the power radiated from 1 sq km of the solar surface, the temperature being 5,800 K, the unit of energy should be 1.16×10^{-4} kWs. The area under the curve is then 6.41×10^{10} kW.

energy scale in Figure 3-2 has been arranged to make the total area under the curve equal to 6.41×10^{10} kW, the power radiated by 1 sq km of the solar photosphere.

With Figure 3-2 available for one particular temperature, it is easy to find the emission for any other temperature. All we need do is to make scale changes of the units in Figure 3-2. The frequency unit simply scales according to the temperature so that a star with a temperature of $2 \times 5,800$ K, for example, would have a frequency unit of 2.42×10^{14} oscillations per second. If we again require the power radiated from 1 sq km of surface, the energy unit is scaled according to the cube of the temperature. A star of temperature $2 \times 5,800$ K would, therefore, have an energy unit eight times larger than Figure 3-2. Otherwise, the form of the curve would stay exactly the same.

Changing the frequency unit of Figure 3-2 changes the distribution of infrared, visible light, and ultraviolet light, as can be seen from the example of a star of 10,000 K shown in Figure 3-3. The black-body curves of Figure 3-2 and 3-3 are often referred to as Planck curves, named for Max Planck (1858-1947), who first found how to obtain them.

An astronomer could easily determine the photospheric temperature of any star simply by studying the frequency distribution of its radiation if the whole of the distribution were available. But severe absorption of radiation in the Earth's atmosphere, for frequencies higher than about 9×10^{14} cycles per sec-

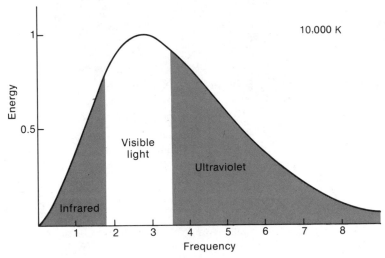

10,000 K

FIGURE 3-3

To obtain the power from a given surface area of a star, the unit for the ordinate must be chosen proportional to the cube of the temperature. Comparing with the solar case, $T = 5,800\,\mathrm{K}$, it is not hard to see that the power emitted from 1 sq km is given for $T = 10,000\,\mathrm{K}$ by the following choice of unit for the ordinate:

$$1.16 \times 10^{-4} \times \left(\frac{10,000}{5,800}\right)^{3} \mathrm{kWs.}$$

ond, seriously truncates the observable distribution for stars with a high surface temperature. This method works very well for stars whose surface temperature is not too high, however. Temperatures obtained in this way are called *color temperatures.*

§3-3. Stellar Spectra

In a diffuse gas, there may not be enough free electrons to produce very much continuous radiation although there can still be enough material to emit significant radiation at discrete frequencies. An example of such a gas has already been seen in Figure 2-22, which showed the Balmer series of hydrogen accompanied by only a weak distribution of radiation of continuous frequencies. The ratio of such discrete "line" frequencies to the continuous frequencies can vary greatly, depending on the density of the gas, its temperature, and its dimensions. By studying this ratio carefully, astronomers are often able to make important inferences about the density, temperature, and content of the gas clouds lying between the stars. Sometimes they can even make inferences about gas clouds in other galaxies.

Consider the situation that arises if a diffuse gas lies in front of a surface that generates a black-body distribution of frequencies corresponding to a certain temperature, T. Within the gas, we have transitions of electrons between discrete states, for example, the transitions of the Balmer series of hydrogen.

If the gas is cooler than the temperature T, it will absorb radiation from the background since the coolness of the gas means that the atoms within it tend to occupy lower energy states than do the atoms that form the hotter background. An observer sees radiation with a *deficit* at certain discrete frequencies determined by the upward transitions of the atoms of the cool gas, as is shown for the example of the Balmer series in Figure 3-4.

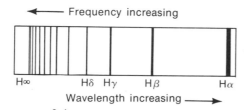

FIGURE 3-4
A schematic representation of the lines of the Balmer series in absorption, with frequency increasing toward the left and wavelengths increasing toward the right. (Compare with the situation of Figure 3-8.)

Conversely, if the intervening gas is hotter than the temperature T, it already has more atoms in the higher energy states than the now cooler background does. For such a hotter intervening gas, the downward transitions dominate the upward ones, and an excess of radiation is added at the discrete frequencies determined by the atoms. The hotter intervening gas thus adds *emission lines* to the background.

Both these situations are illustrated in Figure 3-5, and both situations occur in the gases that lie above the photosphere of the Sun. The gases lying immediately above the photosphere have a lower temperature than the photosphere itself. Consequently, we should find radiation missing at certain discrete frequencies. That we do can be seen from Figure 3-6, which shows the distribution of frequencies emitted by the Sun. These *absorption lines* are produced by many different kinds of atoms, atoms of calcium—notice the H and K lines of calcium—and of iron and other metals playing a particularly prominent role. The temperature, after falling for a while as we progress outward from the

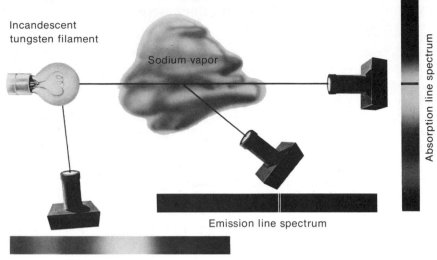

Incandescent tungsten filament

Sodium vapor

Absorption line spectrum

Emission line spectrum

Continuous spectrum

FIGURE 3-5

A gas interposed between the point of observation and a hotter source of continuum light produces *absorption* lines. When the point of observation is such that there is no background of continuum light, the gas produces *emission* lines. (From J. C. Brandt and S. P. Maran, *New Horizons in Astronomy,* Second Edition. W. H. Freeman and Company. Copyright © 1979.)

photosphere, eventually rises with increasing height in the chromosphere of the Sun. With the temperature in the chromosphere higher than that at the photosphere, the chromospheric material adds excess radiation at certain discrete frequencies. Because the chromosphere contains comparatively little material, this effect is not very strong. It does occur, however, as can be seen from Figure 3-7, which shows the Sun photographed by light at the frequency of the K line of calcium.

THE SPECTRUM LINES PRODUCED BY DIFFERENT KINDS OF ATOMS VARY IN THEIR TEMPERATURE SENSITIVITY

The spacings of the levels in energy diagrams are different for different atoms. Atoms whose diagrams have wider spacings have transitions that emit and absorb radiation of higher frequencies than those emitted or absorbed by atoms whose diagrams have smaller spacings. The former atoms produce a greater absorption effect on the spectra of stars of high surface temperature than they do on the spectra of stars of low temperature because there is more high-frequency radiation from the former stars than from the latter, as we can see from a comparison of Figures 3-2 and 3-3.

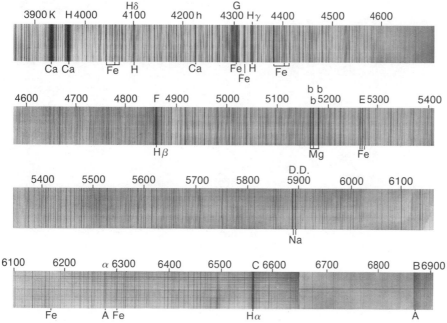

FIGURE 3-6

The many absorption lines (Fraunhofer lines) present in sunlight. This solar spectrum was made with the 13-ft spectroheliograph. The numbers refer to the wavelength of the radiation, measured in a unit known as the angstrom (Å), such that $1 \text{ Å} = 10^{-8}$ cm. The wavelength increases toward the right as in Figure 3-5. The letters below various absorption lines are the chemical symbols of the atoms that produce the lines. (Courtesy of Hale Observatories.)

FIGURE 3-7

Restricting frequency to that of the K line of calcium, the hotter, more diffuse gas lying above the normal solar surface shows distinctive patches of emission. (Courtesy of Kitt Peak National Observatory.)

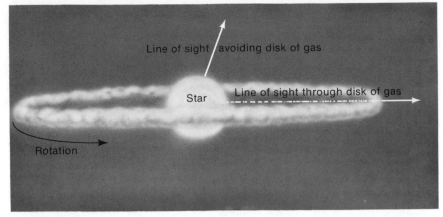

FIGURE 3-8
A star of high surface temperature surrounded by a rotating disk of cooler gas. Absorption lines are seen when the line of sight passes through the disk of gas. (Compare with Figure 3-4).

Of the common atoms, the widest spacings occur for *ionized helium,* helium atoms in which one electron has become free. The next widest spacings occur for *neutral helium,* helium atoms that retain both of their electrons. Next comes hydrogen, followed by ionized metal atoms in which one electron has become free, such as calcium. (Recall that ionized calcium atoms give H and K lines.) The narrowest spacings occur for *neutral* metal atoms, metal atoms with no electrons free, such as iron.

It is possible to classify the stars into categories, or *spectral types,* according to the main features of their frequency distributions—their spectra. The types form a temperature sequence because the different spacings of the energy-level diagrams of different atoms cause the atoms' absorption or emission lines to show prominently only within certain temperature ranges—ranges that happen to match the energy-level diagrams in a suitable way. We expect the appearance of helium to be associated with the hottest stars, hydrogen with less hot stars, ionized metal atoms with even less hot stars, and finally neutral metal atoms with yet less hot stars. The spectral types are catalogued in Table 3-1, and examples are shown in Figure 3-9. The choice of letters to represent the different types is historic, dating from a time when their relation to a temperature sequence was not well understood. It is possible to divide the types into subclasses. In type A, for example, hydrogen absorption lines are stronger in relation to ionized calcium at the upper end of the temperature range than they are at the lower end of the range. Consequently, by estimating the relative strength of hydrogen and ionized calcium for a particular star of type A, it is possible to determine more closely where the star is in the range from 7,200 to 11,000 K. The study of the various subtypes, and the assignment of individual stars to them, forms the subject of *stellar spectroscopy.*

Absorption lines or *bands* caused by molecules appear in cool stars of type

TABLE 3-1
Spectral classification of stars

63

§ *3-3. Stellar Spectra*

Classification symbol	Distinguishing features	Temperature range K
O	ionized helium stronger than neutral helium	Above 30,000
B	ionized helium weaker than neutral helium	11,000–30,000
A	hydrogen at maximum strength ionized calcium appears	7,200–11,000
F	ionized calcium strong hydrogen weakening neutral metals appear	6,000–7,200
G	ionized calcium strong neutral metals strong	5,200–6,000
K	neutral metals strong ionized calcium weakening	3,500–5,200
M	neutral metals strong absorption bands of molecules	Below 3,500

M. Beyond remarking that the most prominent molecule in such stars is tita-nium oxide, we shall not concern ourselves with these bands. Other cool stars have other prominent molecular bands, cyanogen in R stars, carbon in N stars, and zirconium oxide in S stars.

The naming and ordering of the spectrum types can be remembered from the following often-quoted mnemonic aid: Oh Be A Fine Girl, Kiss Me Right Now. Smack!

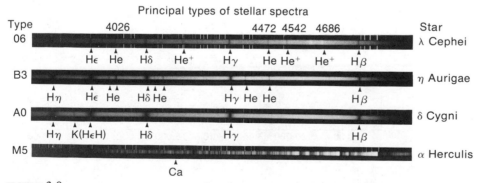

FIGURE 3-9
Examples of the spectra of stars of differing surface temperature. The wavelengths of lines are marked in angstroms (1 Å = 10^{-8} cm). The chemical symbols of the atoms producing the lines are also shown. The digit following the class letter is a subclass designation. (Courtesy of Hale Observatories.)

§3-4. The Hertzsprung-Russell Diagram

Black Bodies, Stellar Spectra, and the H-R Diagram

The Hertzsprung-Russell diagram (H-R diagram) relates the surface temperature of stars to their total luminous output. The unit of luminosity used in Figure 3-10 is the present-day power output of the Sun, 3.8×10^{23} kW. The surface temperatures are in units of 1,000 K and are arranged *to increase toward the left.* Both the luminosity and temperature are displayed *logarithmically.* A

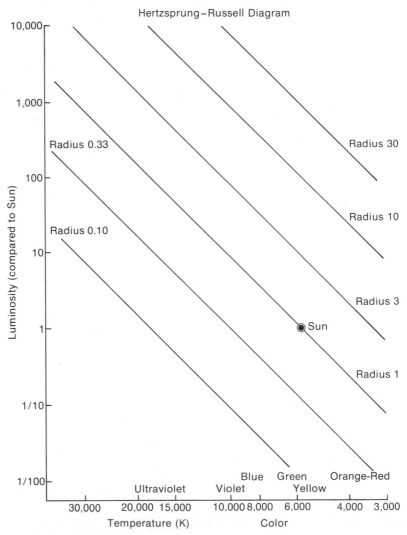

FIGURE 3-10

The Hertzsprung-Russell diagram with lines of equal radius marked (solar radius as unit). Note that the left-hand scale refers to the total emission of radiation of all frequencies by the star. This is the so-called *bolometric* luminosity.

diagram of this kind, with each star represented by a point in the diagram, was first used independently (as its name implies) by Ejnar Hertzsprung (1873–1967) and by Henry Norris Russell (1877–1957). Also shown in Figure 3-10 are lines of equal stellar radius using the present-day solar radius, 6.96×10^{10} cm, as the unit.

THE SURFACE TEMPERATURES AND LUMINOSITIES OF STARS CAN BE MEASURED IN MANY CASES

Astronomers are able to estimate the surface temperatures of stars from either their colors or their spectra. Estimating the total luminous output of a star is a more complex affair, however. For this, it is necessary to know the star's distance. In this chapter, we shall assume distances to be known, postponing a discussion of how they become known until Chapter 9.

Write r for the distance of a particular star and L for its total output of radiation, L kW. Then the rate at which energy from the star is incident on unit area of astronomers' telescopes is $L/4\pi r^2$ because $4\pi r^2$ is the area of the surface of the sphere shown in Figure 3-11. Writing F for this quantity, $F = L/4\pi r^2$, astronomers seek to measure F. Knowing r, the required luminosity L is then immediately calculated from the simple equation $L = 4\pi r^2 F$. It is in this way that the luminosities appearing in the H–R diagram are obtained.

Although the idea behind the use of the equation $L = 4\pi r^2 F$ is simple enough, there is a complication in measuring F that is clearly brought out by Figure 3-12. Only the radiation in the narrow band marked *Visible* reaches astronomers' telescopes. Although this radiation can be measured either photographically or, better still, by the use of a photomultiplier (see Figure 2-31), how are we to allow for the radiation (infrared, ultraviolet, and so forth) that is absorbed in the atmosphere and fails to reach a ground-based observatory? The reason this question can be answered for most stars is that their radiation

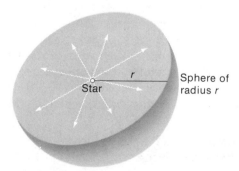

FIGURE 3-11
The rate at which energy crosses a unit area of a sphere of radius r, concentric with the star, is $L/4\pi r^2$, where L is the luminosity of the star.

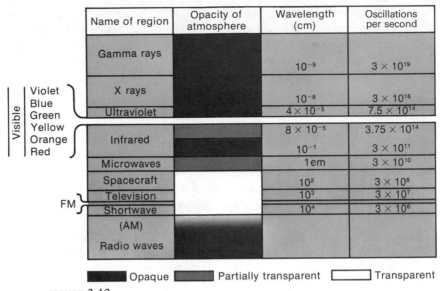

Name of region	Opacity of atmosphere	Wavelength (cm)	Oscillations per second
Gamma rays		10^{-9}	3×10^{19}
X rays		10^{-6}	3×10^{16}
Ultraviolet		4×10^{-5}	7.5×10^{14}
Infrared		8×10^{-5}	3.75×10^{14}
		10^{-1}	3×10^{11}
Microwaves		1em	3×10^{10}
Spacecraft		10^{2}	3×10^{8}
Television		10^{3}	3×10^{7}
Shortwave		10^{4}	3×10^{6}
(AM) Radio waves			

Violet
Blue
Green
Yellow
Orange
Red

Visible

FM

■ Opaque ■ Partially transparent □ Transparent

FIGURE 3-12
Wavelength values have been added to the frequencies of Table 2-1. Radiation in the unshaded areas passes through the Earth's atmosphere from outer space.

approximates the black-body form. We have seen that the black-body relation between E/T^3 and ν/T is always the same, and it is well known. With the surface temperature T of a star also known from its spectrum, we know the relation between E and ν. Hence, it is easy to estimate the fraction of the star's radiation that falls into the visible range of frequencies. Thus, the energy flux, measured for the visible range, can be readily scaled or corrected to what it would have been if the whole frequency range had been observed. This scaling to include the whole of a star's radiation is known as a *bolometric correction*.

Before returning to the H-R diagram, we pause briefly to note that a new quantity appears in the third column of Figure 3-12. The wavelength λ associated with radiation of frequency ν is easily calculated from $\lambda = c/\nu$, where c is the speed of propagation of a pulse of radiation. Outside of matter, the speed c is independent of ν and is determined experimentally to be nearly 300,000 km/second. There are nearly 3.156×10^7 seconds in a year, so that, in a year, a pulse of radiation travels a distance 9.46×10^{12} km. Astronomical distances are often measured in this particular unit, the *light-year*.

PROTOSTARS CONDENSE AT AN APPROXIMATELY CONSTANT SURFACE TEMPERATURE OF 4,000 K

As an example of the usefulness of the H-R diagram, Figure 3-13 shows the evolutionary path of a newly forming star with a mass equal to the Sun. Stars form all the time within interstellar clouds of gas and dust like the Orion nebula

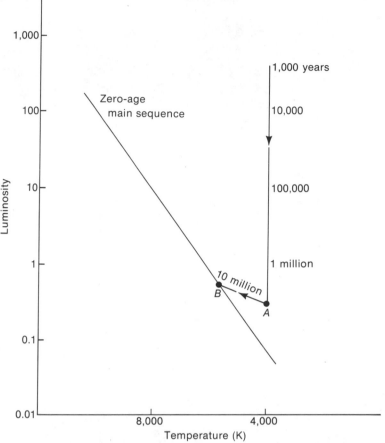

FIGURE 3-13
The path of a condensing solar-type star in the Hertzsprung-Russell diagram.
Note the slowdown in the rate of shrinkage, with the final part of the track
from *A* to *B* taking longer than all the rest of the formation process.

in Figure 3-14 (see also Plate II). Within the main cloud there are smaller,
denser clouds, and within the latter there are still smaller, still denser regions, as
is shown schematically in Figure 3-15. Eventually in this fragmentation cascade,
protostars (indicated by white dots in Figure 3-15) arise. These condense more
and more, with the material inside them becoming hotter and hotter through
compression. Since material deep inside a protostar is hotter than material near
its surface, a flow of heat takes place from the central regions to the surface,
causing the surface to radiate outward into space. As contraction proceeds, the
surface temperature eventually rises to about 4,000 K. This surface temperature
is then maintained steadily until point *A* of Figure 3-13 is reached. The reason
such a steady outer temperature can be maintained is illustrated in Figure 3-16.

FIGURE 3-14
The Orion nebula, a cloud of gas in which stars are now being born.
(Courtesy of Hale Observatories.)

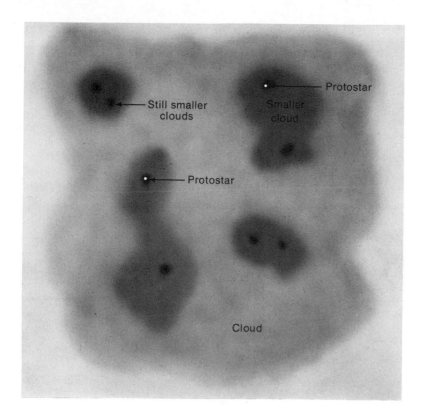

FIGURE 3-15
The fragmentation of a
gas cloud into protostars.

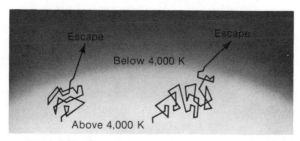

FIGURE 3-16
When the temperature near the surface of a newly forming
star falls below about 4,000 K, the gases are no longer able
to block the escape of radiation in an effective way.

Above a temperature of 4,000 K, the atoms can largely deflect the radiation and
so prevent its free escape. For temperatures below 4,000 K they cannot, and
radiation can stream out into space quite freely. This property requires that the
surface temperature of a protostar must be maintained at about 4,000 K
throughout the later stages of its condensation until a radius comparable to that
of the present-day Sun is reached.

Figure 3-13 gives the time required for the various stages of contraction. The contraction of the protostar is seen to slow down from 100 years to 1,000 years to 10,000 years to 100,000 years. A final, comparatively short, reversed part of the track from *A* to *B* has a special significance that we examine next.

EARLY CONVECTION IN A PROTOSTAR GIVES PLACE TO AN OUTWARD TRANSFER OF ENERGY BY RADIATION

Through the evolution up to point *A*, the surface temperature requirement of Figure 3-16 forces the interior of the protostar to be in convective motion. While the protostar is large in radius, the convection is rapid. Energy is carried swiftly to the surface where it is needed to maintain the temperature of 4,000 K, and it is the swiftness of the outward convective transfer of energy that causes the early, short contraction time of the protostar. The convective motion slows down, however, and disappears entirely deep in the interior as point *A* is reached. It is the disappearance of the interior convection, with outward transfer of energy occurring only by radiation, that causes the reversed part of the track from *A* to *B* in Figure 3-13.

PROTOSTARS BECOME STARS WHEN NUCLEAR REACTIONS ARE TURNED ON

Contraction continues from *A* to *B*, and the central temperature continues to rise until energy in appreciable quantity is delivered by *nuclear reactions*. These reactions change atoms from one kind to another; they involve the strong and weak interactions as well as the electrical interaction between the particles. We will study these reactions in detail in Chapter 8. We are now concerned only with the fact that they deliver energy. Point *B* of Figure 3-13 corresponds to the stage where significant quantities of energy are released by nuclear reactions and where this release establishes a crucial new form of equilibrium.

If within the star we draw a sphere of radius *r*, as in Figure 3-17, the outflow of energy across the surface of this sphere is equal to the nuclear-energy production within its interior. Putting *r* equal to the total radius $R(r = R)$, we see that the loss of energy from the outer surface is then just compensated by the total nuclear-energy production within the whole interior of the star. We thus have a complete energy balance, achieved at the expense of converting one kind of atom into another. If we think of atoms as nuclear fuel, the equilibrium is achieved through the *burning of nuclear fuel*. When this stage is reached, at point *B* in Figure 3-13, the star is said to have reached the *main sequence*. The star is now considered to have formed; it is no longer a protostar.

A similar discussion could be given for protostars with masses different from that of the Sun. The endpoints attained, corresponding to point *B* for the solar case, all lie on the line shown in Figure 3-13. This line is the main sequence itself, and, because it gives the positions of stars when the burning of nuclear fuel first establishes an energy balance, it is sometimes called the *zero-age main sequence*.

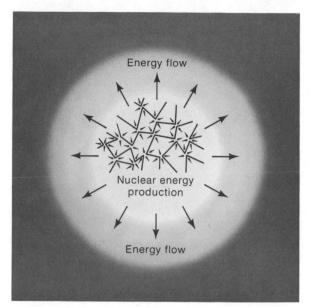

FIGURE 3-17

A condensing star eventually reaches an equilibrium configuration in which the flow of energy across any sphere is equal to the rate of nuclear-energy generation within that sphere.

THE LUMINOSITIES OF STARS ON THE ZERO-AGE MAIN SEQUENCE ARE DETERMINED BY THEIR MASSES

From this point we will use the symbol $\odot$ to denote a mass equal to that of the Sun. We may ask: What is the full range of stellar masses? This question is hard to answer because stars with masses less than 0.3 $\odot$ are intrinsically so faint that they are difficult to observe, whereas stars with masses greater than 50 $\odot$ tend to be confined within the dense, opaque clouds in which they are born, essentially because their lives are very short and because they have inadequate time to move away from their parent clouds. Observations outside the range from 0.3 $\odot$ to 50 $\odot$ are consequently restricted, so the main sequence is usually taken to extend from about 0.3 $\odot$ on the lower side to about 50 $\odot$ on the upper side. We must recognize, however, that stars outside this range exist, probably in large numbers on the low side but rather infrequently on the high side.

The simple proportionality $L \propto M^4$, relating the luminosity L to the mass M, can be used as a reasonable approximation over the mass range from 0.3 $\odot$ to 20 $\odot$. Thus, a star of mass 20 $\odot$ is about 60^4 times brighter than a star of mass 0.3 $\odot$.

Since L must be close to the luminosity of the Sun when $M = \odot$, it follows that the proportionality $L \propto M^4$ gives $L = (M/\odot)^4$ when the solar luminosity is used as the unit of L (as in Figures 3-10 and 3-13). The relation $L = (M/\odot)^4$

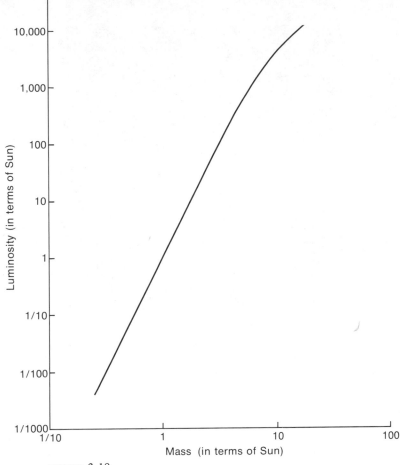

FIGURE 3-18

The general form of the relation between the luminosity of a newly formed star and its mass. The curve tends to lean over at high mass and high luminosity.

is represented graphically in Figure 3-18, except that, as the mass increases above $M = 20 \odot$, the luminosity increases more slowly than the proportionality $L \propto M^4$. This is also illustrated in Figure 3-18.

§3-5. The Stars in the Sky

At this point we can proceed in one of two ways. Having begun a discussion of stars, we could continue and extend the discussion to material inside stars, with particular reference to nuclear reactions. This would require us to drop our consideration of the electrical interaction in favor of the strong and weak interactions. Or we could proceed with the electrical interaction applied to astronomical objects other than stars, objects that give rise to radiation quite different from the black-body form. In a book with a strong astronomical orientation, the former approach would no doubt be preferable. In the present book, with its emphasis on the association of physics with astronomy, it seems better to continue with the application of the electrical interaction to radio astronomy, millimeter-wave astronomy, infrared astronomy, and x ray astronomy. We shall discuss stars again in Chapter 8 when we come to problems associated with the strong and weak interactions.

It may be useful, however, to end this chapter with a short compilation of simple facts concerning stars, including their mapping on the sky and details concerning the nearest and the brightest stars as they appear from a terrestrial observatory. First, we discuss briefly the way stars and other astronomical objects are located on the sky.

Imagine a sphere of large radius concentric with the Earth. Using the center as vertex, project both the Earth's equator and the orbit in which the Sun can be considered to move around the Earth (see Figures 3-19 and 3-20) onto the

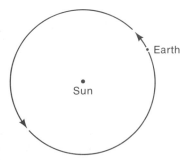

FIGURE 3-19
In many problems, it is useful to think of the Earth as a speck moving in its annual orbit around the Sun. The diameter of the orbit is about 3×10^8 km.

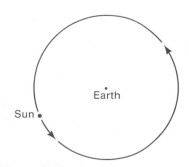

FIGURE 3-20
The relative motion of the Earth and Sun is the same here as in Figure 3-19.

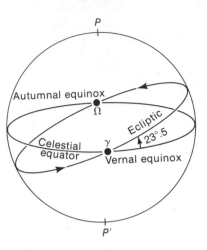

FIGURE 3-21
The apparent path of the Sun in the sky, known as the ecliptic, is inclined at an angle of about 23.5° to the celestial equator. The ecliptic and the equator intersect at the vernal and autumnal equinoxes. The arrows on the ecliptic show the direction of the Sun's apparent annual motion.

sphere, as in Figure 3-21). Both projections are great circles, intersecting at the points γ and Ω. By great circles, we simply mean that a plane through either circle passes through the center of the sphere, as it must from the nature of the construction. During the year, the projection of the Sun goes around one of these circles, the *ecliptic,* being at γ on the day of the vernal equinox and at Ω on the autumnal equinox. The axis of the Earth's rotation intersects the sphere at the points P and P', the *celestial poles.* The sphere itself is called the *celestial sphere.*

A similar projection can be made for every object we see in the sky: stars, planets, comets, and galaxies. When an object appears to us as a point, we obtain a point on the celestial sphere. Where an object is extended—that is, spread out—we obtain a patch on the celestial sphere. The situation is illustrated in Figure 3-22, from which it will be seen that, although the celestial sphere has a large radius compared with the scale of the Earth's orbit around the Sun, the distances of the stars are still larger.

Using the celestial equator and the poles P and P', we can easily set up a system of latitude and longitude on the celestial sphere. Lines of longitude and latitude are drawn analogously to the usual geographical system. For zero longitude, we take the arc of the circle through P, γ, and P'. This arc is known as the *meridian.* Several minor changes from the geographical system are used in practice. Instead of measuring longitudes both east and west, we measure all longitudes in an easterly sense, going through all angles from 0° to 360° (see Figure 3-23).

Longitudes given as angles can be converted into equivalent times in the following way. Divide the range from 0° to 360° into 24 equal steps of 15° each. To each such step, assign 1 hour. To a portion of a step, assign minutes and seconds of time according to the following rules: 1 minute of time ≡ 15 minutes of angle, 1 second of time ≡ 15 seconds of angle. Using these equivalences, we can express longitudes in hours, minutes, and seconds of time (h m s).

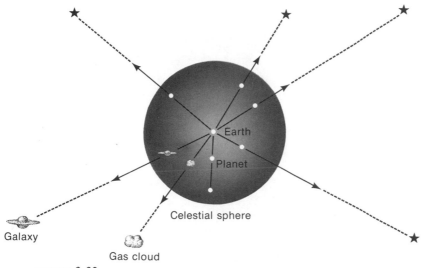

FIGURE 3-22

All manner of astronomical objects are projected on the celestial sphere, which is taken to have a radius that is large compared to the scale of the solar system.

Another minor difference between the geographical and astronomical systems is that, whereas geographical latitudes are given as north or south of the equator, + and − signs are used in the astronomical system. With these small changes, the latitude and longitude of an object on the celestial sphere are referred to as *right ascension* (RA) and *declination* (Dec):

$$\text{right ascension} \equiv \text{longitude}, \qquad \text{declination} \equiv \text{latitude}.$$

Although this situation is straightforward in principle, a vast collection of objects, large both in number and in variety, are thus projected on the celestial sphere. To distinguish different classes of objects, astronomers have constructed a library of catalogues: catalogues of stars (sometimes for separate kinds of stars), of galaxies, of patches of glowing gas, of patches of radio emission known as radio sources, and of sources of unusual forms of radiation—x rays and of infrared—two recently developed branches of astronomy. Indeed, whenever a new variety of object is discovered, one of the first steps taken by the astronomical community is to catalogue them. The listing of the objects in each catalogue is by right ascension and declination. Astronomers, wishing to observe a particular object, will look it up in a catalogue, where they will read the appropriate right ascension and declination. This information determines the point on the sky toward which they must direct their telescopes to make the desired observation.

Because the Earth and the stars are moving, the right ascension and declination of every star change slightly with time. The change caused by the Earth's

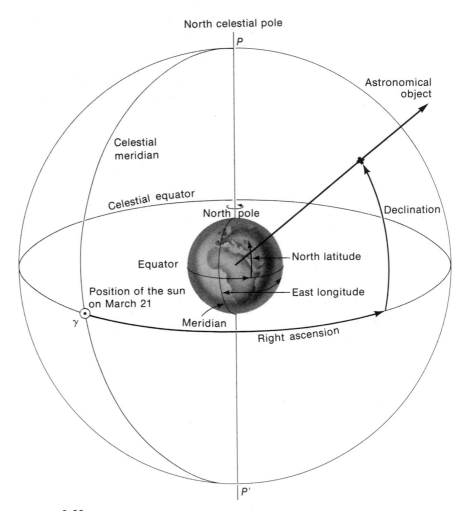

North celestial pole

FIGURE 3-23
The astronomical system of right ascension (RA) and declination (Dec) is similar to the usual geographical system of longitude and latitude. The point from which the RA of an astronomical object is measured is the point γ of Figure 3-21. The scale of the Earth here is obviously grossly exaggerated. (From J. C. Brandt and S. P. Maran, *New Horizons in Astronomy*, Second Edition. W. H. Freeman and Company. Copyright © 1979.)

motion in its orbit around the Sun is termed a *parallax*, and the change due to the star itself is called a *proper motion*. These changes are very small, however, and, to the extent that we neglect them, each star could be given a fixed right ascension and declination if the point γ in Figure 3-23 were fixed in relation to the background of the stars. Unfortunately, this is not the case. The direction from the Earth's center to γ is defined as the line of interaction of the plane of the Earth's equator with the plane of the Earth's orbit around the Sun, and this line of intersection moves slowly with respect to the stars because the Earth is subject to a slow precession like a spinning top. This precession causes γ to make a complete circuit of the celestial equator in about 26,000 years. It follows that, when astronomers specify the right ascension and declination of an object, they must do so for the position of γ as *it was at a particular moment of time*. The moment of time that is used at present is 00.00 h, January 1, 1950. Since 1950, γ has moved by about one-third of a degree. Catalogue positions of stars, given with respect to 1950, must therefore be updated by angles that are of the general order of one-third of a degree.

From the projection of an object onto the celestial sphere, we know its direction but not its distance. The projections of distant objects and near objects can fall adjacent to each other. The Sun, Moon, and planets are close compared to the stars, while the stars are close compared to distant galaxies. We shall consider the determination of distances in Chapter 9.

We can think of the stars and other astronomical objects projected on the celestial sphere in the same way that we think of places on the surface of the Earth. Just as we display places on the Earth in the form of maps, so we can construct maps of the sky, as in Figure 3-24.

The identification of star patterns with quaint terrestrial images has no scientific significance. Yet, there is a continuing attraction in these fanciful constellations, which are listed in Table 3-2. Astronomers still use them in referring to different parts of the sky and in catalogue descriptions of stars. Thus, the brightest stars in the constellations are classified sequentially by letters of the Greek alphabet, α being in general the brightest, β the next brightest, γ the next, and so on. However, the very brightest have specific names, many of Arabic origin. Some of these special cases are marked on the maps of Figure 3-24. The star Vega is also referred to as α Lyrae, for example.

Table 3-3 lists the stars nearest our solar system. The quantity $B - V$ is a color temperature, and the relation between $B - V$ values and temperature values is given at the end of the table. The absolute magnitude values determine the visual luminosity of the stars—the energy output in the visual range of radiation frequencies. Readers not familiar with the somewhat tortuous procedure for determining luminosities from absolute magnitudes will find an explanation in the Glossary. The visual apparent magnitudes in the fourth column are also explained in the Glossary.

Table 3-4 lists stars that appear brightest in the sky. Notice that there is only a very small overlap between Tables 3-3 and 3-4, which means that on the whole the nearest stars do not appear to be the brightest.

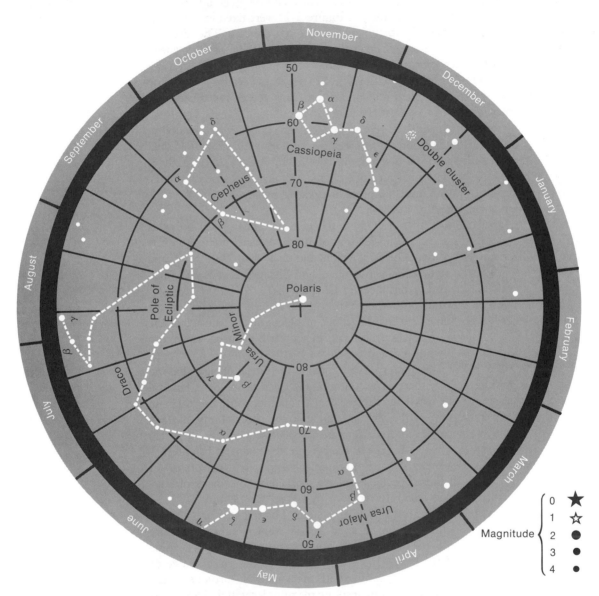

FIGURE 3-24(a)
The constellations of the north circumpolar sky.

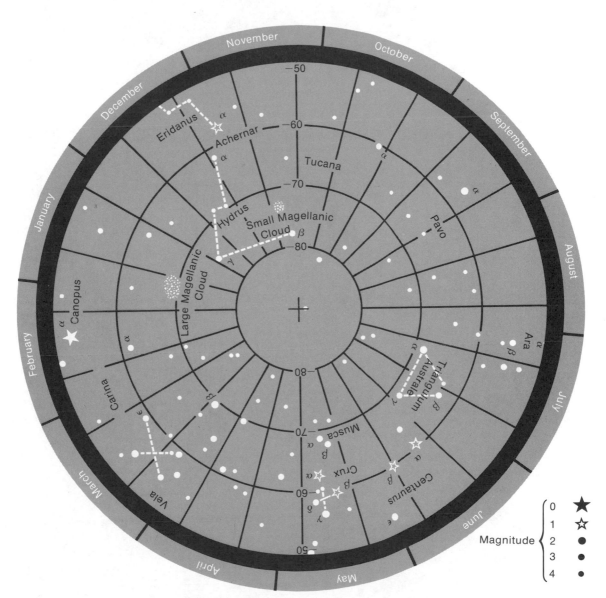

FIGURE 3-24(b)
The constellations of the south circumpolar sky.

80

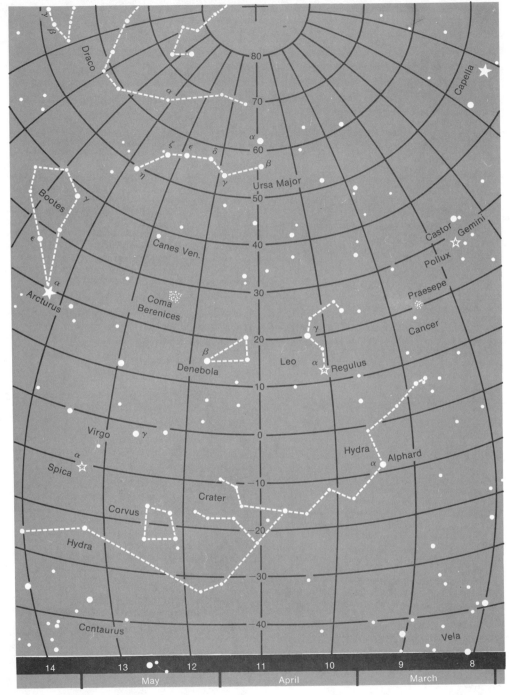

FIGURE 3-24(c)
Night sky in spring (northern hemisphere).

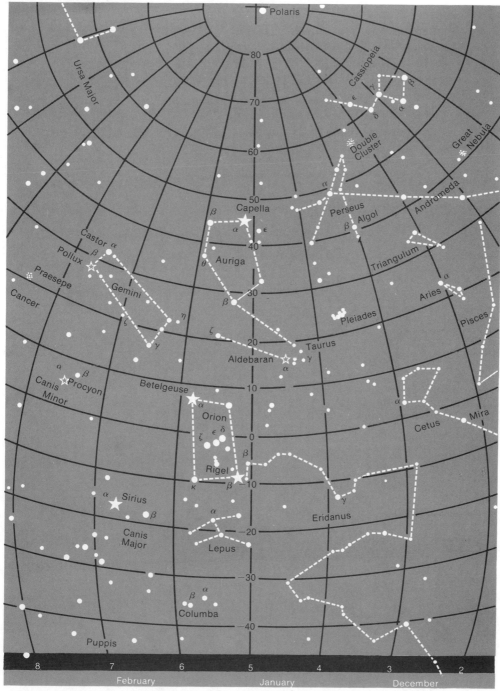

FIGURE 3-24(d)
Night sky in winter (northern hemisphere).

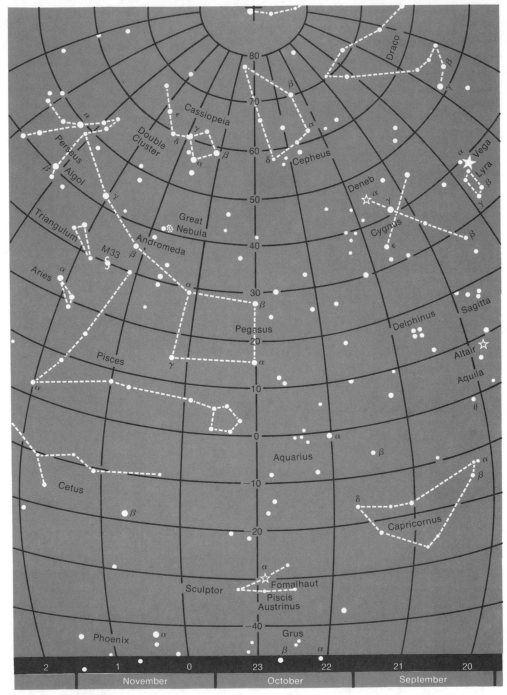

FIGURE 3-24(e)
Night sky in autumn (northern hemisphere).

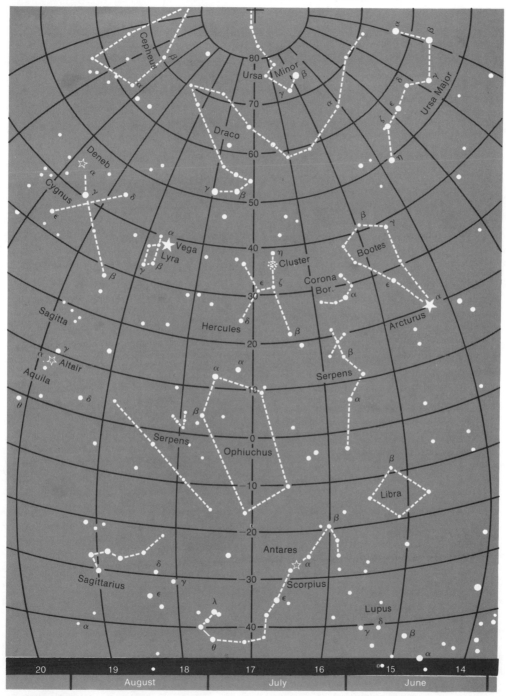

FIGURE 3-24(f)
Night sky in summer (northern hemisphere).

TABLE 3-2
The constellations

Name	Approximate right ascension	Approximate declination	Intended meaning
Andromeda (And)	01	35	Andromeda
*Antlia (Ant)	10	−30	Pump
*Apus (Aps)	17	−75	Bird of Paradise
Aquarius (Aqr)	22	−15	Water Bearer
Aquila (Aql)	20	05	Eagle
Ara (Ara)	17	−55	Altar
Aries (Ari)	02	20	Ram
Auriga (Aur)	05	40	Charioteer
Boötes (Boo)	15	30	Herdsman
*Caelum (Cae)	05	−40	Chisel
*Camelopardus (Cam)	06	70	Giraffe
Cancer (Cnc)	09	20	Crab
*Canes Venatici (CVn)	13	40	Hunting Dogs
Canis Major (CMa)	07	−25	Big Dog
Canis Minor (CMi)	07	05	Small Dog
Capricornus (Cap)	21	−15	Sea Goat
*Carina (Car)	09	−60	Ship's Keel
Cassiopeia (Cas)	01	60	Cassiopeia
Centaurus (Cen)	13	−50	Centaur
Cepheus (Cep)	21	65	Cepheus
Cetus (Cet)	02	−5	Whale
*Chamaeleon (Cha)	11	−80	Chameleon
*Circinus (Cir)	16	−65	Compass
*Columba (Col)	06	−35	Dove
*Coma Berenices (Com)	13	20	Berenice's Hair
Corona Austrina (CrA)	19	−40	Southern Crown
Corona Borealis (CrB)	16	30	Northern Crown
Corvus (Crv)	12	−20	Crow
Crater (Crt)	11	−15	Cup
*Crux (Cru)	12	−60	Southern Cross
Cygnus (Cyg)	21	40	Swan
Delphinus (Del)	21	15	Dolphin
*Dorado (Dor)	05	−60	Swordfish
Draco (Dra)	18	60	Dragon
Equuleus (Equ)	21	10	Small Horse
Eridanus (Eri)	03	−25	River Eridanus
*Fornax (For)	03	−30	Furnace
Gemini (Gem)	07	25	Twins
*Grus (Gru)	22	−45	Crane
Hercules (Her)	17	30	Hercules
*Horologium (Hor)	03	−55	Clock
Hydra (Hya)	10	−15	Water Monster
*Hydrus (Hyi)	01	−70	Water Snake
*Indus (Ind)	20	−50	Indian

Name	Approximate right ascension	Approximate declination	Intended meaning
*Lacerta (Lac)	22	40	Lizard
Leo (Leo)	10	20	Lion
*Leo Minor (LMi)	10	35	Small Lion
Lepus (Lep)	05	−20	Hare
Libra (Lib)	15	−15	Balance
Lupus (Lup)	15	−45	Wolf
*Lynx (Lyn)	09	40	Lynx
Lyra (Lyr)	19	35	Harp
*Mensa (Men)	06	−75	Table (Mountain)
*Microscopium (Mic)	21	−35	Microscope
*Monoceros (Mon)	07	00	Unicorn
*Musca (Mus)	13	−70	Fly
*Norma (Nor)	16	−55	Square
*Octans (Oct)	22	−85	Octant
Ophiuchus (Oph)	17	00	Snake Bearer
Orion (Ori)	05	05	Orion
*Pavo (Pav)	20	−60	Peacock
Pegasus (Peg)	22	20	Pegasus
Perseus (Per)	03	40	Perseus
*Phoenix (Phe)	01	−45	Phoenix
*Pictor (Pic)	07	−60	Easel
Pisces (Psc)	00	10	Fishes
Piscis Austrinus (PsA)	23	−30	Southern Fish
*Puppis (Pup)	07	−35	Ship's Stern
*Pyxis (Pyx)	09	−35	Ship's Compass
*Reticulum (Ret)	04	−65	Net
Sagitta (Sge)	20	15	Arrow
Sagittarius (Sgr)	18	−30	Archer
Scorpius (Sco)	17	−35	Scorpion
*Sculptor (Scl)	01	−30	Sculptor
*Scutum (Sct)	19	−10	Shield
Serpens (Ser)	16	05	Snake
*Sextans (Sex)	10	00	Sextant
Taurus (Tau)	05	20	Bull
*Telescopium (Tel)	18	−45	Telescope
Triangulum (Tri)	02	35	Triangle
*Triangulum Australe (TrA)	16	−65	Southern Triangle
*Tucana (Tuc)	23	−60	Toucan
Ursa Major (UMa)	11	50	Great Bear
Ursa Minor (UMi)	15	75	Small Bear
*Vela (Vel)	09	−50	Ship's Sails
Virgo (Vir)	13	00	Virgin
*Volans (Vol)	08	−70	Flying Fish
*Vulpecula (Vul)	20	25	Fox

*Of modern origin.

TABLE 3-3
The nearest stars[a]

Catalogue designation(s)[b]		1900 Right ascension[c]	Declination	Visual magnitude[d]	B − V[e]	Distance (parsecs)[f]	Absolute magnitude[g]	Mass (☉)[h]	Radius (solar radius)	Remarks[i]
Grm 34 = 43°44	A	00h 13m	43° 27'	8.08	1.55	3.60	10.30			Binary with separation on sky of 38 arc seconds. A is itself an unresolved binary.
(= CC 19)	B		49	11.05	1.78		13.27			
β Hyi		00 20	−77 49	2.79	0.61	6.54	3.71			
η Cas	A	00 43	57 17	3.44	0.57	5.52	4.73	0.85	0.84	Binary with period 490 years and separation 12″.
	B			7.25			8.54	0.52		
v. Maanen = Wolf 28		00 44	04 55	12.34	0.55	4.26	14.19			White dwarf
UV Cet	A	01 34	−18 28	12.41	1.9	2.67	15.28	0.044		Binary with period about 200 years and separation 5″.
	B			12.95	1.9		15.87	0.035		
τ Cet		01 39	−16 28	3.50	0.72	3.64	5.70			
82 Eri = e Eri		03 16	−43 27	4.23	0.71	6.41	5.19			
ε Eri		03 28	−09 48	3.74	0.87	3.30	6.14			
σ² Eri = 40 Eri	A	04 11	−07 49	4.44	0.81	5.00	5.96			Triple system. B is white dwarf. Period of BC is about 250 years.
	B			9.64	0.03		11.16	0.44	0.018	
	C			11.05			12.57	0.21	0.43	
Kapteyn = −45° 1841		05 08	−44 59	8.9		3.98	10.9			
HD 36395 = −3° 1123		05 26	−03 42	7.97	1.48	6.13	9.03			
Ross 47		05 36	12 29	11.58		6.10	12.65			
−21°1377		06 06	−21 49	8.18	1.51	5.88	9.33			
Ross 614	A	06 24	−02 44	11.25		4.02	13.23	0.14		Binary with period 16.5 years and separation 1″.
(= CC 390)	B			14.8			16.8	0.08		

Star		R.A. h	m	Dec. °	′							Remarks
Sirius = α CMa	A	06	41	−16	35	−1.46	0.01	2.67	1.41	2.31	1.8	Binary with period 49.7 years and separation 7″.6. Component B is a white dwarf.
	B	06				8.67	0.04		11.54	0.98	0.022	
Wolf 294		06	48	33	24	10.15		5.92	11.29			
Ross 986		07	03	38	43	11.68		5.81	12.86			
Luyten = 5° 1668		07	22	05	32	9.92		3.76	12.02			
Procyon = α CMi	A	07	34	05	29	0.35	0.40	3.48	2.64	1.75	1.7	Binary with period 40.6 years and separation of 4″.5. Component B is a white dwarf.
	B					10.8	0.5		13.1	0.64	0.01	
L745 − 46	A	07	36	−17	10	13.06		6.10	14.14			Binary with separation of 21″. Component A is a white dwarf.
	B					17.6			18.7			
L97 − 12		07	53	−67	30	14.9		5.88	16.1			White dwarf
Ross 619		08	06	09	11	12.88		6.62	13.78			
LFT 571 = L 674 − 15		08	08	−21	15	13.8		6.02	14.9			
53° 1320	A	09	08	53	07	7.68	1.44	6.13	8.74			Binary with period about 1,000 years and separation of 19″.
53° 1321	B					7.77			8.82			
Grm 1618 = 50° 1725		10	05	49	57	6.60	1.37	4.50	8.33			
AD Leo = 20° 2465		10	14	20	22	9.41	1.55	4.72	11.04			
Wolf 359		10	52	07	36	13.66		2.35	16.80			
Lal 21185 = 36° 2147		10	58	36	38	7.47	1.51	2.51	10.42	0.35		Invisible component of small mass.
44° 2051	A	11h	01m	44°	02′	8.76	1.54	5.81	9.94			Binary with separation of 28″.
(= WX UMa)	B					14.8			16.0			
CC 658		11	40	−64	17	12.5		4.93	14.0			White dwarf

TABLE 3-3 (*continued*)

Catalogue designation(s)[b]		1900 Right ascension[c]		Declination		Visual magnitude[d]	B − V[e]	Distance (parsecs)[f]	Absolute magnitude[g]	Mass (☉)[h]	Radius (solar radius)	Remarks[i]
AC 79° 3888		11	41	79	14	10.92		5.05	12.41			
Ross 128		11	43	01	23	11.13		3.36	13.50			
L 68 − 28	A	12	23	−70	56	15.7		6.58	16.6			Binary with separation of 15".
L 69 − 29	B					17.7			18.6			
Wolf 424	A	12	28	09	34	12.63		4.35	14.44			Binary with separation of 0".5.
	B					12.7			14.5			
15° 2620		13	41	15	26	8.49	1.44	4.98	10.01			
Proxima Cen		14	23	−62	15	10.7		1.31	15.1	0.1		This is the nearest star.
−11° 3759		14	29	−12	06	11.38		6.33	12.37			
α Cen	A	14	23	−60	25	0.00	0.69	1.33	4.39	1.09	1.23	Binary with period of 80.1 years and separation 17".7.
	B					1.38			5.76	0.88	0.87	
−20° 4125	A	14	52	−20	58	5.82	1.12	5.81	7.00			Binary with separation of 20".
−20° 4123	B					8.10			9.28			
−40° 9712		15	26	−40	54	10.1		6.02	11.1			
−12° 4523 = CC 995		16	25	−12	25	10.07	1.60	4.10	12.01			
−8° 4352	A	16	50	−08	09	9.72	1.59	6.58	10.63	0.38		Triple system. C separated from AB by 72".
	B					9.8			10.7	0.34		
(= Wolf 629)	C					11.76		6.25	12.78			AB a close pair with period 1.7 year.

Name	Comp	RA h	RA m	Dec °	Dec ′						Notes
+45° 2505	A	17	09	45	50	9.95	0.85	6.25	10.97	0.31	Triple system of 13.1 years and separation of 0″.7.
(= Fu 46)	B					10.31			11.33	0.25	
36 Oph	A	17	09	−26	27	5.07	0.85	5.68	6.31		Separation of AB is 5″.
(= −26° 12026)	B	17	10	−26	24	5.11	1.14	5.81	6.35		Separation of BC is 12′12″.
(= −26° 12036)	C					6.34			7.52		
−46° 11540		17	21	−46	47	9.4	1.5	4.69	11.0		
−44° 11909		17	30	−44	14	11.1		4.78	12.7		
68° 946		17	37	68	26	9.13	1.52	4.93	10.67		
UC 48		17	38	−57	14	12.9		5.99	14.0		
Barnard = 4° 3561		17	53	04	25	9.53	1.75	1.83	13.21		
70 Oph	A	18	01	02	31	4.22	0.87	5.21	5.64	0.89	Binary with period of 87.8 years and separation of 4″.5.
(= 2° 3482)	B					5.94			7.36	0.68	
59° 1915	A	18	42	59	29	8.90	1.54	3.53	11.16		Binary with separation of 17″.
(= Σ 2398)	B					9.69	1.58		11.95		
Ross 154		18	44	−23	54	10.6		2.86	13.3		
4° 4048	A	19	12	05	02	9.13	1.49	5.85	10.29		Binary with separation of 1′14″.
	B					18.0			19.2		
L 347-14		19	13	−45	42	13.7		5.92	14.8		
σ Dra		19	33	69	29	4.69	0.80	5.68	5.92		
Altair = α Aql		19	46	08	36	0.75	0.25	5.05	2.23		
δ Pav		19	59	−66	26	3.56	0.75	5.88	4.71		
−36° 13940	A	20h	05m	−36°	21′	5.33	0.85	5.81	6.51		Separation 7″.
(= HR 7703)	B					11.5			12.7		
−45° 13677		20	07	−45	28	8.0	1.44	6.29	9.0		

TABLE 3-3 (*continued*)

Catalogue designation(s)[b]		Right ascension[c] (1900)		Declination (1900)		Visual magnitude[d]	$B - V$[e]	Distance (parsecs)[f]	Absolute magnitude[g]	Mass ($\odot$)[h]	Radius (solar radius)	Remarks[i]
61 Cyg	A	21	02	38	15	5.20	1.21	3.42	7.53	0.59		Separation 24″.6. Period 720 years.
	B					6.03	1.40		8.36	0.50		
−39° 14192		21	11	−39	15	6.69	1.42	3.91	8.73			
−49° 13515		21	27	−49	26	8.9		4.57	10.6			
ε Ind		21	56	−57	12	4.73	1.05	3.50	7.01			
Kruger 60 = DO Cep (= 56° 2783)	A	22	24	57	12	9.83	1.63	4.00	11.82	0.27	0.51	Binary with period 45 years and separation 2″.4.
	B					11.37			13.36	0.16		
L 789-6		22	33	−15	52	12.58		3.38	14.93			
−21° 6267	A	22	33	−21	08	9.3		4.57	11.0			Separation 23″.
	B					11.0			12.7			
43° 4305		22	42	43	49	10.05	1.39	5.05	11.53			
−15° 6290 (= Ross 780)		22	48	−14	47	10.16	1.60	4.85	11.73			
−36° 15693		22	59	−36	26	7.39	1.50	3.66	9.57			
56° 2966		23	08	56	37	5.58	1.01	6.58	6.49			

Ross 248	23	37	43	39	12.25	1.8	3.16	14.75
1° 4774	23	44	01	52	8.99	1.49	6.13	10.05
−37° 15492	23	59	−37	51	8.59	1.48	4.57	10.29

[a] Adapted from C. W. Allen, *Astrophysical Quantities*, University of London, 1963. His 1973 (revised) edition gives RA and Dec values for the year 1950.

[b] The same star is sometimes listed in more than one catalogue, and these equivalences are given. Notice the peculiar nature of certain of the catalogue descriptions, for example, σ^2 Eridani = 40 Eri.

[c] The stars have been arranged not by consecutive catalogue entries, but by increasing right ascension. Why? Which of these stars would you expect to be in the night sky during the month of March? The right ascension and declination values are for 0.00h, Jan. 1, 1900.

[d] This magnitude is for light in the visible range.

[e] The value of $B - V$ determines the surface temperature in accordance with the equivalences

$B - V$	Surface temperature (K)
−0.2	18,800
0.0	10,800
0.2	8,190
0.4	6,820
0.6	5,920
0.8	5,200
1.0	4,530
1.2	3,920
1.4	3,480

[f] The distances in parsecs may be converted to light-years by multiplying them by 3.26, since 1 parsec = 3.26 light-years.

[g] This is the visual magnitude that the star in question would have if it were situated at a distance of 10 parsecs instead of at the actual distance given in column 6. These absolute magnitudes are thus standardized with respect to each other. Their differences represent inherent differences between one star and another.

[h] Mass values are available for only a few stars. They are given as a ratio to the Sun's mass. Notice that most of the nearby stars are less massive than the Sun.

[i] Notice the rather large fraction of the stars that belong to binary and to triple systems. Notice also that there are seven white dwarfs.

TABLE 3-4
The brightest stars[a]

Name		Right ascension	Declination	Visual magnitude	B − V	Absolute magnitude	Distance (parsecs)
		1900					
		00h 03m	28° 32'				
Alpheratz	α And	00h 03m	28° 32'	2.07	−0.07	−0.5	31
Caph	β Cas	00 04	58 36	2.26	0.34	1.5	14
Ankaa	α Phe	00 21	−42 51	2.37	1.07	0.2	27
Schedar	α Cas	00 35	55 59	2.20	1.16	−1.3	50
Diphda	β Cet	00 39	−18 32	2.04	1.01	0.8	18
Cih	γ Cas	00 51	60 11	2.15	−0.2	−0.9	40
Mirach	β And	01 04	35 05	2.07	1.62	0.2	24
Polaris	α UMi	01 23	88 46	2.02	0.6	−4.5	200
Achernar	α Eri	01 34	−57 45	0.49	−0.17	−2.2	35
Almach	γ And	01 58	41 51	2.16	1.3	−2.3	80
Hamal	α Ari	02 02	22 59	2.00	1.17	0.3	22
Mira	o Cet	02 14	−03 26	2.00	1.5	−1.0	40
Menkar	α Cet	02 57	03 42	2.53	1.16	−1.0	50
Algol	β Per	03 02	40 34	2.10	−0.05	−0.5	31
Mirfak	α Per	03 17	49 30	1.80	0.48	−4.1	150
Aldebaran	α Tau	04 30	16 19	0.80	1.55	−0.8	21
Capella	α Aur	05 09	45 54	0.09	0.81	−0.6	14
Rigel	β Ori	05 10	−08 19	0.11	−0.05	−7.1	270
Bellatrix	γ Ori	05 20	06 16	1.63	−0.22	−4.1	140
El Nath	β Tau	05 20	28 32	1.65	−0.13	−2.9	80
Mintaka	δ Ori	05 27	−00 22	2.19	−0.21	−6.0	450
Arneb	α Lep	05 28	−17 54	2.58	0.22	−4.8	300
Alnilam	ε Ori	05 31	−01 16	1.70	−0.18	−6.8	500
Alnitak	ζ Ori	05 36	−02 00	1.79	−0.21	−6.2	400
Saiph	κ Ori	05 43	−09 42	2.06	−0.16	−7.1	700
Betelgeuse	α Ori	05 50	07 23	0.4	1.85	−5.9	180
Menkalinan	β Aur	05 52	44 56	1.89	0.04	−0.2	26
Mirzam	β CMa	06 18	−17 54	1.96	−0.23	−4.5	200
Canopus	α Car	06 22	−52 38	−0.72	0.16		
Alhena	γ Gem	06 32	16 29	1.93	0.00	−0.5	30
Sirius	α CMa	06 41	−16 35	−1.44	−0.01	1.41	2.7
Adhara	ε CMa	06 55	−28 50	1.48	−0.17	−5.0	200
Wezen	δ CMa	07 04	−26 14	1.85	0.63	−7.0	600

Name		RA (h)	(m)	Dec (°)	(′)	m	B−V	M	Dist.
Aludra	η CMa	07	20	−29	06	2.42	−0.07	−7.1	800
Castor	α Gem	07	28	32	06	1.56	0.05	0.8	14
Procyon	α CMi	07	34	05	29	0.36	0.41	2.7	3.5
Pollux	β Gem	07	39	28	16	1.15	1.01	1.0	10.7
Naos	ζ Pup	08	00	−39	43	2.23	−0.27	−7.3	800
	γ Vel	08	06	−47	03	1.85	−0.25	−4.2	160
Avior	ε Car	08	20	−59	11	1.94	1.2	−3.1	100
	δ Vel	08	42	−54	21	1.93	0.04	0.1	23
Suhail	λ Vel	09	04	−43	02	2.23	1.7	−4.3	200
Miaplacidus	β Car	09	12	−69	18	1.68	−0.01	−0.4	26
Scutulum	ι Car	09	14	−58	51	2.24	0.18	−4.2	180
	κ Vel	09	19	−54	35	2.45	−0.16	−3.0	120
Alphard	α Hya	09	23	−08	14	2.05	1.43	−0.7	35
Regulus	α Leo	10	03	12	27	1.34	−0.11	−0.8	26
Algeiba	γ Leo	10	14	20	21	2.02	1.2	−0.5	32
Merak	β UMa	10	56	56	55	2.36	−0.02	0.6	23
Dubhe	α UMa	10	58	62	17	1.81	1.06	−0.6	30
Zosma	δ Leo	11h	09m	21	04	2.55	0.12	0.8	23
Denebola	β Leo	11	44	15	08	2.13	0.08	1.6	13
Phecda	γ UMa	11	49	54	15	2.43	0.00	−0.1	32
Gienah	γ Crv	12	11	−16	59	2.58	−0.09	−2.4	100
Acrux	α Cru	12	21	−62	33	0.83	−0.26	−3.7	80
Gacrux	γ Cru	12	26	−56	33	1.68	1.58	−2.5	70
Muhlifain	γ Cen	12	36	−48	25	2.16	−0.01	−1.7	60
Mimosa	β Cru	12	42	−59	09	1.29	−0.25	−4.3	130
Alioth	ε UMa	12	50	56	30	1.78	−0.02	−0.2	25
Mizar	ζ UMa	13	20	55	27	2.12	0.03	0.0	26
Spica	α Vir	13	20	−10	38	0.97	−0.23	−3.1	65
	ε Cen	13	34	−52	57	2.34	−0.23	−3.6	150
Alcaid	η UMa	13	44	49	49	1.86	−0.19	−2.3	70
Hadar	β Cen	13	57	−59	53	0.63	−0.24	−5.0	130
Menkent	θ Cen	14	01	−35	53	2.07	1.02	0.9	17
Arcturus	α Boo	14	11	19	42	−0.05	1.24	−0.2	11
	η Cen	14	29	−41	43	2.39	−0.21	−3.0	120
Rigil Kent	α Cen	14	33	−60	25	−0.27	0.71	4.2	1.3
	α Lup	14	35	−46	58	2.5	−0.22	−2.5	100
Izar	ε Boo	14	41	27	30	2.39	0.93	−0.6	40
Kochab	β UMi	14h	51m	74	34	2.04	1.49	−0.6	33
Alphecca	α CrB	15	30	27	03	2.22	−0.02	0.5	22

TABLE 3-4 (continued)

Name		1900				Visual magnitude	B − V	Absolute magnitude	Distance (parsecs)
		Right ascension		Declination					
Dzuba	δ Sco	15	54	−22°	20'	2.32	−0.14	−4.0	180
Acrab	β Sco	16	00	−19	32	2.52	−0.09	−4.0	200
Antares	α Sco	16	23	−26	13	0.94	1.83	−4.7	130
	ζ Oph	16	32	−10	22	2.56	0.00	−3.4	160
Atria	α TrA	16	38	−68	51	1.93	1.43	−0.4	29
	ε Sco	16	44	−34	07	2.29	1.15	0.6	22
Sabik	η Oph	17	05	−15	36	2.44	0.05	0.8	21
Shanla	λ Sco	17	27	−37	02	1.60	−0.23	−3.2	90
	θ Sco	17	30	−42	56	1.86	0.38	−4.0	150
Ras-Alhagne	α Oph	17	30	12	38	2.07	0.15	0.9	17
	κ Sco	17	36	−38	59	2.39	−0.21	−3.3	140
Eltanin	γ Dra	17	54	51	30	2.21	1.54	−0.8	40
Kans Australis	ε Sgr	18	18	−34	26	1.81	−0.02	−1.7	50
Vega	α Lyr	18	34	38	41	0.03	0.00	0.5	8.1
Nunki	σ Sgr	18	49	−26	25	2.09	−0.20	−2.4	80
	ζ Sgr	18	56	−30	01	2.57	0.09	−0.4	40
Altair	α Aql	19	46	08	36	0.77	0.22	2.3	4.9
Peacock	α Pav	20	17	−57	03	1.94	−0.20	−2.9	90
Sadir	γ Cyg	20	19	39	56	2.22	0.66	−4.8	250
Deneb	α Cyg	20	38	44	55	1.25	0.08	−7.2	500
Gienar	ε Cyg	20	42	33	36	2.46	1.03	0.6	24
Alderamin	α Cep	21	16	62	10	2.43	0.23	1.5	15
Enif	ε Peg	21	39	09	25	2.38	1.56	−4.6	250
Al Na'ir	α Gru	22	02	−47	27	1.75	−0.14	−0.2	25?
	β Gru	22	37	−47	24	2.16	1.62	−2.6	90
Formalhaut	α PsA	22	52	−30	09	1.16	0.09	1.9	7.0
Scheat	β Peg	22	59	27	32	2.50	1.7	−1.4	60
Markab	α Peg	23	00	14	40	2.49	−0.04	0.0	32

[a]Adapted from C. W. Allen, *Astrophysical Quantities*, University of London, 1963. Footnotes c, d, e, f, and g of Table 3-3 also apply to the equivalent columns in this table. It might at first sight be expected that the brightest stars would be the nearest ones, but a comparison of this table with Table 3-3 shows this not to be true. Only four stars appear in both tables. A similar assumption is often made concerning radio sources, and this could also be false. Many of the stars in this table were named by Arabian astronomers, but some were not. Which were not? Which five stars are intrinsically the brightest?

1. How is temperature measured in the absolute (K) scale?

2. Draw the Planck curve for a star of surface temperature 20,000 K, stating the unit you use for the frequency scale, and adjusting the energy scale so that the area under the curve gives the power radiated by 1 square kilometer of surface of the star.

3. After drawing the curve of Problem 2, mark the frequency range of visible light.

4. An experimenter analyzes the radiation of a hot gas with respect to its constituent frequencies. Would you expect emission lines or absorption lines to be found?

5. Radiation from a hot solid surface passes through a cool gas. When analyzed with respect to frequency, would you expect the radiation to contain emission lines or absorption lines?

6. In which spectral class are absorption lines of the Balmer series most prominent?

7. For which spectral classes are the H and K lines of singly ionized calcium atoms (one electron made free) very prominent?

8. In what spectral class are the lines of singly ionized helium most clearly seen?

9. Are stars forming at the present time? If so, in what kind of object?

10. A condensing spherical cloud fragments into 1,000 stars, the mass of each being, on the average, equal to that of the Sun. It does so at the stage where the average density within it is 10^{12} atoms per cm^3. What is the radius of the cloud at this stage? By about how much must each of the resulting 1,000 protostars condense further before normal stars are formed?

11. Once radiation can no longer escape readily from its interior, a condensing gas cloud becomes hotter as it shrinks. Give an example to illustrate this effect.

12. Why do condensing protostars tend to have surface temperatures of about 4,000 K?

13. What physical condition determines that a star has formed?

14. Describe the Hertzsprung–Russell diagram.

15. In a sketch of the Hertzsprung–Russell diagram, indicate the position of the sequence on which newly formed stars are found to lie. What criterion determines where on this sequence a particular star lies?

16. What is the approximate mass range of stars on the zero-age main sequence? Give the approximate relation between luminosity and mass for such stars.

Chapter 4
Radio Astronomy

§4-1. A Matter of History

In the previous chapter we met the idea of a bandwidth, the small range of frequencies between the adjacent values v and $v + dv$ so that dv was the amount of the variation over the range. We saw that when a temperature could be assigned to matter, and when there were sufficient absorptions and emissions of quanta with such frequencies, then the energy distribution of the radiation emerging from the matter into space had the form of a black-body curve. At the low frequencies of radio waves, there is not much energy in a black-body curve unless the temperature is enormously high. For this reason, astronomers thought for a long time (until the late 1940s) that it would not be worthwhile to attempt to observe the universe by means of radio waves even though the Earth's atmosphere was transparent for radio waves (see Figure 3-12). Consequently, most of the fundamental discoveries of radio astronomy were made from outside astronomical observatories. Before World War II, they were made more or less accidentally by those engaged in electronic investigations of a commercial nature. After the war, they were made in physics laboratories by scientists who had been engaged in the invention and production of radar equipment. What astronomers had overlooked was the fact that situations such as the one mentioned in connection with Figure 2-8 might exist, situations with low densities of fast-moving electrons. With the electrons spiraling in magnetic fields, very many quanta would be emitted in each radio bandwidth dv, permitting them to be detected by ordinary radio methods (see Figure 2-2).

Such situations are far removed from the closed-box arrangement discussed in Chapter 3, the fast-moving electrons being able to radiate freely into space. It is curious that even after this mistake had become apparent, the astronomical world could not rid itself of its preoccupation with black bodies. The authors of papers on radio astronomy continued to write about a so-called "brightness temperature," the temperature that a fictitious black body would need to possess in order to emit as much radio intensity as a real astronomical object was observed to do over some particular bandwidth. This meaningless practice has been discontinued since it has no physical significance except in very rare cases.

Suppose we write $I\,dv$ for the energy flux of radiation in the bandwidth dv that is received from an astronomical object, which might be a cloud within our galaxy, a different distant galaxy, or a quasar. By *flux* we mean the rate at which energy crosses unit area of a sphere centered at the object, with a radius equal to our distance from the object (see Figure 3-11). In general, the value of I will depend on the particular frequency v at which the bandwidth is located. This dependence can be denoted by $I(v)$; we can see that I changes when v changes. The detailed form of $I(v)$ plotted as a graph against v is still called the *spectrum* of the object, just as it was in the black-body case. For many objects, $I(v)$ is approximately proportional to a power of v, $v^{-\alpha}$, with α a constant. The minus sign appears so that α is positive for most sources (at any rate, for most of the sources discovered in the beginning). When $I(v)$ can be described in this way, the source is said to have a power-law spectrum with *index* α. With α positive, the intensity declines toward the higher frequencies. For those sources in which α turns out to be negative, the intensity increases toward the higher frequencies. Such sources are said to have *inverted spectra*. More and more sources of radio waves with inverted spectra are now being discovered.

To return to our earlier discussion, in 1930 the Bell Telephone Laboratories at Holmdel, New Jersey, were concerned with building ship-to-shore communications equipment at a frequency of 2×10^7 cycles per second. The question arose as to the sources and the nature of external interference that might prove a nuisance for the operation of this equipment. Radio static from thunderstorms was an obvious example of the interference that might be expected. Karl Guthe Jansky was to investigate the sources of interference, and for this purpose he built a rotating antenna 30 m wide and 4 m high. Through 1931, Jansky continued his work. By 1932, he was able to report that, in addition to the loud static of local storms and the persistent cracklings of distant storms, there was a steady hissing sound on his headphones. The hiss was variable as the antenna rotated. It so happened, when the experiments first began, that the Sun lay more or less in the direction from which the hiss was loudest. It seemed natural, therefore, to think the Sun must be the source of the hiss. Instead of becoming fixed in this point of view, Jansky continued to listen. As the months went by, the Sun moved gradually away from the region of the strongest hiss, which therefore came, not from the Sun, but from an area fixed on the sky. That is to say, the hiss came from an area fixed on the star maps of Figure 3-24. The question was: From where on the maps did it come? By 1933, Jansky had satisfied himself that the region of strongest hiss lay in the direction of the center of our galaxy. The source was therefore on an awesome cosmic scale that not even the resources of the mighty American Telephone and Telegraph Company could remove. *The New York Times* carried a front page report on Jansky's discovery under the headline, "New Radio Waves Traced to Center of Milky Way." One reporter, after listening to the radio waves, wrote that they sounded "like steam escaping from a radiator." To the scientific world of 1933, still immersed in the aftermath of the upheaval caused by quantum mechanics, and even then anticipating the rise of nuclear physics, Jansky's discovery of a hiss that sounded like steam escaping from a radiator seemed but a small thing. Yet it proved to be one of the major discoveries of the twentieth century.

It is interesting that a second major discovery came in 1965 from the Bell Laboratories at Holmdel, again from research for a commercial enterprise. A. A. Penzias and R. W. Wilson were asked to investigate sources of noise interference, this time for ground-to-satellite equipment. As in Jansky's case, they found a source that was neither local nor associated with their experimental antenna or radio receiver. Once again, the source came from the sky. This time, unlike Jansky's case, the source was smoothly distributed over the sky (which made it harder to detect). These smoothly distributed radio waves are now known as the *microwave background*. The significance of this second discovery was immediately appreciated, a significance that goes to the very roots of theories of the structure of the whole universe, as we shall discuss in detail in Chapters 12 to 14.

In retrospect, it is strange that no observatory or university followed up on Jansky's discovery. The next step was taken by a young radio engineer, Grote Reber, who built his equipment at his own expense in his own back garden. Without the resources available to Jansky, Reber could not construct a rotating or a steerable antenna, so he built a larger one. With it, by the year 1944, he had clearly demonstrated that radio waves were coming from the plane of the Milky Way, with the greatest concentration toward the center in the constellation of Sagittarius, but with subsidiary concentrations in Cygnus and Cassiopeia. The remarkable map that Reber constructed is shown in Figure 4-1.

A strange aspect of radio astronomy to this point is that the discoveries had all come when they were least expected. Yet no one had been able to detect the Sun as a source of radio waves, even though everyone had expected for 50 years or more that it was such a source. How the Sun came to be detected in wartime Britain is an interesting story.

A large fraction of Britain's professional physicists had been drafted in 1939 into the radar field. With such a heavy concentration of intellect on one topic, it was natural that most of the possible ideas for using radar had been conceived by 1942. (Techniques were sometimes too crude in 1942 for these ideas to be made operational.) One simple idea involved the use of a profusion of thin metal strips

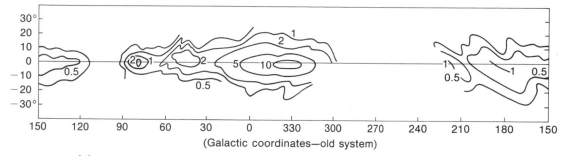

FIGURE 4-1

Grote Reber's first map of galactic radio emission. The map demonstrates the concentration of the emission toward the galactic plane, that is, toward latitude 0°. (After *Astrophysical Journal*, 100, p. 279, 1944).

to produce ghost radar echoes, thereby simulating the effect, when distributed from a single airplane, of a large group of attacking planes. This elementary concept became the subject of an intense power struggle between those who were concerned with attacking Germany and those who were responsible for the defense of Britain. The attack group wanted to use the metal strips (*Window* was the strips' code name) to confuse the German radar, whereas the defense was acutely worried that, if the idea ever became known, the Germans would instantly use it against the British defensive screen. Against this background, reports of strong ghost echoes from Army radar field stations started a panic at the London headquarters of all the armed services. Maybe the Germans had *Window* already?

J. S. Hey was the young man sent by the Army Operations unit to investigate. Fortunately, Hey had the right idea from the beginning. He quickly noticed that the radar ghost echoes had coincided in time with the appearance of an exceptionally large and active sunspot close to the center of the solar disk. Thus, by the time other scientists told him that the Sun did not produce radio waves (at the operating frequencies of the Army radar equipment, about 5×10^7 cycles per second), Hey knew the correct answer to this criticism: there had been no suitably placed large, active sunspots when the earlier attempts to detect radio waves from the Sun had been made.

It is worth a short digression here to discuss the nature of sunspots. They wax and wane, both in their sizes and numbers, in a somewhat irregular cycle of about 11 years. Figure 4-2 shows a typical situation. Sunspots were discovered by Galileo Galilei, (1564–1642) using his newly constructed telescope (see Figure 4-3). Under favorable conditions, large sunspots are easily visible to the naked eye, so it is surprising that their discovery had to await the dawn of modern science. On two occasions one of the authors has observed such naked-eye spots, even from the astronomically unfavorable climate of the United Kingdom. Naked-eye spots are best seen near sunset, when absorption in the atmosphere cuts the normal glare of the Sun down enough so one can look

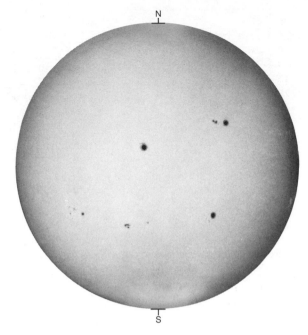

FIGURE 4-2
Viewed in full sunlight, spots can be seen on the Sun, contradicting the ancient belief that the face of the Sun must be perfect. The diameter of the Sun is about 1.4×10^6 km. (Courtesy of Dr. N. Sheeley, NOAA Space Environment Laboratory, Boulder, Colorado.)

FIGURE 4-3
Galileo's telescope. (Courtesy of Prof. ssa. Dott. ssa. Maria Luisa Righini-Bonelli, Istituto e Museo di Storia della Scienza, Firenze.)

directly at the solar disk. Although this condition is realized quite often, usually the atmosphere destroys clarity of definition for low angles of elevation, leaving the setting Sun as a "boiling" image. The essential condition for large sunspots to be seen by the naked eye is that the atmosphere be absorptive but without distortion, and this condition is encountered only rarely. Nevertheless, naked-eye spots must have been seen long before Galileo, probably by thousands of people. The surprise is that their existence does not seem to have been recorded by monks or court chroniclers.

Sunspots are dark only in contrast with the brilliance of the surrounding regions of the solar surface. The temperature within them is actually about 4,000 K, considerably hotter than most industrial furnaces. The surrounding regions have temperatures of about 5,800 K, however. Since the emission from a hot surface goes as the *fourth power* of the temperature (T^4), the surrounding regions are appreciably brighter to the eye.

To return to radio waves from the Sun, a few months after Hey's investigation of the British Army ghost echoes in 1942, G. C. Southworth at the Bell Telephone Laboratories also obtained Jansky's characteristic hiss from the Sun, using radio equipment that operated at a hitherto unprecedentedly high frequency of about 10^{10} cycles per second. Whereas sunspots had been the causative agent in Hey's case, Southworth detected the normal solar black-body emission for a surface temperature of about 6,000 K. Black-body emission is much stronger at Southworth's high frequency than it is at the lower frequencies that had been previously used. While this further detection of the Sun as a source of radio waves was an additional landmark in radio astronomy, especially in its use of a high radio frequency, Hey's work had the more interesting implication by demonstrating unequivocally that radio sources could emit intensities vastly higher than black-body calculations suggested.

Four years later, in 1946, Hey made another fundamental discovery. Hey, together with S. J. Parsons and J. W. Phillips, set out to examine the emission of radio waves along the Milky Way in more detail than Reber had been able to do. They found a fluctuating source in the constellation of Cygnus. This source came from a small intense patch of the sky, and it became known as Cygnus A. Unfortunately the antennas then available could not pinpoint the position of Cygnus A with anything like the precision needed to make an identification with an optically visible object. It was not until 1951 that F. G. Smith at Cambridge obtained the radio position with sufficient accuracy for Walter Baade at the Mount Wilson and Palomar Observatories to make an identification of Cygnus A. Baade's historic photograph is shown in Figure 4-4. Cygnus A was a distant galaxy, a galaxy that we would describe today as being disturbed, but which Baade then thought to be a collision of two galaxies. What Hey, Parsons, and Phillips had discovered was the first *radio galaxy*.

Perhaps the first person in the recognized astronomical community to take Jansky's discovery seriously was Jan H. Oort, Director of the Leiden Observatory, Holland. Yet, like other astronomers, Oort felt that a general hiss that was not characteristic of any particular frequency could never be of much help in the

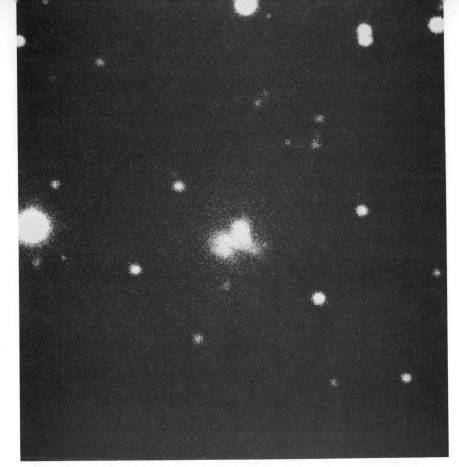

FIGURE 4-4
The radio source Cygnus A, a very intense source. (Courtesy of Hale Observatories.)

study of stars and gas clouds. But, if radio emission belonging to a particular frequency could be found, the situation would be different. Oort gave the problem of making a search for possible radio spectrum lines to a young student, H. C. van de Hulst, as the topic for his doctorate. By 1944, van de Hulst had concluded that a transition of hydrogen atoms at a frequency of 1.42×10^9 cycles per second (the transition noted in Figure 2-19) was the most likely possibility. In 1945, he concluded further that a search for this emission from clouds of hydrogen in our galaxy would be technically feasible.

After an agonizing period as a hostage in the hands of the Gestapo, Oort returned in 1945 to his position at the Leiden Observatory, and, together with C. A. Muller, he set about testing van de Hulst's prediction. The project moved slowly to begin with because of the shortage of high-grade electronic equipment in postwar Holland. Then the work was delayed seriously by a disastrous fire. Thus, it was not until 1951 that Oort and Muller at last succeeded in detecting the so-called 21-cm line of atomic hydrogen (21 cm is the wavelength $\lambda = c/\nu$

corresponding to $\nu = 1.42 \times 10^9$ cycles per second). A week or two earlier, H. L. Ewen and E. M. Purcell had also detected the 21-cm line from gas clouds in the galaxy. The results of these two independent groups were published contemporaneously in Volume 168 of *Nature*. Shortly afterward, there was another detection of this radio emission line by W. N. Christiansen and J. V. Hindman in Australia.

This work caused a sensation in the astronomical world. For the next decade or so, much attention was payed to delineating, with the aid of the 21-cm radiation, the distribution in a spiral arm pattern (Figure 4-5) of the atomic hydrogen in our galaxy. In one important sense, results like those shown in Figure 4-5 were misleading because the unsubstantiated view that *all* of the un-ionized hydrogen had thereby been detected became widespread. This mistaken view was a serious impediment to studies of the cloud structure of interstellar gas and to the study of star formation within clouds. The true importance of the 21-cm line was that it foreshadowed the observation of a wide range of radio spectrum lines having frequencies mostly higher than 3×10^{10} cycles per second—that is, with wavelengths in the millimeter range. As we shall find in the next chapter, the new topic of millimeter-wave astronomy now almost rivals all the rest of radio astronomy in importance.

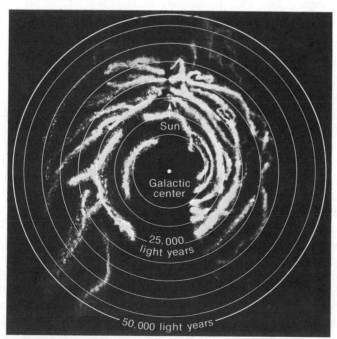

FIGURE 4-5
The lanes of gas in our galaxy, determined by radio astronomy.
(Courtesy of Leiden Observatory.)

Immediately following World War II, two important university research groups in radio astronomy were started in Britain, one at the Cavendish Laboratory of Cambridge University under M. Ryle and the other at Manchester University under A. C. B. Lovell. In Australia, the Radiophysics Department of the Commonwealth Scientific Industrial Research Organization (CSIRO) under E. G. Bowen devoted a large fraction of its time and effort to radio astronomy. In the United States, a start was made at the Naval Research Laboratory, and Harvard University under B. J. Bok was an early entrant in the field, but it was not until almost a decade later that radio astronomy began to gather momentum in the United States. Indeed, it was not until the early 1960s, 30 years after Jansky's discovery, that the National Radio Astronomy Observatory (NRAO) became a dominant organization in the field.

The first identifications with optically visible objects in the galaxy were made in 1948 and 1949. In Cambridge, F. G. Smith and M. Ryle identified the radio source that had become known as Cassiopeia A with a network of gaseous filaments, while at CSIRO J. G. Bolton, G. J. Stanley, and O. B. Slee identified the source Taurus A with the Crab nebula (Figure 2-7). The latter identification has turned out to be crucial to all subsequent astronomy; the Crab, as it is now familiarly called, will be discussed later in a special section. Bolton and his colleagues made a second epoch-making identification, that of the radio source Centaurus A with a galaxy having the catalogue number NGC 5128—the galaxy shown in Figure 4-6. Two years later, R. Hanbury Brown and C. Hazard detected radio waves from the nearest large galaxy to our own, galaxy M31 in the constellation of Andromeda, shown in Figure 4-7. The central regions of M31 can easily be seen by naked eye. Assuming that our own galaxy were at the distance that M31 is, our galaxy would be detectable at much the same level of intensity as M31 itself, a level typical for a normal galaxy. The situation was quite different for NGC 5128, however; it was both 10 times more intense as a source and 10 times more distant than M31, making NGC 5128 intrinsically 1,000 times more powerful than M31 or our own galaxy.

By 1951, about 100 radio sources had become known. A radio source is a small patch on the sky from which the radio waves are more intense than they are in the surrounding sky. It is thus a kind of island of high emission intensity on the sky. A few dozen of the sources were positioned along the Milky Way, and these were clearly sources in our own galaxy. Examples are the Crab nebula and the network of wisps that constitutes Casseopeia A; these sources are now known to be the remnants of supernovae (see Chapter 8). The other sources were distributed more or less uniformly over the sky, and because of this uniformity it was generally agreed that these other sources had to be either very close to the solar system, like nearby stars, or very distant, far outside our galaxy. Opinion among astronomers favored heavily the former possibility, and these sources were thus known as radio stars. This opinion was challenged by T. Gold, who maintained (with apparently the sole support of one of the authors) that the sources were distant galaxies. The argument against distant galaxies, put forth primarily by those who were beginning to catalogue the radio

FIGURE 4-6

The system of NGC 5128. This is a giant elliptical galaxy with a diameter of about 100,000 light-years, out of which a vast cloud of gas and dust seems to have emerged. (Courtesy of Hale Observatories.)

sources, was that the catalogue would not be understandable unless there were exceptional galaxies having radio emissions on the order of a million times greater than a normal galaxy like our own. But since NGC 5128 was already a thousand times greater, this objection did not seem overriding. Nor was it correct. When, a year or so later, Baade identified the source Cygnus A as the distant galaxy of Figure 4-4, it was clear that Gold's point of view was correct. This particular galaxy was then seen to be some 10 million times more intense in its radio emission than our own galaxy, a circumstance that is evidently related to its disturbed condition.

By the middle 1950s, radio astronomy was emerging from its golden age, an age in which almost every intelligently designed observation yielded an important discovery. There were still two major discoveries to be made: (1) quasars by C. Hazard and M. Schmidt in 1963 (although A. R. Sandage and R. Hanbury Brown had been close to this discovery a year or two earlier) and (2) pulsars by S. J. Bell and A. Hewish in 1967.

FIGURE 4-7
The galaxy M31, about 100,000 light-years in diameter, situated in the constellation of Andromeda. This galaxy, sometimes referred to as the Andromeda nebula, is the nearest of the large galaxies. The center regions are visible to the naked eye. (Courtesy of Hale Observatories.)

Radio astronomy was now enlivened by a spirited debate as to the form that new equipment should take. There was a split between two groups, a split foreshadowed by the different antenna systems built by Jansky and Reber, the one smaller and movable, the other larger and fixed. The split was well exemplified in Britain, with Ryle at Cambridge moving to larger and larger fixed systems and Lovell at Manchester seeking to build huge, steerable parabolic metal mirrors. The justly famous 250-ft radiotelescope at Jodrell Bank, completed around the end of 1957, set the stage for all the large, steerable radiotelescopes of the future. By a happy choice, Lovell had hit on just about the largest size that can be achieved without incurring the penalty of a rapid escalation of engineering problems. To this day, only the 100-m steerable telescope at Bonn, Germany, shown in Figure 4-8, exceeds the Jodrell Bank instrument in scale. A similar dichotomy existed in Australia, with a group led by B. Y. Mills preferring large fixed instruments, and others including J. G. Bolton, preferring steerable radiotelescopes.

FIGURE 4-8
The 100-m radiotelescope of the Max Planck Institut für Radioastronomie, Bonn, Germany. This is the largest fully steerable radiotelescope yet constructed. (Courtesy of Max Planck Gesellschaft.)

Two arguments were pressed for steerable instruments, only one of which was to prove correct. The incorrect argument was that steerable instruments would be better able to determine the exact positions of sources on the sky. This had seemed to be so during the golden age, but the large fixed systems, developed particularly by Ryle at Cambridge, eventually proved capable of the most precise positions for radio sources that rivaled or even exceeded the accuracy of measurement of the optical position of a star. The correct argument was the one pressed most strongly by Bolton, namely, that steerable instruments would be better adapted than large fixed ones to operate at very high frequencies. However, the technology of building radio receivers at high frequencies was then (in the middle 1950s) rather crude. Thus, it was argued that any advantage gained on the antenna would be lost on the receivers. In reply to this, Bolton argued that receiver technology would inevitably improve, as indeed it did, dramatically. The shift toward high frequencies was initially intended to reach about 3×10^9 cycles per second, but, as time went on, the highest frequencies attainable by the radiotelescopes of the 1960s were about 10^{10} cycles per second. This was just sufficient for radio astronomy to cross the threshold into *millimeter-wave astronomy*, which we discuss in the next chapter. It was not quite in the true millimeter range of wavelengths, but it was adequate to give ample hints of the rich field that awaited the first investigators to build equipment that would range up to a frequency of 1.5×10^{11} cycles per second (wavelength $0.2 \, \text{cm} = 2 \, \text{mm}$). It was to be the entry into a new golden age.

§4-2. The Crab Nebula

The Crab nebula, the blue inner portion shown in Plate I, sets the basic theoretical problem of radio astronomy. What is the source of the intense radio waves and of the light that comes from this inner part of the nebula? The ideas we discuss here were contained in a book by G. A. Schott (*Electromagnetic Radiation,* Cambridge University Press, 1912). After World War II, nuclear physicists became interested in the construction of a laboratory accelerator known as the *synchrotron* and found that fast-moving electrons circulating in a magnetic field lose a good deal of their energy through the emission of light. The problem was restudied (although it had been treated by Schott) by L. A. Artsimovich and I. Y. Pomaranchuk in the USSR (1946) and later by J. Schwinger (1949) in the United States. Because papers in Soviet scientific journals were not then translated into English, Schwinger's work had the main impact on radio astronomers in Europe and America. The work led to the suggestion in 1950 by H. Alfvén and N. Herlofson that the process first considered by Schott might be responsible for the origin of radio waves from stars. (Recall that in 1950 astronomers believed most radio sources were stars.) Almost contemporaneously, K. O. Kiepenheuer in Germany suggested that the Schott process, or synchrotron process as it had become called, might be responsible for the origin of radio waves from the Milky Way, that is, for the hiss discovered by Jansky.

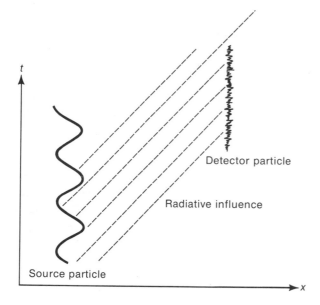

FIGURE 4-9
The motion of a detector particle is complex, its oscillations being much more frequent than those of the source particle.

FAST-MOVING PARTICLES IN A MAGNETIC FIELD
EMIT RADIATION OVER A WIDE RANGE OF FREQUENCIES

Let us begin our discussion of the synchrotron process by recalling that the simple radiation effect of an oscillating electric particle (as we studied it in Chapter 2, Figure 2-2) was always contingent on the source particle moving much more slowly than the speed of light (300,000 km/second). When this is no longer true, we have the more complex situation of Figure 4-9, in which the detector particle no longer has a well-defined frequency equal to that of the source particle. How, then, do we cope with the situation of Figure 4-9?

One hundred and fifty years ago, the French mathematician Jean B. J. Fourier (1768–1830) proved that the complex path of the detector particle in Figure 4-9 could be regarded as a mixture of simple oscillations. Actually, Fourier determined the mixture of oscillations, each with a definite frequency ν, that, when *added together,* would give precisely the same motion as in Figure 4-9. This mathematical problem is important because an electron with the complex path of Figure 4-9 has the same electrical effect as the sum of the oscillations into which it can be separated by the Fourier method.* Since we know how to cope with a simple oscillation, we can cope with the motion of Figure 4-9 by the addition of many such oscillations.

---

*Notice that it is the physical behavior of radiation that makes Fourier's mathematical problem useful. An electric particle in complex motion behaves like the summation of simple oscillatory motions. A similar method *cannot* be used, however, for the gravitational effect of a particle with complex motion. Thus, in Einstein's theory of gravitation, the effect of a particle with a complex path is *not* the same as the sum of the oscillatory motions into which it can be resolved by the method of Fourier. Thus, the radiative and the gravitational effects of a moving particle behave quite differently; gravitation is said to be *nonlinear,* whereas radiative effects are *linear.*

The situation is therefore that fast-moving source particles, as in Figure 4-9, cause detector particles to have oscillations (in the sense of Fourier) that cover a wide range of frequencies. As with the radiation from a black body, it is not possible to use a tunable receiver to detect a unique oscillation frequency of the detector particles. The strength of the receiver reponse will change slowly as we vary the tunable frequency ω. In other words, we will have the impression of a continuum of radiation, as we do for a hot gas, even though the oscillation frequency ν of the source particles is well defined.

At first sight, we might expect to obtain the strongest response in the receiver when ω is reasonably close to ν, and we do when the speed of the source particle is only moderately close to the speed of light. On the other hand, when the speed of the source particle becomes very close to that of light, the strongest response occurs in the receiver at values of ω that are *very much larger than ν*. It is possible for ν to be a low frequency, say, 1,000 oscillations per second, and yet for the stongest receiver response to occur at ω of order 10^{15} oscillations per second—that is, at the frequencies of visible light. A situation of this kind is believed to occur in the Crab nebula, shown in color in Plate I. Even though ν is estimated to be about 1,000 oscillations per second, the blue light from the inner part of the Crab is generated in this way (but not the outer red light, which is *line radiation* from neutral hydrogen atoms). Indeed, response signals are observed from the Crab even at frequencies in the x-ray and γ-ray ranges, from 10^{18} to 10^{21} oscillations per second.

Particles moving at speeds close to that of light are not easily deflected from their paths by collisions with other particles. Thus, at first we might expect the high-speed electrons in the Crab nebula to move in straight paths. This would mean no oscillation and therefore no radiative interaction. So what causes these electrons to oscillate? The answer is that the Crab nebula contains a magnetic field that forces the electrons to move along helical paths of the form of Figure 4-10. Notice that in Figure 4-10 we have returned to a plot of the electron path with respect to the spatial dimensions x,y,z, and that the magnetic field is considered to have the direction of the z dimension. If we now follow a simple method whereby the behavior with respect to the time t is specified in three separate diagrams, one of x versus t, one of y versus t, and the third one of z versus t, we have the situation of Figure 4-11. Evidently both the x and y dimensions have the required oscillatory behavior, both with the same oscillatory frequency. Collapsing the x,y dimensions into a single dimension, we thus arrive at the picture of Figure 4-8. The behavior with respect to the z dimension, corresponding simply to uniform motion, has no effect on the radiative problem and does not need to be considered in any detail.

Hence, we have answered the basic and crucial question of the source of the blue light from the Crab nebula and also of the source of γ rays and x rays from the Crab. It is the effect of a magnetic field within this remarkable object that leads to the electron oscillations that then generate these radiative interactions. From detailed calculations, it is found that a magnetic field with an intensity of about one-thousandth of the Earth's magnetic field will suffice to produce the required electron oscillations. At first this might seem a rather weak field, but it

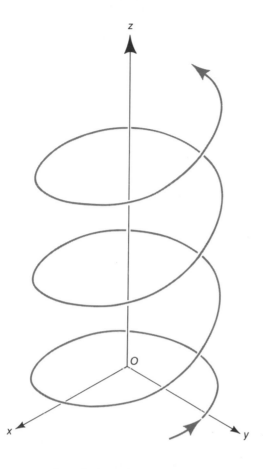

FIGURE 4-10
Electrons in the Crab nebula, moving at speeds close to that of light, follow helical paths because of the presence of a magnetic field.

must be remembered that the magnetic field of the Crab is maintained throughout a region with a diameter of the order of 10 light-years, and this dimension is enormous compared with the scale of the Earth's magnetic field, with the Sun's magnetic field, or indeed with the scale of the magnetic field of any ordinary star.

In Chapter 8 we shall discuss exploding stars known as supernovae. The Crab nebula is believed to be the remains of a supernova that occurred nearly a thousand years ago. The outer material, the red filaments of Plate I, are observed today to be in rapid outward expansion. When traced backward in time, this outward expansion suggests that an explosion from some centrally located object occurred about a thousand years ago. Such an argument does not, of course, give a precise date for the explosion of the supernova. Yet, a precise date is available from the records of Chinese astronomers, who observed the light that accompanied the explosion. The supernova was seen on July 4 in the year AD 1054 as a "guest star visible by day like Venus." It remained as bright as Venus until July 27. Thereafter, it faded gradually, becoming invisible to the naked eye by April 17, 1056, almost two years after the outburst. Moreover, the day-by-day variations estimated by the Chinese agree very well with the behavior of

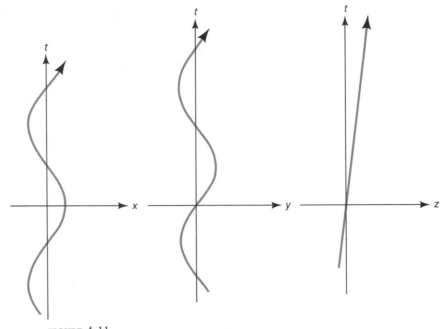

FIGURE 4-11
The (x, t), (y, t), (z, t) diagrams for a magnetic field aligned in the z direction.

supernovae observed today. The position in the sky given by the Chinese astronomers for their guest star agrees, within expected uncertainties, with the present-day position of the Crab nebula.

It is curious that no similar accounts have been found in European or Arabic records. An apparition in the sky at least as bright as Venus can scarcely have been missed. It remained exceedingly bright for about a month, and it is hard to believe that European, North African, and Arabian skies could have been cloud-covered for as long as this. We must suppose that the supernova was seen by millions of people. Most would be illiterate and therefore unable to record it. What the specialized class of chroniclers chose to record seems to have turned on their beliefs. In China, it was thought that terrestrial events could be foretold from occurrences in the sky, and, therefore, peculiarities in the sky were avidly searched for and noted. In Europe, on the other hand, the monkish chroniclers believed the heavens to be the personal handiwork of God, and therefore perfect and not subject to change. Any record to the contrary would have been heretical and would have provoked an angry response from theologians and philosophers, just as did Galileo's discovery of spots on the Sun more than five centuries later.

Two pictographs shown in Figure 4-12 have been reported by William C. Miller. They were discovered at different sites during exploration of the cave dwellings of the Pueblo culture of the North American Indians, a culture that spanned the year AD 1054. Miller believes the two pictographs refer to a moment

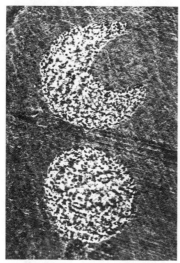

FIGURE 4-12
Two Pueblo Indian pictographs. The one on the left is from Navajo Canyon, and the one on the right is from White Mesa. (Courtesy of William C. Miller.)

when the Moon, in its monthly motion across the sky, approached close to the guest star of the Chinese. Another interpretation might be that the object close to one of the horns of the crescent is the planet Venus, but, as Miller points out, such a close approach of Venus to the Moon occurs every two or three years. The fact that only two among very many pictographs have ever been discovered with an undoubted astronomical connotation, namely, the two reproduced in Figure 4-12, suggests that the Pueblo Indians were not usually interested in astronomical events and that only a quite uncommon circumstance would have impelled them to make these remarkable visual records. What else but the supernova of AD 1054? But did the Moon really approach so close to the supernova, and did it do so for an observer in Arizona, but not for an observer in China? Otherwise, the Chinese astronomers would almost surely have recorded the event. Miller found the answers to both questions to be affirmative, thereby providing strong confirmation that these two pictographs really do refer to the supernova of 1054. The chroniclers of Europe were too prejudiced to record the event, but not so the Indians of North America. Unable to write, they nevertheless managed to leave a record.

You will note that the two pictographs of Figure 4-12 are different, being reflections of each other in a vertical axis. Miller notes that the artists chipping out the separate pictographs would have had their backs turned to the sky and would thus have experienced a left-right ambiguity. Does one mentally invert the left-right sense or not? Imagine yourself looking over your shoulder at the horns of the Moon. Would you draw the Moon with the actual left-right sense, or would you draw it as if you turned around full face toward the sky?

§4-3. A Personal Recollection

The ideas discussed here concerning the nature of the Crab nebula were formulated in the decade from 1950 to 1960. Although astronomers became convinced of their correctness, a big mystery remained. Radiative interactions cause source particles to lose energy, an effect illustrated in Figure 4-13. Both the amplitude and the frequency of oscillation become less. This energy loss is particularly marked for electrons moving at speeds close to that of light, as the electrons in the Crab nebula are thought to do. It can be calculated that the electrons responsible for the blue light of Plate I would lose most of their energy in only a few years; yet the Crab nebula was about 900 years old. High-speed electrons produced in the explosion observed by the Chinese in 1054 would long ago have lost their energy and hence their ability to radiate visible light. So where did the electrons responsible for Plate I come from, if not from the actual explosion of the supernova?

The late Walter Baade, who, up to the time of his death, was responsible for much of the optical work on the Crab nebula, had identified a peculiar yellow star near its center, which he believed might have been responsible for the explosion of 1054. He had also detected moving ripples in the nebulosity itself. These and other matters connected with the Crab were topics of discussion at a conference of astronomers in Brussels in the summer of 1958. In an outside-the-conference-room talk with Jan Oort from Leiden, Holland, it was speculated that Baade's star might be showing fast light variations, and one of the authors

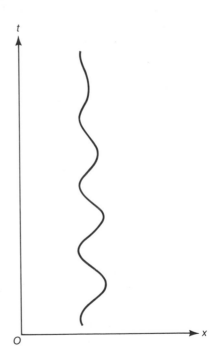

FIGURE 4-13
A radiating particle loses energy. For an electron moving in a magnetic field, both the amplitude and the frequency of oscillation become less. Only the (x, t) diagram is shown here (see Figure 4-11). Actually, many more oscillations than the ones shown here are needed for the particle to lose appreciable energy.

suggested that such variations might be connected with the continuing source of the high-speed electrons that we knew were needed to supply the blue light of the Crab. We were sufficiently impressed by the idea to refer it to Baade, asking if it could be checked observationally. Baade immediately wanted to know how fast the light variations might be, and the author replied that they might occur in a few seconds, basing the estimate on the concept of a pulsating white-dwarf star. Baade seriously considered the possibility of taking a series of photographs at about one-second intervals, which was the only feasible method at that time. In any event, however, he made no attempt in this direction, presumably because the photographic method proved too insensitive. It was only when more refined electronic devices replaced the photographic plate that the problem could be successfully tackled. It then turned out that variability is in fact present, as can be seen in Figure 4-14, but not in the one or two seconds that we had estimated. The oscillation frequency was found to be about 30 per second, that is, each

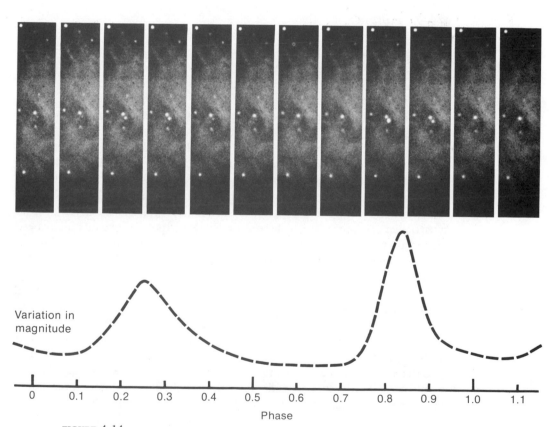

FIGURE 4-14
The oscillating starlike object in the Crab nebula (NP 0532). The whole of the Crab nebula was shown in Figure 2-7; find NP 0532 in that figure. (Courtesy of S. P. Maran, Kitt Peak National Observatory.)

oscillation being performed in only 0.033 second. This much shorter pulsation period implies that the object responsible for the light variations must be much smaller than a white-dwarf star. The object is now thought to be a *neutron star,* whose properties we discuss in Chapter 8. It belongs to a class of objects known as *pulsars,* which were first discovered in 1967 by S. J. Bell, A. Hewish, and their collaborators in Cambridge, England.

The point of this anecdote is to emphasize once again the importance of the technical resources available to astronomers. Having the right idea is not in itself sufficient. The idea must be supported by instruments sophisticated enough to enable the idea to be applied and developed. Otherwise, the idea remains sterile, as it did following the 1958 conference. It was only some 10 years later, when new methods were available, that a further step could be made. By then Baade was dead, and the investigation had passed to a younger generation.*

§4-4. *Pulsars*

A pulsar, as its name implies, emits a regular sequence of pulses. Although pulses of light are detected from the Crab, the pulses are best discovered by radio observations. Two hundred or so of these remarkable objects have now been found. The spacing between two pulses is known as the pulsar *period.* Typically, pulsar periods lie in the range from 0.25 second to about 2 seconds, although much shorter periods can occur, only 0.033 second for the Crab. The spacing between two pulses is usually much longer than the duration of an individual pulse, as can be seen from the example shown in Figure 4-15.

The pulses are thought to arise from rotation, the radiation being emitted in the fashion of a lighthouse beam, as is shown schematically in Figure 4-16, with the radiation coming from a particular spot on the surface of the object or from a particular place in an atmosphere around it. Since pulsars generate high-speed electrons, which escape outward and in the Crab light up the whole nebula, it is reasonable to suppose that high-speed particles are also present in the immediate surroundings of the objects themselves. The radiative emission giving rise to the lighthouse beam of Figure 4-16 is thought to be connected with such high-speed particles and with their behavior in strong magnetic fields. The details of the emission are not straightforward, however.

It is clear that any object going through a complete rotation in only 0.033 second, like the Crab pulsar, must be small in size compared to ordinary stars like the Sun. If spun even at a rate of one rotation per 1,000 seconds, the Sun would immediately fly apart under the influence of powerful rotary forces. Even much more compact stars, white dwarfs, could not be spun as fast as the Crab

---

*The discovery that regular light variations come from the pulsar in the Crab nebula was made by W. J. Cocke, M. J. Disney, and D. J. Taylor, "Discovery of Optical Signals from Pulsar NP 0532," *Nature, 221* (1969), p. 525.

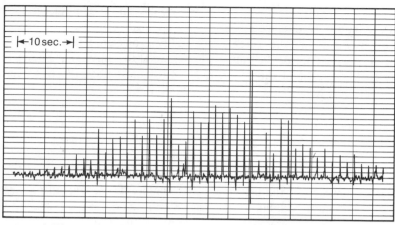

FIGURE 4-15

A pulsar record. (Courtesy of Commonwealth Scientific and Industrial Research Organization.)

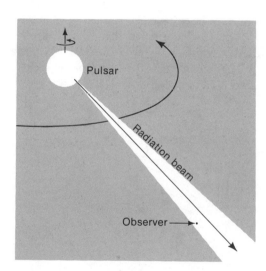

FIGURE 4-16
Pulsars are thought to behave like the rotating beam of a lighthouse.

pulsar. The only known stable objects that can be spun as fast as this are the neutron stars.

Neutrons in bulk can be stable provided there are neither too many of them nor too few. About as many as are in the Sun, 10^{57}, can exist in a stable configuration, but if there were appreciably more, they would shrink into what is called a *black hole*. (The properties of black holes are discussed in Chapter 11.) Stable neutron stars could also exist with appreciably less than 10^{57} neutrons. However, most of the observed pulsars are thought to have just about this number. What is so special about it? What is magic about 10^{57}?

Here is something to think about. Consider the universe on the scale of the most distant observed galaxies, explicitly, on a scale of 10^{28} cm. This is some 10^{22} times greater than the radius of a typical neutron star. The total number of particles contained in such a grand-scale picture of the universe is known to be about 10^{79}. The ratio of this to our magic 10^{57} is also 10^{22}. Hence we have a remarkable equality

$$\frac{\text{Scale of universe}}{\text{Scale of neutron star}} = \frac{\text{Number of particles in universe}}{\text{Number of neutrons in star}}.$$

This is the first of a number of remarkable coincidences. We shall meet others in our later discussion of the gravitational attraction. These coincidences are at present little understood, but they appear to imply a deep relationship between local physics and the large-scale structure of the universe.

PULSARS ARE EXAMPLES OF THE SYNCHROTRON PROCESS

One of the authors, together with W. A. Fowler, pointed out in 1960 that the residue from a supernova explosion must be either a neutron star or a black hole. F. Pacini, before the discovery of pulsars, considered that such neutron-star remnants would be endowed with rapid rotation. After the discovery of pulsars, this idea was further developed by T. Gold in what after 10 years still seems the simplest explanation of the lighthouse effect of Figure 4-16.

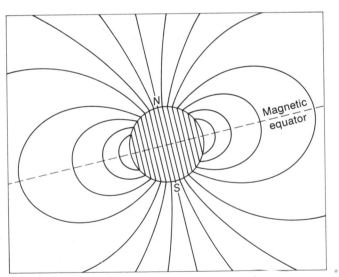

FIGURE 4-17
The magnetic field of the Earth gives a schematic representation of the inner closed region of the magnetic field of a pulsar. Note that the magnetic axis *NS* does not coincide with the Earth's axis of rotation.

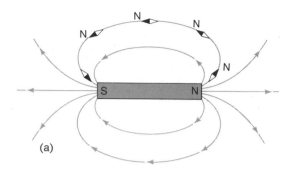

(a)

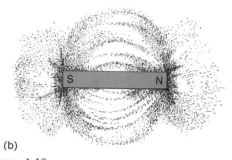

(b)

FIGURE 4-18

The field around a magnet has an effect both on a
compass needle (a) and on iron filings (b), showing that
magnetic fields can cross a gap. (Diagram (a) is from
G. Christiansen and P. Garrett, *Structure and Change.*
W. H. Freeman and Company. Copyright © 1960.
Diagram (b) is from A. Baez, *The New College Physics.*
W. H. Freeman and Company. Copyright © 1967.)

Figure 4-17 shows the *lines of force* of the Earth's magnetic field. This pattern
is very like that of the lines of force of a slightly inclined bar magnet placed at
the center of the Earth. Iron filings around a bar magnet arrange themselves
in a pattern with respect to the lines of force, as is shown in Figure 4-18. The
question arises as to what happens to the picture of Figure 4-17 as the Earth
spins. Does the system of lines of force also turn? If we answer affirmatively, the
motion of rotation around the Earth's axis of spin becomes faster at points along
lines of force that go further and further away from the Earth's center. Indeed,
in principle there would be a line of force that would go so far out that, at its
equator, the motion would have to rise to the speed of light in order to complete
a rotation once each day. There would even be other lines of force emerging
from near the Earth's magnetic poles, at places on which the turning motion
exceeded the speed of light. As long as we think in the abstract about lines of

force and not about material particles, this conclusion is no embarrassment. As A. S. Eddington once remarked, there is no barrier to our thinking faster than the speed of light; we can *think* ourselves a million light-years away in a fraction of a second. What is forbidden by basic physics is any material particle moving faster than the speed of light. A problem only arises, therefore, when we consider material particles (like the iron filings of Figure 4-18) to be embedded in the lines of force of a rotating magnetic field. What happens then is that, at any particular moment, the material particles and the magnetic lines of force can be considered to co-rotate out to a distance dependent on the density of the particles. Beyond that distance, the material particles are accelerated so violently as to break open the lines of force, preventing them from remaining closed across the equator as in Figure 4-17. (By closed we mean going from one of the Earth's magnetic poles to the other in a loop across the equator.) When the lines of force are thus broken, the material particles slide away from the Earth along the opened lines.

Neutron stars are believed to possess magnetic fields similar in form to the magnetic field of the Earth, but on the order of 10^{12} times more intense. The great strength of such a field makes it much harder for a diffuse gas composed of electrons and ionized atoms to break open the loops of the magnetic field. Indeed, it is only by coming extremely close to the speed of light that the particles of low-density gas could break open the lines of force and prevent them from crossing the magnetic equator of the star. Yet, this will happen close to the speed of light, just as it does for the Earth. Exceedingly energetic particles can thus be generated near a neutron star, and, after they break open the lines of force, they then leak away from vicinity of the star, thus becoming capable of injecting fast-moving electrons and ionized atoms into the world outside. It may well be that the high-energy electrons of the Crab nebula arise in this way; perhaps all high-energy electrons and cosmic rays in space generally arise this way. The breaking open of the lines of magnetic force occurs at a distance d from the neutron star, given by the simple equation $d = cP/2\pi$, where c is the speed of light (300,000 km/second) and P is the period of rotation of the neutron star. For the Crab pulsar, d is about 1,600 km, about one-quarter of the radius of the Earth. In comparison, the neutron star itself has a radius of no more than about 10 km.

Now let us consider in a little more detail the lines of force as they emerge from the neutron star. Those lines of force that emerge in low magnetic latitudes close in a loop across the equator without going far out from the star and create no problem. For lines of force emerging at higher and higher latitudes, the closure across the equator goes outward to greater and greater distances, however, eventually reaching the critical value of $cP/2\pi$. Thereafter, the lines of force are broken open, and they do not cross the equator. We can thus think of a magnetically closed region surrounded by a magnetically open region. The latter feeds high-energy particles to the outside world, whereas the closed region contributes the radiation from the pulsar itself. Figure 4-19 represents the

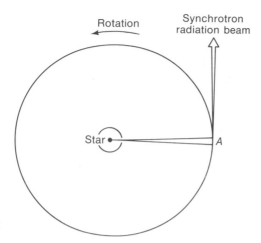

FIGURE 4-19
Equatorial section of the magnetically closed region generated by a pulsar, showing the direction of the radiated beam.

equatorial section of both the star and the inner closed, rotating region. The outer line of the diagram represents the places on the magnetic equator at which the outermost lines of force of the closed region cross from the North magnetic hemisphere to the South magnetic hemisphere. This line is drawn as a circle, and it refers to lines of force that emerge from the star *at a particular magnetic latitude*. In a perfectly symmetrical situation, the star would be concentric with the circle, but, in a real situation, there will inevitably be some small displacement between the center of the star and the center of the circle formed by the magnetic lines of force. On the circle there will be a greatest distance from the star, shown here as a region close to point A. It will be at a region such as this where the critical condition $d = cP/2\pi$ is first reached as the magnetic latitude of emergence from the star is increased. Thus, in the neighborhood of A, particles have much higher energy than they have elsewhere on the circle of Figure 4-19, and this higher energy causes synchrotron radiation from A along the beam marked in the figure. This is the *lighthouse effect*.

At first, it might seem as if we can obtain the critical condition $d = cP/\pi$ at longitudes other than A, simply by considering the lines of force to emerge from the star at still higher magnetic latitudes. But such a system of lines of force would inevitably be broken open at the longitude of A, and this breaking open provides an escape route for *all* the particles at the higher magnetic latitude in question. This is due to the fact that, in an irregular situation, particles are able to *precess* (to change slowly) their magnetic longitude, and they do so until they find the open region of escape at the longitude of A. Particles are therefore absent at higher latitudes of emergence of the magnetic field from the star.

This theory provides a simple and elegant explanation of the main features of pulsars, invoking nothing further than processes that are known to take place in the Earth's magnetic field and in the accelerators that physicists construct in the laboratory.

RADIO GALAXIES ARE ALSO EXAMPLES OF THE SYNCHROTRON PROCESS

Events in galaxies sometimes cause explosions of exceedingly great violence. Such explosions are believed to occur in the very central regions of the galaxies. Particles with speeds approaching that of light are shot out, usually in two opposite directed streams, as shown schematically in Figure 4-20. Should such a stream impinge on a cloud of gas that extends throughout, or even beyond, the galaxy, the high-speed outward motion of the particles will be checked by the magnetic field that always exists in such a cloud. An equally rapid motion of the particles around the direction of the magnetic field is then set up, as in Figure 4-21. Such a circling motion causes electrons to emit radio waves by a process similar to that which takes place in the Crab nebula. The effect is to produce two patches of radio emission, as illustrated in Figure 4-22. Radio galaxies frequently show this double pattern of emission, as shown in Figure 4-23.

The outbursts from radio galaxies are much more energetic than those of supernovae, even though a supernova can temporarily become as bright as a galaxy. The explosive energy in a supernova is on the order of 10^{51} ergs. Yet, the outbursts of radio galaxies release a still vaster amount of energy, ranging from about 10^{54} ergs on the low side to about 10^{60} ergs in the most extreme outbursts.

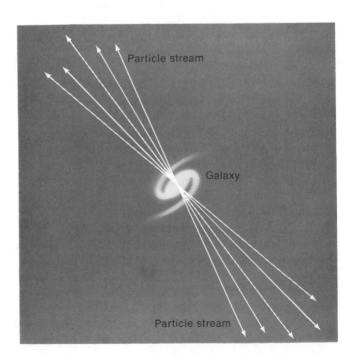

FIGURE 4-20
Particles are shot out with speeds approaching that of light in violent outbursts from galaxies, usually in two oppositely directed streams.

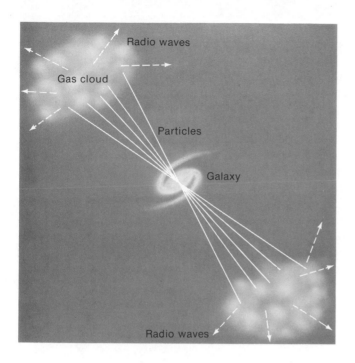

FIGURE 4-21
If the streams in Figure 4-20 impinge on an external gas cloud, radio waves are emitted in the presence of a magnetic field.

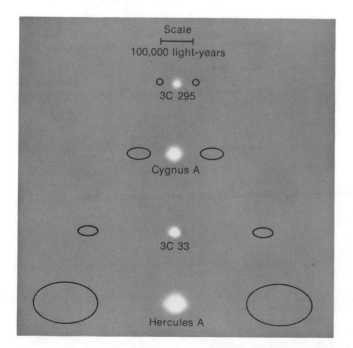

FIGURE 4-22
The effect of the situation in Figure 4-21 is to produce two patches of radio emission. The names here refer to explicit radio sources.

So great is the energy that one may wonder where it can come from. Some astronomers have speculated that the energy resident in a mass millions of times the Sun's mass may be released by completely annihilating matter, but it has proved difficult to make such a theory work satisfactorily when one considers the details. It is more usually supposed that a large mass in the central regions of a galaxy becomes highly compact, hundreds of millions of solar masses being confined in a region not much larger than the solar system. Because of the very strong gravitational fields, the speeds of motion of such compact masses would be expected to approach the speed of light. Sudden changes in these motions may lead to the explosive ejection of enormous streams of particles, which give rise to the radio galaxies.

These galaxies are, of course, exceptions; perhaps one galaxy in a thousand is the kind that astronomers term a radio galaxy. Evidence of explosion shows clearly in Figure 4-24, where a remarkable jet emerges from the center of the galaxy M87. The light emitted by this jet comes from high-speed electrons moving in helices of the form of Figure 4-10 about the direction of a magnetic

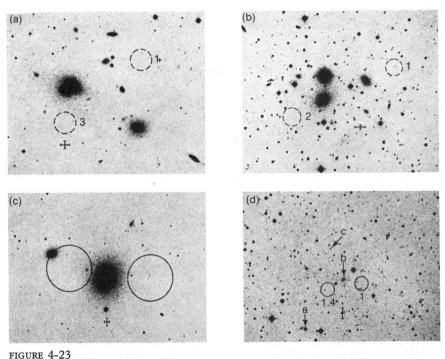

FIGURE 4-23
Examples of radio sources with double patches of emission: for (a), 3C 40; for (b), 3C 66; for (c), 3C 270; and for (d), Hercules A. These are *negatives* of the usual photographs with the galaxies appearing dark against a light sky. (Reprinted courtesy of P. Maltby, T. A. Mathews, and A. T. Moffet and *The Astrophysical Journal*, published by the University of Chicago Press; © 1963 The American Astronomical Society.)

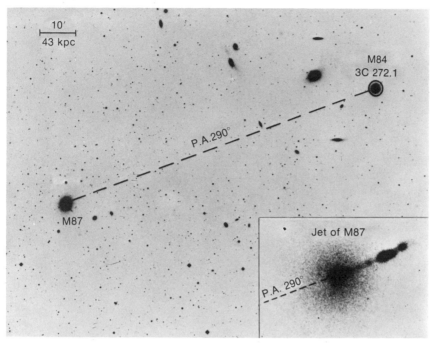

FIGURE 4-24
The jet of M87 points almost directly toward the galaxy M84, which is also a radio source. Note that 1 kpc equals 3,260 light-years. (Courtesy of Prof. G. R. Burbidge, University of California, San Diego.)

field. Thus, the optical light from the jet of M87 is synchrotron radiation, not starlight. The galaxy M87 also emits radio waves by the same synchrotron process. It is interesting and curious that the nearest radio galaxy to M87 is the system of M84, shown in Figure 4-24, the direction joining M87 to M84 being almost the same as the direction of the jet. Is this mere happenstance, or does it imply a connection between M84 and M87? According to the usual ideas, the coincidence of the directions must be ascribed to chance, since most astronomers feel the implication that M84 has been shot out of M87, or that both galaxies have been shot out of some invisible intermediate object, to be too strange to merit serious attention.

Another curious case is shown in Fig. 4-25. This is the system of NGC 7603, with a satellite galaxy apparently connected to the main galaxy by both a clearly marked arm and a fainter arch. A Doppler-shift technique (described in detail in the next section) has shown that, whereas the main galaxy is moving away from us at a speed of about 8,000 km/second, the satellite galaxy is moving away at a speed of about 16,000 km/second. Unless this apparent association is a remarkable fluke, the least radical explanation of these results is that the satellite

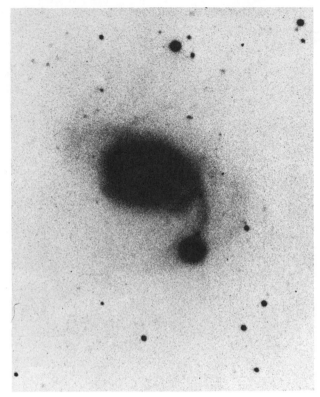

FIGURE 4-25
The strange case of NGC 7603. The smaller galaxy appears to be
connected to the main galaxy not only by a clearly marked arm
but also by a fainter arch. Yet the Doppler shift for the smaller
galaxy (16,000 km/second) is twice that for the main galaxy
(8,000 km/second). (Courtesy of H. Arp, Hale Observatories.)

has been shot out of the main galaxy at a speed of at least 8,000 km/second. The
concept that one galaxy can disgorge another in this way is highly unorthodox,
however, and it is not accepted by most astronomers.

§4-6. Quasars

Quasars were discovered by M. Schmidt in 1963 through the detection of the
Balmer lines of hydrogen in the spectrum of the radio source 3C 273, a spectrum
already shown in Figure 2-25. An interesting history in which radio astronomy
played a crucial role preceded Schmidt's discovery. By 1959, R. Hanbury Brown
had noticed an unusual class of radio source, later found to be quasars. Whereas
other radio sources were determined at Jodrell Bank, Cheshire, to be patches of

measurable extent on the sky, the unusual class had patches so small as to evade precise measurement by the instruments available at that time. A further serious technical limitation in the early 1960s was that the positions of radio sources on the sky could not generally be found with sufficient accuracy for satisfactory identifications with visible objects. All radio astronomers could then do was to position their sources within what were known as *error boxes*. Unfortunately, the error boxes were usually so large as to contain many visible objects.

Even so, some two years before the definitive discovery, A. R. Sandage had found a starlike object with a remarkable optical spectrum present within the error box associated with a source of the Jodrell special list, the source 3C 48 shown in Figure 4-26(a). The spectrum was awkward to interpret, however, and its nature eluded those astronomers who examined it, although a remark made to one of the authors by J. G. Bolton would soon have led to the correct interpretation had it been pursued.

The difficulty presented by 3C 48 was psychological as well as spectroscopic. The highly compact sources of the Jodrell special list seemed to be uniformly distributed over the sky, so that yet again there was a choice between thinking of them as very distant—far outside our own galaxy—or being local objects close by the solar system. Moreover, each was a very small patch on the sky, and it seemed natural to most astronomers to think that the special list was comprised of the local radio stars believed in so earnestly during the early days of radio astronomy. Attempts to interpret the spectrum of 3C 48 in terms of a local star, even an unusual star, were doomed to failure, however.

In 1962, one of the authors, together with W. A. Fowler, suggested that radio sources were compact objects of unprecedentedly large masses, upward of a million times the mass of a typical star. At first we had thought of nuclear energy as the source of radio outbursts, but, before the end of 1962, we had come to the view that gravitation was a more promising energy source. (This idea is discussed in Chapter 11 in connection with black holes.) While these substantially correct suggestions implied that the compact sources of the Jodrell special list were not local radio stars at all but distant sources outside our galaxy, this work was insufficient to break the psychological logjam. The break came about in a different way.

Because of its orbital motion around the Earth, the Moon completes a circuit of the sky in each lunar month of 27.3 days. The Moon's orbit around the Earth varies slightly from month to month, and this variation causes the Moon's path on the sky to vary correspondingly. From time to time, the Moon comes in front of a star, producing an eclipse or *occultation* of the starlight. Because of the monthly changes of the Moon's orbit, stars in a certain band on the sky are subject to occasional occultation. This would also be the case for the radio emission of sources of the special Jodrell list if they happen to lie in the appropriate band on the sky. C. Hazard noticed that one or two sources in the list did in fact lie in the lunar occultation band. He also noticed that, by examining the details of an occultation, particularly with respect to its onset and duration, and by using the information available from the error boxes, it would

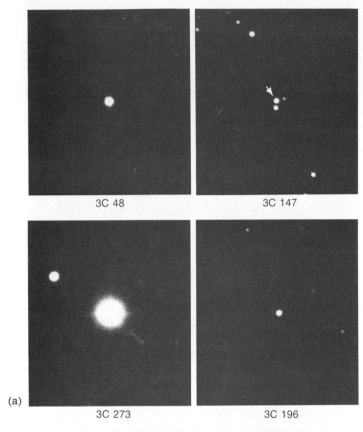

3C 48   3C 147

(a)

3C 273   3C 196

FIGURE 4-26

(a) When photographed directly, the quasi-stellar objects appear like ordinary stars. (b) A long exposure showing the jet of 3C 273. According to the usual interpretation of the redshift of quasi-stellar objects, the jet extends to about 300,000 light-years from the center of the system. (Courtesy of Hale Observatories.)

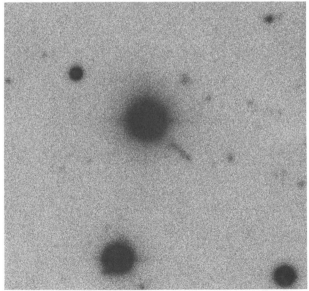

(b)

be possible to obtain positions of sources on the sky with great accuracy, to within seconds rather than minutes of arc. This was due to the fact that the position of the Moon is known accurately at all times. The source 3C 273 met the required conditions.

Although this technique was first tested successfully at Jodrell Bank, Hazard happened to be working in Australia in 1962 as the time for the occultation of 3C 273 approached. He was able to continue the investigation of 3C 273 in collaboration with A. Shimmins using the 210-ft CSIRO radiotelescope at Parkes, NSW, under the direction of J. G. Bolton. The investigation was successful, and the position of 3C 273 was accurately obtained. Bolton communicated the position by letter to Schmidt at the California Institute of Technology. Figure 4-26(b) shows the object that Schmidt found at the Hazard–Shimmins position, an object bright enough for an optical spectrum to be readily obtained. The spectrum turned out to contain the Balmer lines of hydrogen, which soon made unequivocal the nature of the object. (The redshift property discussed later became evident then.) With the information thus available from 3C 273, it was possible to return to 3C 48 and interpret the spectrum of that source correctly. Within a few more months, Schmidt, together with J. L. Greenstein, gave an extensive discussion of quasar spectra that was to prove a model for all subsequent investigations. Quasars had not only been discovered, but their nature as very massive, highly condensed bodies was also understood. E. E. Salpeter soon added a further correct idea, that radio-source outbursts were connected with the infall of new material into such compact bodies. The basic steps necessary for progress along the lines discussed in Chapter 11 had been taken.

The optical light of quasars displays a continuum of frequencies that is thought to be produced by the synchrotron process, as in the jet of M87. There are also line radiations from atoms; hydrogen, helium, carbon, nitrogen, oxygen, neon, magnesium, silicon, sulfur, and iron have all been detected in quasars. Usually, but by no means always, these line radiations are in emission lines. Typical radiation distributions (spectra) are shown in Figure 4-27. These spectra are characteristic of a moderately diffuse gas, say, with a density of the order of 10^6 atoms per cm^3. They are highly distinctive and can readily be distinguished from the spectra of other kinds of astronomical objects. It is generally believed that, whereas the continuum light is generated in the compact central region of a quasar, the line radiations come from a much more extended cloud of gas with dimensions in the range from 10 to 100 light-years.

When photographed directly against the sky in the ordinary way, quasars appear like faint, rather blue stars, as can be seen from the examples shown in Figure 4-26. After examining Figure 4-26, one may wonder how, given all the ordinary stars that are scattered over the sky, quasars are ever discovered. The first method uses radio observations. Radio astronomers can now determine with great accuracy the positions in the sky from which radio emissions are detected; such patches of emission are called radio sources. Sometimes the patches of emission are large and extended, especially for radio radiation generated by

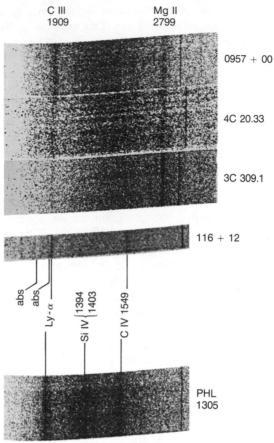

FIGURE 4-27
The spectra in negative form of several quasi-stellar objects (quasars). Such spectra are characteristic of a moderately diffuse gas. They show emission lines of common atoms—hydrogen, carbon, magnesium, silicon. (Courtesy of Dr. R. Lynds, Kitt Peak National Observatory.)

rapidly moving electrons in our own galaxy. Other patches of emission on the sky are highly localized, especially the quasars. With their positions accurately known, it is a comparatively straightforward task to search optical photographs of the relevant parts of the sky. When an optical object is found to be coincident with a radio position, there is a presumption that the optical object is associated with the source of the radio emission. To decide whether this is so, the light from the object is analyzed to discover its frequency distribution. With the spectrum thus available, it becomes immediately clear whether the object is a quasar. Should the spectrum be of the characteristic quasar type, the initial presumption of a connection with the radio emission is taken to be confirmed. The source is then said to be identified.

A second method of discovering quasars is simply to search a restricted area of the sky for faint blue starlike objects and then to obtain their spectra. Those that are quasars are readily identified by their spectra. The disadvantage of this second method is that only a very limited area of sky can be searched in this way within any reasonable amount of time. There is the advantage, on the other hand, that quasars can be discovered in this second way whether or not they happen to be sources of radio waves. The surprise was that most quasars are *not* sources of radio waves. Although the radio method first led to quasars being discovered, it has turned out that the property of radio emission is a comparative rarity, just as intense radio emission is a rarity for galaxies. Only about one galaxy in a thousand is a radio galaxy.

We can now see that the word quasar used in this sense is a misnomer. Quasar is a contraction of *quasi-stellar radio source,* which came into widespread usage at a time when it was thought that all such objects were radio sources. It is, therefore, quite meaningless to use the word *quasar* for objects that are not radio sources. For this reason many astronomers prefer the more general description *quasi-stellar object,* which is applicable whether or not the object happens to be a source of radio waves. This more logical name is usually contracted simply to QSO.

THE SPECTRA OF QSOS ARE REDSHIFTED

How do we know that QSOs are appreciably brighter than galaxies? The first step in answering this question is to understand the so-called redshift phenomenon. The line radiations in the light from a distant galaxy or a QSO have precisely defined frequencies that can be measured. As examples, we might have the Balmer lines of hydrogen or the H and K lines of the singly ionized calcium atom. The ratios of the measured frequencies of the lines to each other are always the same in a single spectrum, whether from an extraterrestrial source or in the terrestrial laboratory. However, the absolute frequency of a line in a spectrum from a distant galaxy or QSO is always found to be less than the corresponding frequency measured in the laboratory. By analogy, suppose we think of the line radiations as corresponding to the notes of a piano. Although the "piano" that constitutes a distant galaxy or QSO is in perfect *relative* tune, the absolute pitch of such a piano is lower than that of a similar piano here on the Earth. A quantity z is used by astronomers to express this lowering of pitch, namely,

$$1 + z = \frac{\text{A pitch on Earth}}{\text{Pitch in distant object}}.$$

If we write ν_{lab} for the frequency of a certain line radiation as measured in the laboratory and ν_{obj} for the frequency of the same line observed in the light from a distant object, we have $1 + z = \nu_{lab}/\nu_{obj}$.

Now that we understand what z means and how it can be measured, we can return to the problem of determining the relative brightness of QSOs and galaxies. For galaxies, the measured z values are found to increase systematically with the distances d, the distances being determined by the method discussed in Chapter 9. This systematic behavior shows itself when a diagram like Figure 4-28 is constructed, with the values of z and d plotted logarithmically. One remarkable result emerges from the measured z and d values: galaxies that are structurally similar to each other (for example, the brightest galaxies in similar clusters) tend to fall on the straight line of Figure 4-28.

Turning to the QSOs, *the assumption is made that the z and d values are related in the same way as for galaxies.* However, most QSOs turn out to have z values appreciably greater than those of the measured galaxies, so that an immediate comparison with galaxies is not possible. To cope with this difficulty, the line of Figure 4-28 is extended until it reaches large enough z values, and the

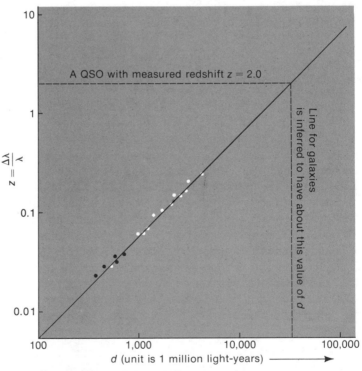

FIGURE 4-28
Schematic illustration of the observed relation between the redshift z and the distance d for galaxies that are structurally similar to each other. An extrapolation of this relation is used, together with the measured z values for quasi-stellar objects, to infer distance values for the latter.

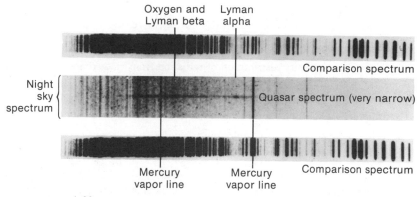

FIGURE 4-29

The spectrum of a quasi-stellar object of very large redshift, OH 471 with $z = 3.40$. (Courtesy of R. F. Carswell and P. A. Strittmatter, Steward Observatory.)

QSO distances are then read off from the line, as indicated schematically in Figure 4-28. With the d values for the QSOs determined in this way, it is possible to compare their intrinsic luminosities with those of the galaxies. As we have already stated, it is by this method that we find most QSOs are brighter than the most luminous galaxies, some QSOs very much so. Some are indeed at least 100 times more luminous than the brightest galaxies.

The measured values of z for galaxies, whether ordinary galaxies or radio galaxies, have rarely turned out to be greater than 0.3. On the other hand, values of z up to about 3.5 have been observed for quasars, as shown in Figure 4.29. It is reasonable to consider whether the straight-line relation of z and d, applicable at small z, can be extended to values as large as 3.5. Should the relationship of z to d become curved at large z? This is an important question that we will consider in Chapters 12 to 14, but the answer to it does not significantly affect the issues we are discussing here.

It is more appropriate here to ask *why* should QSOs be appreciably brighter than galaxies? Start with an ordinary galaxy and suppose that intense emission of radiation by a synchrotron process becomes operative in the compact central nucleus. Indeed, let the nucleus become so inordinately bright that, when seen from a great distance, the nucleus overwhelms the normal starlight of the galaxy. Seen from a distance, only a central brilliant point of light can then be distinguished. Many astronomers believe this to be a QSO. In short, a QSO is a galaxy in which the nucleus has temporarily become exceedingly bright. The physical nature of the process responsible for the brightening and for the intense outbursts of radio sources in general is discussed in Chapter 11.

General Problems and Questions

1. Write an essay on the early development of radio astronomy.

2. Writing the intensity distribution of radiation from an astronomical object as $I(\nu)\, d\nu$, explain the meaning of $I(\nu)$ and of $d\nu$. What form does $I(\nu)$ take for a radio source of index α?

3. Why does a sunspot with a temperature of about 4,000 K appear dark?

4. What discoveries were made with the aid of the 21-cm line of atomic hydrogen?

5. Discuss the emission of radio waves by the synchrotron process.

6. In what important respect does the emission of radiation from an oscillating electric particle differ for fast motion (comparable to the speed of light) and slow motion?

7. Write an essay on the Crab nebula.

8. Discuss pulsars. What is the significance for these stars of a distance $cP/2\pi$, where P is the pulsar period and c is the speed of light?

Chapter 5
Millimeter-Wave Astronomy

§5-1. Molecules

In the preceding chapter, we saw that hydrogen atoms emit radio waves with an intensity adequate for the waves to be detected readily with radiotelescopes. The frequency of the emission is 1.42×10^9 cycles per second, and this frequency is centrally located in the waveband to which radio techniques can be applied. The corresponding wavelength λ, determined from $\lambda = c/\nu$, is close to 21 cm.

Observation at this frequency lets us estimate how much hydrogen gas there is in atomic form in the interstellar spaces within our galaxy. With modern large radiotelescopes, we can make similar measurements for other nearby galaxies. The results range from about 1% to about 10% of the total mass of a galaxy. The larger proportions of atomic hydrogen tend to be found among the smaller galaxies, those of comparatively low total mass, although not all galaxies of small mass possess much gas. The results of these 21-cm radio observations for our own galaxy were seen in Figure 4-5. Results obtained in Holland for the galaxy M51 (NGC 5194) are shown in Figure 5-1. Notice the radio picture has two bright spots that do not appear on the optical picture.

It used to be thought that most of the interstellar hydrogen existed as individual atoms and was thereby detectable by the 21-cm radio observations.

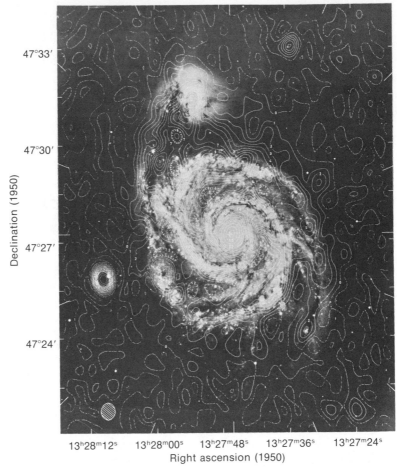

47°33′

47°30′

47°27′

47°24′

Declination (1950)

13ʰ28ᵐ12ˢ   13ʰ28ᵐ00ˢ   13ʰ27ᵐ48ˢ   13ʰ27ᵐ36ˢ   13ʰ27ᵐ24ˢ
Right ascension (1950)

FIGURE 5-1
Here the radio intensity is presented as a contour map with the optical
picture of M51 superimposed on it. The beam size of the radio equipment is indicated at the lower left. The diameter of the galaxy is
∼ 100,000 light-years. (From D. S. Mathewson, P. C. van der Kruit,
and W. N. Brouw, *Astronomy and Astrophysics*, Vol. 17, 1972.)

Molecules of hydrogen, H_2, do not emit this particular radiation, and consequently they cannot be detected by this radio technique, whereas monatomic
hydrogen can be. It is now known from other (nonradio) data that as much
interstellar hydrogen may exist in molecular form as it does in atomic form.
Molecular hydrogen exists in compact dense clouds with perhaps as many as a
million particles per cubic centimeter, whereas atomic hydrogen is much more
uniformly spread at a low density of only one or two atoms per cubic centimeter.

It is very possible that even more H_2 than H may exist, so that considerably more interstellar gas may well exist in our galaxy than has hitherto been supposed. If it does, most of the gas would be in dense clouds of the kind we considered in Chapter 3 in discussing problems of star formation.

THE ELEMENTS HAVE PERIODICITIES IN THEIR CHEMICAL PROPERTIES

Before continuing this discussion of molecules in space, we will consider the reason atoms form themselves into molecules at low enough temperatures. In Chapter 2, we saw that the behavior of electrons in atoms is highly organized. This organization controls the chemical properties of the various kinds of atoms.

Early in the second half of the nineteenth century, an English chemist, John Newlands, pointed out that, when the elements are arranged in the order of Figure 5-2, every eighth element seems to have similar properties. This idea was developed by a Russian, Dmitry Ivanovich Mendeleev (1834–1907), by setting out the elements in an array like the one shown in Figure 5-2. The distinctive feature of this array is that the elements in each column have similar chemical properties.

Not all the elements set out in Figure 5-2 were known in Mendeleev's time. Consequently, his table had gaps in it. From the dates of discovery of the various elements, given at the outset in Chapter 1, the reader will notice that, whereas most of the elements discovered in the twentieth century are heavier than lead, five elements lighter than lead were discovered in our century. They are: technetium at number 43, found in 1937; promethium at number 61, found in 1947; lutetium at number 71, found in 1907; hafnium at number 72, found in 1923; and rhenium at number 75, found in 1925. By noting the position of the gaps, Mendeleev was able to predict what elements would eventually be discovered. Today, no gaps are left. New elements will be found in the future only by extending Figure 5-2 to still heavier atoms.

CHEMICAL PROPERTIES ARE RELATED TO THE ARRANGEMENTS OF ATOMIC ELECTRONS IN SHELLS

In Figure 5-2, the numbers attached to the elements ordered them with respect to increasing proton numbers in their nuclei, but for electrically neutral atoms we could also have ordered the elements according to the number of electrons present in each kind of atom. If we had done so, we would have obtained exactly the same arrangement. We can therefore think of Mendeleev's periodic table as a classification of atoms with respect to electron number.

Working from the top of the periodic table and counting the number of elements in each row, we have 2, 8, 8, 18, 18, 32 (that is, 18 plus the 14 in rare earths), with the final row incomplete. The elements in the rightmost column are helium (order number 2), neon (order number $10 = 2 + 8$), argon (order

The Periodic System

IA	IIA	IIIB	IVB	VB	VIB	VIIB		VIII		IB	IIB	IIIA	IVA	VA	VIA	VIIA	0
1 H																	2 He
3 Li	4 Be											5 B	6 C	7 N	8 O	9 F	10 Ne
11 Na	12 Mg											13 Al	14 Si	15 P	16 S	17 Cl	18 A
19 K	20 Ca	21 Sc	22 Ti	23 V	24 Cr	25 Mn	26 Fe	27 Co	28 Ni	29 Cu	30 Zn	31 Ga	32 Ge	33 As	34 Se	35 Br	36 Kr
37 Rb	38 Sr	39 Y	40 Zr	41 Nb	42 Mo	43 Tc	44 Ru	45 Rh	46 Pd	47 Ag	48 Cd	49 In	50 Sn	51 Sb	52 Te	53 I	54 Xe
55 Cs	56 Ba	57* La	72 Hf	73 Ta	74 W	75 Re	76 Os	77 Ir	78 Pt	79 Au	80 Hg	81 Tl	82 Pb	83 Bi	84 Po	85 At	86 Rn
87 Fr	88 Ra	89† Ac	104	105	106												

*Lanthanides (rare earths)

58 Ce	59 Pr	60 Nd	61 Pm	62 Sm	63 Eu	64 Gd	65 Tb	66 Dy	67 Ho	68 Er	69 Tm	70 Yb	71 Lu

†Actinides

90 Th	91 Pa	92 U	93 Np	94 Pu	95 Am	96 Cm	97 Bk	98 Cf	99 Es	100 Fm	101 Md	102 No	103 Lr

FIGURE 5-2
The periodic system of the elements. This is a scheme of classification adopted in the nineteenth century by D. I. Mendeleev.

number $18 = 2 + 8 + 8$), krypton (order number $36 = 2 + 8 + 8 + 18$), xenon (order number $54 = 2 + 8 + 8 + 18 + 18$), and radon (order number $86 = 2 + 8 + 8 + 18 + 18 + 32$). Except in certain very special circumstances, these six elements are chemically inert, remaining in gaseous form as individual atoms except at very low temperatures, whence they are known as the *noble gases*. This property of chemical inertness can be explained in terms of the preceding electron numbers that represent closed shells that rarely interact with other atoms. Helium, with 2 electrons, has the first closed shell. Neon has two closed shells, one of 2 electrons, the other of 8 electrons. Similarly, radon has six closed shells, of 2, 8, 8, 18, 18, and 32 electrons. Other atoms complete the closed shells in the same order, but, since they do not have the same number of electrons as the noble gases (that is, 2, 10, 18, 36, 54, or 86), some electrons are inevitably left over after all the possible closed shells have been filled. Thus, sodium (Na), with 11 electrons, fills two shells of 2 and 8 electrons with 1 electron left over. Similarly, potassium (K), with 19 electrons, fills three shells of 2, 8, and 8 electrons with 1 electron left over. Omitting, for simplicity, the two subsidiary series of 14 elements given at the foot of Figure 5-2, it is easy to see that all the elements in each column of this figure have the *same number of electrons left over after all the possible complete shells have been filled*. In this connection, the shells must always be taken in the same order—2, 8, 8, 18, 18, 32—whatever the element. It is this property of the electron structures of the elements that determines which of them will be similar in their chemical behavior.

Although there is no way all the electrons of the sodium atom can arrange themselves into closed shells, a sodium atom, by interacting with another suitable atom, can use its extra electron to form a hybrid closed shell. The chlorine atom (Cl) is one electron short of being able to form a complete set of closed shells. Thus, if the sodium atom lends its extra electron to a chlorine atom, a hybrid closed shell is formed. In the terminology of Chapter 2, the excess electron of the sodium then has paths of high probability that remain close to the chlorine atom, causing the two atoms to form a composite particle written as NaCl. Such composite particles are called *molecules*, and NaCl is the molecule of sodium chloride, common salt.

It is possible for more than two atoms to be involved in this process of electron sharing. In the molecule of ammonia, for example, the 3 electrons of three hydrogen atoms join the 5 extra electrons of a nitrogen atom to form a hybrid shell of 8 electrons, the molecule in this case being written in the form NH_3.

Although atoms with one or two spare electrons, such as Na, Mg, K, and Ca, can be thought of as lending atoms, and atoms such as O or Cl, which require only one or two electrons to fill a shell, can be considered receiving atoms, this concept is not entirely clear-cut. Carbon has 4 spare electrons, but it also requires 4 electrons to complete a shell of 8 electrons. Thus, carbon can be a lending atom, as in the molecule of carbon tetrachloride, CCl_4, or a receiving atom, as in the molecule of methane, CH4. The ability of carbon to operate both ways is a reason why carbon is able to join other atoms in a most prolific way, and thence why carbon forms the basis for the complex chemical processes that occur in biological material.

Hybrid closed shells are not as mutually standoffish as the closed shells of the noble gases. Two molecules, both entirely closed in the present hybrid sense, may well be capable of joining into a composite molecule. For example, CO_2 and CaO are both completely closed in a hybrid sense, and yet they can combine to give the molecule of calcium carbonate, $CaCO_3$. Water, H_2O, and sulfur trioxide, SO_3, combine similarly to give sulfuric acid, H_2SO_4. Such associations are usually not very strong, however, and they can be dissociated without undue difficulty, often simply by heating. Thus, calcium carbonate is dissociated into CaO and CO_2 in a lime kiln. Much of industrial chemistry is concerned with the associations and dissociations of such closed molecules.

It is possible for molecules to be formed in such a way that, although one or more hybrid shells are produced, some spare electrons nevertheless remain. In the simple molecule, CO, 4 electrons from the carbon, together with 6 electrons from the oxygen, give a hybrid shell of 8 plus 2 spare electrons. These spare electrons are able to join with a second oxygen atom to give a second hybrid shell in the molecule of carbon dioxide, CO_2. Thus, CO will always tend to acquire O, a property that produces a kind of suffocation if we breathe too much CO gas. Almost every molecule with electrons left over is similarly poisonous, some very markedly so.

Usually, in any given mixture of atoms and molecules, all the hybrid shells that are possible are already formed; spare electrons occur only when there is no way for them to form further shells. Occasionally, however, a solid or liquid can be prepared in such a way that possible hybrid shells still have not been formed. Such a situation occurs for a *chemical explosive*. Detonation of an explosive consists of starting up hybrid shell formation. Once started, shell formation proceeds rapidly and spontaneously to completion, accompanied by a sudden energy release. It is this energy release that constitutes the explosion itself. Any sudden energy release produces the phenomenon of explosion, whether or not the energy is generated chemically. A chemically inert stone dropping at very high speed from the sky, that is, a meteorite, produces an explosion when it hits the ground.

Molecules can form even without a single hybrid shell being formed. A hydrogen atom will lend its electron to an oxygen atom, for example, giving the molecule OH. Or a hydrogen atom will lend its electron to a carbon atom to give the molecule CH. These very simple molecules are exceedingly active, so much so that they can hardly be studied at all in the laboratory where they immediately join with some other atom or molecule to produce a closed shell. Such exceedingly active molecules are given the special name of *free radicals*. They are widespread in astronomy in gas clouds like the Orion nebula, shown in Plate II.

MOLECULES ARE DETECTED FROM THEIR ROTATIONS

Molecules emit spectrum lines that are sharply defined in their frequencies due to transitions in the state of their rotation. In Chapter 2, we saw that the states of an atom could be denoted by ψ_1, ψ_2, and so forth. In a similar way, we can denote the possible states of rotation of a molecule by ϕ_1, ϕ_2, . . . , arranging them according to energy, so that a jump to the left corresponds to emission of radiation and a jump to the right corresponds to absorption of radiation. Just like the energy diagrams for atoms we studied in Chapter 2 (for example, Figure 2-17 for hydrogen) are the energy diagrams for molecules, with the state ϕ_1 at the bottom, ϕ_2 as the first raised level, ϕ_3 as the next raised level, and so on. An important difference between the rotational states of molecules and atoms lies in the frequencies associated with transitions in the energy diagram. Frequencies for atoms are usually on the order of 10^{15} oscillations per second, whereas, for jumps among the rotational states of molecules, they are usually on the order of 10^{11} oscillations per second. And although it is necessary to use techniques appropriate to visible light to detect the emission of atoms, we must use the techniques of radio antennas and receivers to detect the emission of molecules. Receivers can be tuned to the precise frequency associated with the particular transition of a particular molecule and so detect the presence of the molecule in a cosmic gas cloud.

The construction of equipment to detect molecules is thus a natural extension of the methods of radio astronomy, especially of the methods used for the 21-cm line of atomic hydrogen. Therefore, it is curious that, with the exception of the National Radio Astronomy Observatory and the CSIRO installation at Parkes, N.S.W., the established centers of research in radio astronomy have been slow to adapt themselves to this new field, a field still very much in its golden age. Once again, the Bell Telephone Laboratories at Holmdel (under Arno A. Penzias) have been in the forefront of the new work.

Table 5-1 lists the 40-odd molecules detected to the time of writing. The first theoretical prediction concerning molecules in interstellar space was made in 1940 by R. A. Lyttleton, together with one of the authors. Very shortly thereafter, Andrew McKellar detected CH and CN by using an optical technique and not by radio methods. Nothing much happened from then until 1963 when OH was discovered by S. Weinrab, A. H. Barrett, M. L. Meeks, and J. C. Henry. This discovery led to a burst of activity in what was still the normal frequency band of radio astronomy (1.662×10^9 cycles per second). Five years later, in 1968, A. C. Cheung, D. M. Rauk, C. H. Townes, D. C. Thompson, and W. J. Welch, at Berkeley, California, found NH_3 and H_2O at wavelengths of 1.26 cm and 1.35 cm, respectively. The latter discoveries were the outcome of a deliberate search for these molecules, occasioned by much earlier (1955) predictions by Townes.

TABLE 5-1
Observed interstellar molecules in dense molecular clouds

Inorganic		Organic
H_2 hydrogen OH hydroxyl SiO silicon monoxide SiS silicon sulfide NS nitrogen sulfide SO sulfur monoxide	DIATOMIC	CH methylidyne CH^+ methylidyne ion CN cyanogen CO carbon monoxide CS carbon monosulfide
H_2O water N_2H^+ H_2S hydrogen sulfide SO_2 sulfur dioxide	TRIATOMIC	CCH ethynal HCN hydrogen cyanide HNC hydrogen isocyanide HCO^+ formyl ion HCO formyl OCS carbonyl sulfide HNO nitroxyl
NH_3 ammonia	4-ATOMIC	H_2CO formaldehyde HNCO isocyanic acid H_2CS thioformaldehyde C_3N cyanoethynyl
	5-ATOMIC	H_2CHN methanimine H_2NCN cyanamide HCOOH formic acid HC_3N cyanoacetylene H_2C_2O ketan
	6-ATOMIC	CH_3OH methanol CH_3CN cyanomethane $HCONH_2$ formamide
	7-ATOMIC	CH_3NH_2 methylamine CH_3C_2H methylacetylene $HCOCH_3$ acetaldehyde H_2CCHCN vinyl cyanide HC_5N cyanodiacetylene
	8-ATOMIC	$HCOOCH_3$ methyl formate
	9-ATOMIC	$(CH_3)_2O$ dimethyl ether C_2H_5OH ethanol HC_7N cyanotriacetylene C_2H_5CN ethyl cyanide
	11-ATOMIC	HC_9N cyano-octa-tetrayn

Also in 1968, Carl Heiles made the important observation that certain localized regions, while strongly emitting the 18-cm line of OH, were *without* emission of the 21-cm line of atomic hydrogen. Here was a strong hint (emphasized by P. M. Solomon and N. C. Wickramasinghe) that, within these regions, now known as molecular clouds, the atomic hydrogen has been converted to molecular hydrogen, H_2. Two further critical discoveries then opened the whole subject of molecular astronomy to its current phase of rapid development.

To this point (1968), it had been thought that interstellar molecules would be essentially confined to species containing only two or three atoms. The discovery in 1969 of formaldehyde (H_2CO) by P. Palmer, B. M. Zuckerman, D. Buhl, and L. E. Synder first brought this opinion under suspicion. The discovery in 1970 by R. W. Wilson, K. B. Jefferts, and A. A. Penzias of strong radiation at a wavelength of 2.6 mm from carbon monoxide (CO) provided the crucial working tool for investigating the interstellar gas in fine and exquisite detail. This was the consummation of the vision of Jan Oort in the early 1940s. After the discovery of strong radiation from CO, further discoveries followed thick and fast. Here we mention only the detection of HC_3N by Barry Turner in 1971 and the detection of HCN by L. E. Synder and D. Buhl, also in 1971. Since 1971, molecular astronomy, conducted now largely at millimeter wavelengths, has emerged as an important branch of astronomy in its own right.

§5-3. *Giant Molecular Clouds*

MILLIMETER-WAVE ASTRONOMY DETERMINES THE DETAILED STRUCTURE OF THE INTERSTELLAR GAS

Figure 3-15 is a schematic representation of a giant molecular cloud (GMC) with its several hierarchical stages of condensation, while the Orion nebula in Figure 3-14 and in Plate II is an actual example of a GMC. Clouds can be observed with respect to the radiation emitted in the rotational transitions of many of the molecules of Table 5-1, with each molecule yielding information about its own distribution within a cloud. Since the different molecules are distributed differently, each one gives its own picture. The various pictures bear family relationships to one another, however, with the more common molecules displaying the widest ranging structures. Of all the molecules, carbon monoxide (CO) gives the largest scale picture. Indeed, the CO picture of the Orion nebula extends well beyond the photograph of Figure 3-15, as we can see in Figure 5-3. Carbon monoxide is a particularly stable molecule built from two common atoms. The transition of CO at a wavelength of 2.6 mm displays the general distribution of gas in our galaxy far more effectively than the 21-cm line of atomic hydrogen does. We can see this fact from the exquisitely detailed CO map by P. M. Solomon, D. B. Sanders, and N. Z. Scoville shown in Plate III.

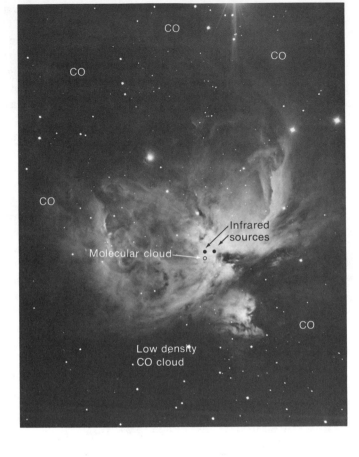

FIGURE 5-3
The Orion nebula, a cloud of gas
in which stars are now being born.
(Courtesy of Hale Observatories.)

Carbon monoxide has also been detected from other galaxies, although not in
the remarkable detail of Plate III.

The small region marked "Molecular cloud" in Figure 5-3 is a region of
particularly high gas concentration in which the less common molecules are
found. Whereas the enveloping diffuse material marked "CO" in Figure 5-3 has
a density of only a few hundred hydrogen atoms per cubic centimeter, the small
molecular cloud region has on the order of 500,000 H_2 molecules per cubic
centimeter. Such small, much denser regions typically have diameters of about
a light-year, whereas the whole complex has a diameter some 100 times larger
than this, about 100 light-years. As we already indicated in Figure 3-15, the new
stars are being born in highly dense compact regions of the GMCs.

Distinguishing new stars by direct optical methods—that is, with an optical
telescope—is, unfortunately, not feasible for a reason that will emerge in more
detail in Chapter 6. (Myriads of small solid particles with dimensions comparable to the wavelength of visible light are contained within the gas, and these

particles produce an exceedingly dense fog throughout such regions thereby obscuring the activity within them from optical analysis.)

An intensive recent study of pictures like Plate III has shown that, within our galaxy, there are about 4,000 GMCs, with typical masses about 5.10^5 times the mass of the Sun, giving a total of about 2.10^9 ⊙ for the mass of molecular gas, most of it H_2. This mass is about 1 percent of the mass of the whole galaxy, the rest of the mass being overwhelmingly in the form of stars.

STARS FORM SPORADICALLY INSIDE THE GIANT MOLECULAR CLOUDS

Star formation is now proceeding rather slowly in the GMCs, whereas, in the early history of the galaxy, star formation must have proceeded far more rapidly. This somewhat mysterious circumstance has caused much current discussion and controversy among astronomers. Several explanations have been put forth. Perhaps the present-day clouds are less massive than former clouds, hence less able to pull themselves together by gravitational forces. Perhaps magnetic fields within the clouds impede star formation more than they used to. Or perhaps the stars that are born now are more massive and much more luminous than they used to be.

Of these alternatives, the third is the most intriguing. We saw at the end of Chapter 3 that the luminosities of stars increase with their masses according to about the fourth power, so that even a comparatively small quantity of gas can temporarily make a big splash if it condenses into a few stars of exceptionally large masses. Such stars would, by their emission of intense radiation, violently stir up the gas cloud in which they happened to be born, with the likely effect of stopping (for a while) any further star formation. Thus, star formation would become a stop-and-go phenomenon. If there were no new massive stars, a few of them would quickly condense. Once a few were condensed, their intense emission of light and heat would set the gas into a strong internal turbulence that would stop further star formation until the few had passed through their evolution and were laid to rest in their appropriate graveyards (Chapter 8). Then the cycle would begin once more.

The late Walter Baade was strongly attracted by this stop-and-go idea. It would provide a possible simple explanation he insisted, of the fact that the two Magellanic Clouds (Figures 9-11 and 9-12) are so very different from each other, the so-called Small Cloud with no massive bright stars and the Large Cloud with massive very bright stars scattered throughout its whole structure. According to the stop-and-go view, the two Clouds would be in opposite phases of their star-formation cycle.

We might seek to explain the spiral structures of whole galaxies in a similar way, as in Figures 9-14 and 9-16, in terms of star-formation cycles operating inside the GMCs of these galaxies. The problem lies in understanding why the cycles should be locked in phase from one GMC to another. How could all the

GMCs be in the go phase together, as they seem to be in the Large Magellanic Cloud, or in the stop phase together, as they seem to be in the Small Magellanic Cloud?

If we are to proceed successfully with such a theory, we must suppose that some galaxy-wide event has locked the individual GMC cycles together. There have indeed been several proposals as to what such an event might be. A major explosion occurring in some central highly massive object, such as we considered in the previous chapter in connection with radio galaxies, appears to be the most plausible of these suggestions. A weaker effect of a gravitational nature—a so-called *density wave*—seems to be a less attractive explanation, since it is hard to see what significant effect a modest gravitational disturbance could have on an object with as loose an internal structure as a GMC. A small disturbance in the outer regions of Figure 3-15 would not be of much consequence to the inner, far more dense regions. For the theory to work, the go portions of the GMC cycles must all be triggered by some galaxy-wide flash, a flash propagated throughout the galaxy at high speed. Since events in radiogalaxies are certainly propagated at high speeds, this phenomenon is not impossible.

In support of the stop-and-go idea, the gases inside GMCs are certainly being stirred into turbulence, usually with motions of about 3 km/second, by some internal agency. Giant molecular clouds have rotation in periods that are typically about 30 million years. They are often strung together, one after another, like beads on a string. It is likely that the stringing occurs with respect to the lines of force of the magnetic field that pervades our galaxy.

This whole picture of GMCs, widely distributed in the galaxy, is one that theoreticians conjectured many years ago. Support for the conjecture has come only very recently, however, and it has come from maps like that of Plate III, which have been made possible through the development of radio amplifier techniques at exceedingly short wavelengths in the millimeter range. This development has come as part of the general modern revolution in electronics.

General Problems and Questions

1. Describe the periodic classification of the elements introduced in the nineteenth century by D. I. Mendeleev.

2. Explain the relation of Mendeleev's periodic table to the arrangements of atomic electrons in shells.

3. How are molecules in space identified?

4. What molecules have been found in space?

5. What has been learned from the CO molecule concerning the distribution of the interstellar gas?

6. Outline the general problem of star formation.

Chapter 6
Interstellar Grains and
Infrared Astronomy

§6-1. The Birth of a New Science

Infrared astronomy is still in its golden age, and it challenges all other branches of astronomy in its cost effectiveness. A great deal of interest and importance has come from modest expenditures. Reference to Table 2-1 and to Figure 3-12 shows that, whereas the infrared extends over a wide frequency range from about 3×10^{11} cycles per second at the low end to about 3.75×10^{14} cycles per second on the high side, infrared radiation does not penetrate the Earth's atmosphere freely over most of this range. There are, however, a number of restricted frequency bands, known as *windows,* through which radiation penetrates comparatively freely. These windows are shown in detail in Figure 6-1, where it will be seen that astronomers have given the letter designations J, H, K, L, M, N, and QZ to these bands. The scale used in Figure 6-1 is the *micron,* a length equal to one-millionth part of a meter. This is the wavelength λ of the radiation, related to the frequency ν by $\lambda = c/\nu$, where c is the speed of light.

We talk conversationally about the light and heat that we receive from the Sun, the light being the radiation that produces a structural change in the retinal molecules of the eye and the heat being the sensation of warmth we receive when radiation is incident on the skin. It is not clear from this crude subjective experience that light and heat are both forms of the same thing, namely radiation. That heat is indeed a form of radiation was proved long ago by W. Herschel. Figure 6-2 shows the effect of a prism of glass on an incident beam of white light. The prism separates the band of frequencies in the beam into its constituent colors. Herschel used a beam of sunlight, not white light. If sunlight contained radiation of higher frequency than visible blue-violet light, such radiation would be deflected onto the lower part of the viewing screen; this would be "ultra" violet light. And, if sunlight contained lower frequency radiation, it would be deflected onto the upper part of the screen, beyond the visible red light; this would be "infra" red light. Herschel had available only a very crude detector for such invisible radiations, a simple thermometer. How-

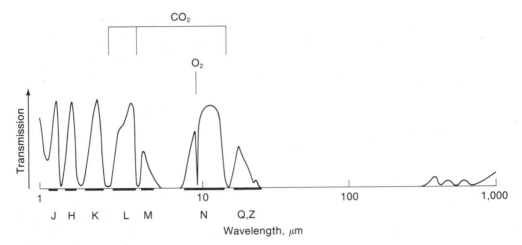

FIGURE 6-1
The atmospheric windows accessible from the earth's surface.
(After D. A. Allen, *Infra-red, the New Astronomy,* David & Charles, Devon, England, 1975.)

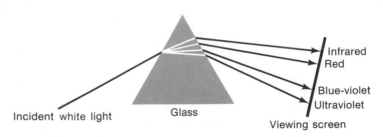

FIGURE 6-2
We can use a glass prism to separate light of mixed colors, and we can project the separate colors on a viewing screen.

ever, by putting the bulb of a mercury thermometer in the infrared position of the spectrum, he found an easily detectable effect on the length of the column of mercury. A similar attempt on the ultraviolet side of the viewing screen was not so successful. The reason is made clear by Figure 3-12; the Earth's atmosphere does not have windows for ultraviolet radiation so the solar ultraviolet radiation was not getting through the terrestrial atmosphere in Herschel's experiment.

Little further progress was made in infrared astronomy until the past two decades because no suitable method for detecting and measuring infrared radiation was available before the advent of modern electronics. (Modern electronics was itself an outcome of the practical application of the ideas of quantum mechanics.) Table 2-1 lists the "effect on crystals" as the modern method for detecting infrared.

Crystals, when heated, always change their physical properties to some degree. The problem in detecting infrared radiation is to find a particular crystal that changes its electrical properties in a highly sensitive way, even for comparatively slight heating. No one kind of crystal that is capable of spanning all infrared wavelengths has been found. For the range from 1μ to 4μ (1μ is 1 micron), crystals of lead sulphide (PbS) are used. The method used is very similar to the operation of a cadmium sulphide crystal in an ordinary photographic exposure meter. Light falling on the crystal causes the electrical resistance of the cadmium sulphide to change, with the change showing itself on an electric current driven through the crystal by a battery.

Infrared astronomers have some formidable advantages over astronomers in other fields. Because of the longer wavelengths of infrared radiation, their telescopes do not need to be made with a precision as high as those of astronomers working with visible radiation, and sunlight scattered by the Earth's atmosphere does not interfere as drastically with their observations as it does for visible light. Consequently, infrared astronomers can often make useful observations during the daytime as well as at night. Against these advantages is the disadvantage that the crytals must be kept very cold indeed to gain adequate sensitivity for their task. Liquid nitrogen at a temperature of 77 K ($-196°C$) is used for PbS crystals, whereas the temperature has to be much lower still for the germanium crystals used for the wavelength range from 5μ to 15μ. Thus, the temperature must be no higher than 2 K ($-271°C$) for germanium crystals deliberately doped with a gallium impurity (invented by Frank J. Low in 1961). Such temperatures demand the boiling of liquid helium at very low pressure, and the low pressure has to be obtained by the use of a powerful vacuum pump.

Table 6-1 gives the technical details of the crystal detectors used for the various windows shown in Figure 6-1. The column headed "sky brightness" brings out another problem for infrared astronomers, namely that the atmospheric gases themselves emit infrared radiation—by night as well as by day. This radiation from our own atmosphere enters the telescopes and contributes to the heating of the detector crystal, thereby producing an unwanted signal. To minimize this difficulty, infrared astronomers like to site their telescopes on high

TABLE 6-1
The infrared atmospheric windows[a]

Waveband (μ)	Crystal	Designation	Sky transparency	Sky brightness
1.1–1.4	PbS	J	high	low at night
1.5–1.8	PbS	H	high	very low
2.0–2.4	PbS	K	high	very low
3.0–4.0	PbS/InSb	L	3.0–3.5μ fair 3.5–4μ high	low
4.6–5.0	InSb/Ge:Ga	M	rather low	high
7.5–14.5	Ge:Ga	N	8–9μ and 10–12μ fair, rest low	extremely high
17–40	Ge:Ga	17–25μ (Q) 28–40μ (Z)	very low	very high
330–370	Ge:Ga/Si		extremely low	low

[a]After D. A. Allen, *Infra-red, the New Astronomy,* David & Charles, Devon, England, 1975.)

mountains to reduce the amount of atmosphere (and particularly the amount of atmospheric water vapor) that lies above the telescopes. On the island of Hawaii, Mauna Kea, with an altitude of about 14,000 ft, is a particularly excellent site in this respect, although working at such a height is not to everybody's taste.

Even with their telescopes on mountains, it is essential for infrared astronomers to use a technique known as *beam chopping*. During the observation of an astronomical object, the area of the object is alternated with an equal area of sky *off* the object, and the *difference* of the infrared radiation from the two areas is measured electronically. This process is known as *filtering out the sky*. Yet the infrared radiation from the sky is not entirely uniform, and sky brightness gradients cause difficulties that call for still further refinements of technique.

All these problems are precisely the kind in which physicists revel, for they are problems in which ingenuity is of greater importance than massive finance. Mention has already been made of the doped germanium detector of F. J. Low. There have also been remarkable developments in low-temperature techniques by Edward P. Ney and his colleagues at the University of Minnesota. The pioneering work of Robert B. Leighton and Gerry Neugebauer at the California Institute of Technology opened the modern field of infrared astronomy. It is described next.

INFRARED ASTRONOMY REVEALS THE PRESENCE OF NEWLY FORMED STARS

Leighton and Neugebauer undertook to survey the three-quarters of the whole sky accessible from California, about 30,000 square degrees, at an infrared wavelength of 2.2μ for all sources of radiation above the sensitivity limit of their equipment. For this work, they constructed their own telescope with an aperture of 62 inches, the mirror being produced by a method that goes back to the time of Isaac Newton. If one spins a liquid contained in a dish around a suitable vertical axis, the surface of the liquid takes on the shape of a paraboloid, which is just the shape to which the glass mirrors of many telescopes are carefully ground. What Leighton did was to spin liquid epoxy resin in an aluminum dish, and the epoxy set itself into the required paraboloidal shape as it cooled and solidified. From the point of view of the astronomer working with visible radiation, this simple inexpensive process does not produce a sufficiently accurate surface, but, at the longer infrared wavelength of 2.2μ used by Leighton and Neugebauer, it was good enough.

A catalogue of 5,612 sources of infrared radiation published by Leighton and Neugebauer in 1969 came as a surprise to most astronomers. Astronomers had expected the survey to show a number of very red stars, but what appeared were sources that were not connected with any visible object, as, for instance, the two sources marked in Figure 5-3. At first, it was thought that sources like those of Figure 5-3 were clouds of gas and dust that had become heated to temperatures of about 500 K through condensation, that is, through compression by gravitation. It was not long, however, before it became clear that the amounts of energy involved in many such objects were far too large for the heating of the clouds to be gravitational in origin. Only objects with access to *nuclear energy,* such as those considered for stars in Chapter 8, could emit energy as prodigally as many of the sources of infrared radiation discovered by Leighton and Neugebauer. The inference seems to be that enormous nuclear reactors are tucked away inside the clouds. These nuclear reactors are probably newly formed stars, high toward the upper left of the H–R diagram, heating the gas and dust that surround them. Indeed, it seems likely that what Leighton and Neugebauer found was the birthplace of stars, stars still embedded in primeval cocoons of cloud material that prevent us from seeing the stars directly as visible objects. We shall return later, toward the end of this chapter, to the detailed forms of infrared spectra of sources such as those in Figure 5-3, when we consider what the spectra may tell us of the detailed chemical composition of the fine grains of solid material that exist within the clouds. Suffice it for the moment to say that the most energetic of these cloud sources possess infrared luminosities on the order of a million times greater than the energy output of a typical star like the Sun. Plainly, something very exceptional and unusual had been discovered.

THE INFRARED EMISSION FROM GALAXIES HAS
ANALOGIES TO THE RADIO GALAXIES

In infrared astronomy, just as in radio astronomy, there were early investigations of the nearest large galaxy to our own, the Andromeda nebula (M31 in Figure 4-7). The Andromeda nebula was first observed at infrared frequencies by Harold L. Johnson in 1965; it was observed in more detail in 1969 by A. R. Sandage, E. E. Becklin, and G. Neugebauer. As in radio astronomy, it was soon found that, whereas the Andromeda nebula was not greatly different from our own galaxy in its infrared output, other more distant galaxies exceeded our

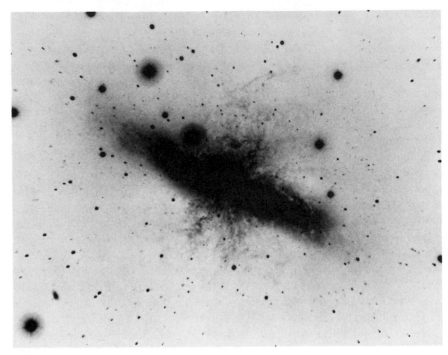

FIGURE 6-3
The galaxy M82. (Courtesy of Hale Observatories.)

output very substantially. Early among such studies was an investigation in 1970 by D. E. Kleinmann and F. J. Low. The galaxy M82 shown in Figure 6.3 (see also Plate IV) is an example, as are two galaxies that will be familiar from Chapter 4, the comparatively nearby NGC 5128 in Figure 4-6 and the classic distant radio galaxy Cygnus A in Figure 4-4. The same galaxies that had been found two decades earlier to be very strong emitters of radio waves now turned out to be strong emitters of infrared radiation. Just as the region of intense emission of radio waves has been eventually tracked down as small central cores of radio galaxies, so E. E. Becklin, J. A. Frogel, D. E. Kleinmann, G. Neugebauer, E. P. Ney, and D. W. Strecker found in 1971 that the strong infrared emission from NGC 5128 (Figure 4-6) came from a small central core. Also in 1971, similar results were found for the galaxy NGC 1068 by G. Neugebauer, G. Garmire, G. H. Ricke and F. J. Low.

Taken together, these discoveries immediately suggested that infrared radiation might be generated by the same synchrotron process that gave rise to the intense emission of radio galaxies. It is a feature of the synchrotron process, as we emphasized in Chapter 4, that it generates a wide range of frequencies, extending in such objects as the Crab nebula (Plate I) from radio waves at the low frequency to the visible light of Plate I, and thence, as we shall see in the next chapter, to the still higher frequencies of x rays and γ rays. Consequently, it seemed natural to many astronomers to think that the synchrotron process might also be responsible for the strong infrared emission of galaxies such as NGC 5128 and NGC 1068.

There were difficulties with this apparently straightforward suggestion. The intensities with which both NGC 5128 and Cygnus A emit radio waves fall off with increasing frequency, that is, they have radio spectra of the form $\nu^{-\alpha}\,d\nu$ with α positive. This falling away of the energy output with increasing frequency would need to undergo a dramatic reversal at the still higher frequencies in the infrared range (at about 10^{14} cycles per second) if synchrotron radiation were responsible for the infrared emission. In addition, the output would also have to reverse itself to fall away steeply through the visible frequencies and into the ultraviolet and x-ray ranges. Another problem was that the central infrared cores of NGC 5128 and NGC 1068, although very small compared to the sizes of the galaxies themselves, were yet much larger than the radio cores. Whereas the latter were at most only a few light-years in diameter, the infrared cores had diameters of about 200 light-years.

DUST PLAYS A ROLE IN THE EMISSION OF INFRARED

The several points just mentioned led to an alternative suggestion for the source of the emission from strong infrared galaxies, namely that it was heat radiation by fine dust particles densely concentrated in the central cores of the galaxies. An argument against this proposal was that the infrared emission appeared variable in time, like the emission of light from quasi-stellar objects discussed in Chapter 4. A dust cloud 200 light-years in diameter certainly could not show the variability that infrared galaxies were claimed to exhibit. Then an investigation by W. A. Stein, F. C. Gillett, and K. M. Merrill cast serious doubt on this supposed time variability, and now it is generally thought that dust is indeed the most likely generator of infrared radiation from most galaxies. It is likely that synchrotron radiation is still involved, but only indirectly, with the dust in the central cores of galaxies absorbing both visible light and higher frequencies that may be generated by the synchrotron process at the very center of the galaxies. The dust simply degrades the higher frequencies into the infrared radiation, which it reemits, and which infrared astronomers observe.

INFRARED ASTRONOMY HAS APPLICATIONS TO THE SMALL (ASTEROIDS) AS WELL AS THE LARGE (GALAXIES)

Turning from the very large to the very small, from galaxies to asteroids, we have an interesting example of how new astronomical methods can often solve problems that seemed quite insoluble by older methods. Asteroids are small bodies that move around the Sun like planets; for this reason, they are often referred to as minor planets. For the most part, they lie in the region between Mars and Jupiter, and they have sizes that range from a few meters to a few hundred kilometers. The problem at issue was the determination of the diameters of the larger asteroids. This was very hard to do by direct optical methods because even the largest asteroids show only a tiny disk in an optical telescope. It is not possible to distinguish this tiny disk from the natural twinkling of optical images caused by movements of air in the Earth's atmosphere, such as the twinkling of stars.

Because we know the distance of an asteroid from the Sun, we know the intensity of the sunlight that is incident upon it. Some of the sunlight is reflected, say, a fraction A. It is this fraction, known as the *albedo*, that gives a visible brightness to the asteroid, permitting it to be observed by optical astronomers. By measuring the brightness of an asteroid at visual frequencies, and by doing so for different orientations of the asteroid with respect to the Earth and the Sun (thus eliminating angular effects from the problem), optical astronomers can determine a relation between A and the diameter d of the asteroid. Such measurements will not give A and d separately, since both together affect the observations. However, what infrared astronomers can observe is the *absorbed* fraction $(1 - A)$ of the incident sunlight. They can observe the absorbed fraction because absorbed sunlight heats the asteroid until just as much energy is emitted in the infrared as is absorbed from the sunlight. What happens now is that infrared astronomers find a relation between $(1 - A)$ and the diameter d. Like their optical colleagues, infrared astronomers cannot separate $(1 - A)$ and d. But if infrared and optical astronomers pool their results, they can find A and d separately.

Using this idea, D. A. Allen in 1971 obtained 540 km for the diameter of the asteroid Vesta, and this was some 50 percent larger than the diameter had previously been thought to be. Table 6-2 compares the new infrared diameters of the four largest asteroids with the older estimates, and the new values are seen to be systematically larger.

TABLE 6-2
Diameters of the largest asteroids

Asteroid	Infrared diameter (km)	Old diameter (km)
1. Ceres	1,040	740
2. Pallas	570	480
3. Juno	250	200
4. Vesta	540	380

The new infrared diameters have an interesting effect on the *densities* that are calculated for the asteroids. Treating Ceres as an example, the old diameter of 740 km gave a volume of 2.12×10^{23} cm^3. The mass of Ceres has been estimated to be about 6.10^{23} g, so the density was formerly thought to be in the region of 3 g/cm^{-3}, a density typical for ordinary rock. The infrared diameter of

1040 km gives a volume of 5.89×10^{23} cm³, however, and the density is then close to 1 g/cm³, that is, close to the density of water-ice. Thus, the new infrared diameter would imply that Ceres and the other larger asteroids are really snowballs rather than chunks of rock. The values obtained for A from the observations require the surfaces of the asteroids to have dark, dirty coverings of dust; they are not clean, shining snowballs.

Again we come back to the ubiquitous dust that we encountered repeatedly in the molecular clouds of Chapter 5. We found it in the centers of galaxies, in huge clouds surrounding whole galaxies, as for instance around NGC 5128 (Figure 4-6), and now we have it in our own solar system. In the next section, we attempt to track down the composition of this dust, using observations both at visual frequencies and in the infrared.

§6-2. *Interstellar Dust*

Interstellar dust is visually the most conspicuous component of the matter between stars. It shows up as dark patches and striations against the background of distant star fields. The light from individual stars is dimmed, as well as reddened, by its passage through interstellar clouds, due to interaction with dust particles. The reddening occurs by a process similar to the reddening of a street lamp through a fog, or to the reddening of sunlight in the evening sky. Blue light is scattered, as well as absorbed, more strongly by dust than is red light. For natural light, which is comprised of a mixture of all colors, blue light is preferentially removed compared to red light, which is transmitted. The result is that the original source—Sun, star, or street lamp—appears reddened. Blue light is also scattered more strongly than red light by atoms and molecules; this phenomenon is the reason the sky is blue.

INTERSTELLAR GRAINS HAVE DIMENSIONS OF THE SAME ORDER AS THE WAVELENGTH OF LIGHT

Early investigations of interstellar dust were restricted mainly to a study of the way the dust absorbs and scatters starlight. The first attempts to obtain quantitative estimates of the dimming, or *extinction* as it is often called, were made as long ago as the 1930s. It was shown that, at a single photographic wavelength close to 4,500 Å ($1 \text{ Å} = 10^{-8}$ cm), the extinction of starlight amounted to a halving of the energy flux for every 2,000 light-years of passage of the light along the plane of the Milky Way. From this result, it was easy to infer that interstellar dimming could be reasonably attributed only to solid particles that have dimensions comparable to the wavelength of the light itself, about 5×10^{-5} cm. Other kinds of absorbers and scatterers—for example, electrons, atoms, or small molecules—could be shown to require implausibly large density values if they were to produce the required amount of extinction.

With the advent of photoelectric techniques in astronomy, it became possible to study the way interstellar extinction varies with wavelength. The study is done by comparing the spectra of two stars that are intrinsically similar, but one of which is more dimmed than the other. The relation between extinction and wavelength obtained from such a comparison is called the *interstellar extinction curve,* and it has an important bearing on the chemical nature and properties of the interstellar dust.

ABSORPTION IN THE ULTRAVIOLET IS A CLUE TO THE COMPOSITION OF INTERSTELLAR GRAINS

The first stellar extinction curve, obtained in the 1930s by J. Stebbins, C. M. Huffer, and A. E. Whitford, covered the wavelength range from 3,500 Å to 5,500 Å (that is, 3.5×10^{-5} cm to 5.5×10^{-5} cm). Over this range, a rule known as the $1/\lambda$ law was established. It stated that the *logarithm* of the factor by which starlight was extinguished was proportional to the reciprocal of the wavelength of the light, $1/\lambda$. Remarkably, this rule was found to hold good from one star to another over wide areas of the sky. In recent years, the interstellar extinction curve has been extended to shorter wavelengths, into the ultraviolet as well as into the infrared. The form of the curve, as it is now known, is shown in Figure 6-4. Note that it is $1/\lambda$, with λ expressed in microns, that is plotted in

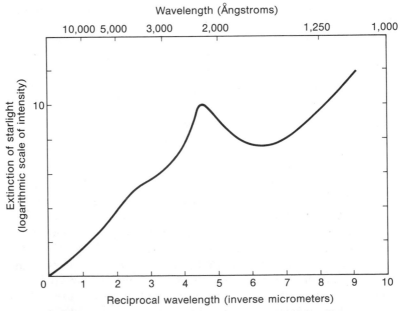

FIGURE 6-4
Dimming—also known as extinction or absorption—of starlight in its passage through cosmic dust clouds.

Figure 6-4. Thus, the old $1/\lambda$ law is simply the straight bit of the curve to the left between about 2 and 3 in this scale for $1/\lambda$. The data in the ultraviolet—that is, for $1/\lambda$ greater than about 3—were obtained by T. P. Stecher, R. C. Bless, and A. D. Code using equipment carried on rockets and Earth satellites. The most conspicuous feature of the whole curve is the broad hump centered on the wavelength 2,200 Å in the ultraviolet.

Many astronomers have come to see this 2,200-Å feature as indicative of the presence of carbon particles in the form of graphite, following calculations made in 1965 by N. C. Wickramasinghe and one of the authors. Indeed, the observed extinction hump at 2,200 Å does have an uncanny resemblance to the effect of spherical carbon particles with diameters less than about 500 Å. But this explanation raised a difficulty because one would expect graphitic carbon to be in the form of small plates rather than spheres. Faced with this difficulty, both Hoyle and Wickramasinghe, as well as groups in Japan, noticed that organic molecules with the chemical formula $C_8H_8N_2$ have ultraviolet absorptions close to 2,200 Å. The point here is that $C_8H_8N_2$ can be broken up to produce organic molecules appearing in Table 5-1,

$$C_8H_8N_2 \longrightarrow HCN + HC_7N + 3H_2$$

and

$$C_8H_8N_2 \longrightarrow HC_3N + HC_5N + 3H_2,$$

so that an interesting link between the ultraviolet measurements and millimeter-wave astronomy might be established in this way. However, the need for carbon particles to be small spheres rather than larger plates may not be the difficulty it was thought to be at first. If one begins with a long organic polymer instead of with carbon as graphite, degradation of the polymer through the loss of the more weakly attached atoms would cause an initially needle-shaped particle to ball up like wool into a spherical form. Thus, starting with a needle-shaped polymer, say 5.10^{-5} cm in length, we would arrive at specks of "soot" only some 1 to 200 Å in diameter, in good agreement with the requirement of the 2,200-Å hump of the extinction curve.

INTERSTELLAR GRAINS MAY BE A COMPLEX MIXTURE

Perhaps the most dramatic discovery of infrared astronomy concerning the nature of the interstellar dust was made in 1969 by E. P. Ney, D. A. Allen, W. A. Stein, J. E. Gaustad, F. C. Gillett, and R. F. Knacke. They found a broad feature appearing in the spectra of certain highly luminous stars lying in the giant region of the H–R diagram in the wavelength range from 8μ to 12μ. This feature was attributed to the presence of mineral silicates that are condensed within material flowing from the stars into space, and it was suggested, as a deduction from this observation, that mineral silicates might well constitute a dominant feature of the interstellar dust. Silicate particles, if they were of the right size (and for certain other reasons, if they were rod-shaped), could explain the behavior of interstellar dust over the visual range of frequencies. In particular, silicate particles of the right size could explain the $1/\lambda$ rule already mentioned. But silicate particles cannot explain the 2,200-Å hump in the ultraviolet, and it seems that this hump must still be attributed to carbon particles or to organic molecules.

The silicate hypothesis received a further boost from results such as those in Figure 6-5, all of which indicate powerful infrared sources where no visible

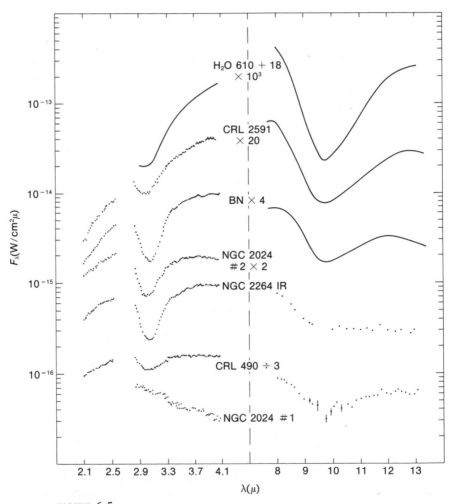

FIGURE 6-5

The 2.1–4.1μ and 8–13μ spectra of infrared objects associated with molecular clouds. All objects, except NGC 2024 No. 1, show absorptions at 3.08μ and around 9.7μ, attributed to ice and silicate absorption, respectively. (Reprinted courtesy of K. M. Merrill, R. W. Russell, and B. T. Soifer, and *The Astrophysical Journal*, published by the University of Chicago Press; © 1976 The American Astronomical Society.)

exciting star is seen. These results, obtained with increasingly refined instrumentation by K. M. Merrill, R. W. Russell, and B. T. Soifer, show absorption occurring from 8μ to 12μ and also from 2.9μ to 3.3μ. The former absorption was attributed to silicate particles and the latter to small crystals of water-ice. Neither of these forms of particles could be responsible for the main infrared emission, however, and the main emission may be due to carbon. There would thus be three types of particle: carbon, responsible for the overall emission; water-ice, responsible for absorbing the emission from the carbon over the wavelength range from about 2.9μ to 3.3μ; and silicate particles, responsible for absorbing the emission from the carbon over the wavelength range from 8μ to 12μ. The difference from one source to another in Figure 6.5 was explained partly by differences in the temperature of the carbon particles and partly by the assumption that silicate and water-ice grains are present in different proportions.

AN ORGANIC COMPONENT MAY BE PRESENT AND ALSO DOMINANT

An attempt to simplify this picture of interstellar dust was recently made by Hoyle and Wickramasinghe. They were looking for a single material that could be responsible for all the emission and all the absorption. These men found that a single material with infrared properties remarkably similar to those measured in the laboratory for the commonest of all biopolymers, the well-known material cellulose (present in its purest form in cotton), gave results in reasonable agreement with the sources of Figure 6-5, as can be seen from the selection of cases shown in Figure 6-6.

Most astronomers believe this association with the known properties of cellulose to be fortuitous because they think it unlikely that a biomolecule as complex as cellulose could be present in very large quantity in the interstellar material. Even so, the cellulose molecule has some interesting properties that are worth mentioning. A single cellulose strand has the shape of a long plank of timber, giving it the rodlike property we know the interstellar dust must have.* Many planks can be firmly stacked together as they can in a timber merchant's yard. Indeed, it is just this stacking property that gives cellulose its importance in providing the structural strength of plants and trees. Tubular arrangements of the strands, such as those that occur in wood and in the stalks of ground plants, permit other substances to fit conveniently within hollows between the strands, a property that could be significant in an astronomical context.

Another point of relevance to the composition of interstellar dust is the amount of it. To explain the observed extinction of visual starlight, dust must make up from 1 to 2 percent of the total mass of the interstellar clouds. And this makeup is very similar to the amount of carbon, oxygen, and nitrogen present in the clouds, whereas it is at least three times too large for silicate materials.

---

*The reddening of starlight shows so-called polarization properties that require a rodlike shape for the grains.

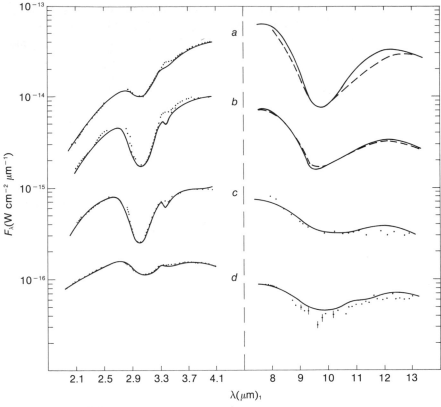

FIGURE 6-6

Cases *a* through *d* represent observational data (points) from Figure 6-5 and theoretical calculations (solid curves) for an absorbing cloud of cellulose. The theoretical curves are normalized to match the observations at one wavelength. (From *Nature*, 268, 610, 1977).

Moreover, the space observations of ultraviolet light from bright hot stars are subject to absorption by the interstellar gas at the particular frequencies of the quantum transitions of the atoms that make up the gas. This has the effect of introducing absorption-line characteristics of common atoms into the ultraviolet light from the stars; from the amount of this absorption, it is possible to infer the abundances in the gas of common atoms such as carbon, oxygen, and nitrogen. What has emerged from such investigations is that C, N, and O atoms are *underabundant* compared to hydrogen atoms. Since these underabundances are *for the gas only,* the inference has been made that the bulk of interstellar carbon, oxygen, and nitrogen must be tied up in solid particles, as indeed may also be the case for other atoms such as magnesium, silicon, and iron. But, among all such atoms, carbon, oxygen, and nitrogen have the higher abundances, and consequently they would dominate the composition of the interstellar dust. It seems quite likely, therefore, that some form of carbonaceous polymer forms an important component of the interstellar dust.

We have reserved the discussion of the origin of the organic molecules set out in Table 5.1 to this point because two of the three theories that have been suggested involve interstellar dust. The theories are:

1. That the molecules are formed within the interstellar gas by *ion-molecular chemistry,* a term that will be explained in a moment.
2. That the molecules are formed by *surface chemistry,* the surfaces being provided by the grains of dust.
3. That the molecules are break-up fragments from the dust itself, an important component of the dust being an organic polymer.

Atoms within the gas are constantly being ionized by high-speed particles—high-speed protons and electrons—the electrons being the kind that produce radio waves from the Milky Way by the synchrotron process, the radio waves first discovered by Karl Jansky. When an atom A, say, has been ionized (that is, when it has lost an electron), we denote it by A^+. Now, whereas molecule formation in a gas by a radiative reaction between two ordinary electrically neutral atoms, A and B, say, by

$$A + B \longrightarrow AB + \text{radiation},$$

takes place only very slowly indeed, the reaction

$$A^+ + B \longrightarrow AB^+ + \text{radiation}$$

occurs much more readily. The latter fast reaction is of the ion-molecule type from which the term ion-molecule chemistry is derived. Next, we can have

$$AB^+ + C \longrightarrow AB + C^+$$

also taking place very rapidly, yielding the molecule AB and also the ionized form C^+ of some other atom, which can itself indulge in a further rapid reaction, say,

$$C^+ + D \longrightarrow CD^+ + \text{radiation}.$$

Then we could have

$$CD^+ + E \longrightarrow CD + E^+,$$

and so on, in a rapid cascade that began with A^+.

This idea for promoting molecule formation within the diffuse interstellar gas was suggested in 1972 by P. M. Solomon and W. Klemperer, who proposed that organic molecules are formed by cascades that begin with ionized atoms of hydrogen and helium (that is, $A \equiv$ hydrogen or $A \equiv$ helium in the preceding reactions). Undoubtedly, this process must occur, and a strong indication that it does is the fact that it correctly predicts that the methylidyne ion, CH^+, should be the most common molecule after H_2.

What ion-molecule chemistry does not predict, however, is that the more complex molecules of Table 5-1 are quite comparable in their abundances to simpler molecules. Thus, while we should expect

$$AB + C^+ \longrightarrow ABC^+ + \text{radiation}$$

and

$$ABC^+ + D \longrightarrow ABC + D^+$$

to lead to three-atom molecules, and similarly to four-atom molecules and so on, we would expect the abundances of larger molecules generated by the more complex cascades to be markedly lower than the abundances of the smaller molecules generated by the simpler cascades. Yet this has not been found to be so. Thus, the sequence HC_3N, HC_5N, HC_7N, HC_9N has the following approximate abundance ratios:

$$10 : 5 : 1 : \tfrac{1}{3}.$$

While the abundances certainly decrease along the sequence, the amounts of the decrease are not nearly as large as the calculations of ion-molecule chemistry seem to require.

It is hard to assess the potentiality of grain-surface chemistry at the time of writing because many uncertainties enter the calculations that have been made for this theory and because laboratory experiments in surface chemistry do not simulate the conditions of interstellar space with adequate fidelity. Some workers have claimed that all the molecules of Table 5.1 can be generated on grain surfaces, but, even when the most favorable assumptions are made, the resulting abundances for the various molecules that have been calculated differ from the observed abundances in very crucial respects.

If the more complex molecules of Table 5-1 are fragments from much larger organic polymers, it is rather easy to see why their abundances are roughly comparable with one another. Thus, the third theory explains in a natural manner the otherwise puzzling feature that the abundances do not fall off sharply as the complexities of the molecules increase. The tentative opinion of the authors is that a combination of the first and third theories is mainly responsible for Table 5-1, with ion-molecule chemistry dominant for the small molecules and with the break up of polymer grains responsible for the larger molecules. These studies are at an early stage of development, however, and it is still premature to make any firm statement on the matter. Just as the existence of the molecules of Table 5-1 came as a great surprise to astronomers, so the eventual resolution of their origin may contain ideas that have not yet been suggested.

General Problems and Questions

1. What atmospheric windows in the infrared are accessible from the Earth's surface?

2. How did William Herschel prove that the sun limits infrared radiation?

3. Describe the infrared survey of Leighton and Neugebauer. What results did it achieve?

4. Compare the emission of infrared radiation from galaxies with the emission of radio waves.

5. Describe the properties of interstellar grains.

6. Write an essay on the chemical nature of the interstellar grains.

7. Discuss the theories that have been proposed to explain the origin of interstellar molecules.

Chapter 7
X-Ray Astronomy

§7-1. Techniques

Reference to Figure 3-12 shows that the Earth's atmosphere is opaque to x rays. Unlike the situation in infrared astronomy, where, by ingenious methods, it is possible to deal with a situation in which the atmosphere absorbs a great deal of the radiation, the atmosphere is impenetrable for x rays. Hence, to observe x rays, the detecting instruments must be above the atmosphere, and they must be carried there either by rockets or by satellites. This inevitable need immediately puts x-ray astronomy in a different financial bracket from other newly developing branches of astronomy. With infrared astronomy, it is simply necessary to convince the head of a university physics department (such as E. P. Ney at the University of Minnesota) that infrared astronomy is worth doing; for x-ray astronomy, it is necessary to convince governments. This difference in scale made the task of pioneers of x-ray astronomy more difficult.

We saw in Chapter 2 that a telescope is a device for concentrating radiation, as, for instance, radio waves are brought to a concentrated focus by the huge metallic mirrors used in radio astronomy (Figure 4-8). While telescopes for x rays have been constructed in recent years, the pioneering work in x-ray astronomy did not make use of telescopes for a reason that can be appreciated by

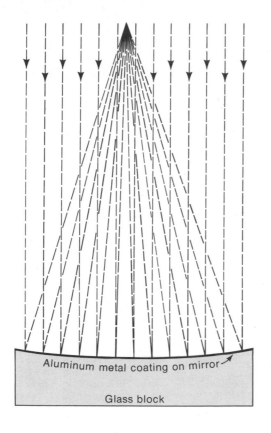

FIGURE 7-1
Parallel rays of light from a distant source are brought to a focus by a mirror with the shape known as a paraboloid. In practice, the focal point would be about three times farther from the mirror than it is shown to be here.

a comparison of Figures 7-1 and 7-2. Figure 7-1 illustrates the method by which visual light is brought to a focus, the mirror for visual light being an appropriately shaped block of glass with a surface coating of aluminium. The essential feature of Figure 7-1 is that the mirror is able to reflect the light rays full face on, that is, at normal incidence. For x rays, however, the use of mirrors requires the radiation to be incident at a *grazing angle* as in Figure 7-2. The lower half of Figure 7-2 illustrates the way in which geometrical shapes are combined directly for x rays rather than being separated spatially as in an optical telescope that has distinct primary and secondary mirrors.

Without telescopes, the early x-ray observations were rather like looking at the sky with the naked eye. When people look at the sky with the naked eye, the eye and the brain can separate different bits of the sky even though the whole sky is exposed to our view, but x-ray equipment cannot separate bits of sky in this way. To concentrate on a particular bit of the sky, x-ray equipment looks through a tube as we ourselves might look through a hollow tube if we wanted to study a particular area of the sky with the naked eye without being distracted by the rest of the sky. In fact, early astronomers actually made their observations in much this way. Obviously, if we use a tube with a circular cross section, we should see a circular patch of the sky. If we use a tube with a rectangular cross

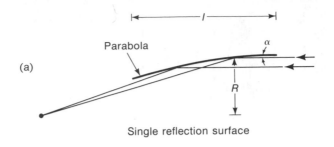

Single reflection surface

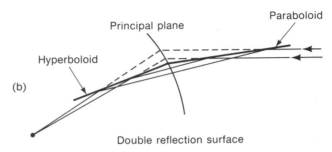

Double reflection surface

FIGURE 7-2
Focusing by a parabola line element. For small incident angles, the focal length is $R/2\alpha$, and the mirror intercepts a cross-sectional length αl of the beam. (a) The parabola is translated a height h perpendicular to the plane of the figure to form a plate that will focus in one dimension to a line of length h. The collecting area is αlh. (b) A hyperboloid following a parabola of revolution will produce a two-dimensional focus, with focal length $R/4\alpha$. The collecting area is $2\pi R\alpha l$. (From R. Giacconi and H. Gursky, eds., *X-Ray Astronomy*. Hingham, Mass.: D. Reidel, 1974.)

section, we should see a rectangular section of the sky. X-ray astronomers have found it more convenient to use rectangular tubes than circular ones because the detecting equipment that must be mounted at the end of the tube is more conveniently constructed inside a flat rectangular box than inside a circular box. An example of a detecting box, known as a *proportional counter,* is shown in Figure 7-3. The box is closed on all sides by metal surfaces impenetrable by x rays, except for one window made from a material through which the x rays can penetrate to the interior of the box. Beryllium sheet, aluminium foil, and plastic films are commonly used for this window.

Methods of x-ray detection are all based on the photoelectric effect in which an x-ray quantum of radiation ejects a fast-moving electron from an atom. The fast-moving electron then knocks other electrons out of other atoms, thus producing a trail of free electrons in its wake. In the ordinary visual frequency range, something of the same idea is used in the photomultiplier shown in Figure 2-31, where the incident quantum of radiation ejects an electron from a metal cathode. In a proportional counter, the electrons do not, however, come from metal surfaces as they do in Figure 2-31. Instead, they come from a gas contained within the detector box. A noble gas is used, usually xenon (He, Ne, A, Kr, Xe, and Rn are the noble gases). The anode of Figure 7-3 can be given a

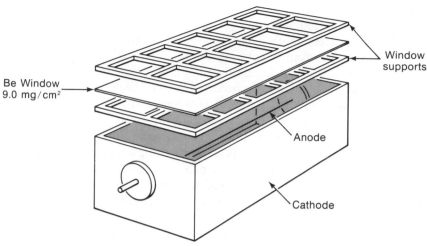

Window
supports

Be Window
9.0 mg/cm²

Anode

Cathode

FIGURE 7-3
Schematic layout of a thin window gas proportional counter. The Be window is cemented between a supporting "sandwich" that in turn is hermetically sealed to the cathode to preserve the gas integrity. The anode is usually kept under tension by a spring. The charge sensitive preamp and high voltage power supply are ideally mounted as close as possible to the anode feedthrough. (From R. Giacconi and H. Gursky, eds., *X-Ray Astronomy*. Hingham, Mass.: D. Reidel, 1974.)

positive electric charge by a battery, and this positive electric charge draws the trail of negatively charged electrons toward the anode wire. This trail is the one started by the x-ray quantum. As the electrons approach close to the anode wire, they can actually be speeded up to produce still more electrons, again by knocking electrons out of atoms of gas near the anode. Thus, a veritable pulse of electrons eventually arrives at the anode wire, a pulse that is then detected electronically. In this way, the box of Figure 7-3, plus electronic equipment and a battery to supply power, replaces the human eye at the end of the tube.

While this x-ray detection is simple in principle, there are many complicating issues in practice. There are also cosmic rays incident on a rocket or a satellite—mostly highly energetic electrons and protons coming to the Earth from along the plane of the Milky Way—as well as the wanted x rays. These cosmic rays also cause ionization of gas atoms within a proportional counter. Fortunately, there are two ways to distinguish the wanted x rays from the unwanted cosmic rays.

Cosmic rays penetrate the detecting box from all angles, not merely from the tube open to the sky. Those that come in from the side, not down the tube, can be detected separately by surrounding the outside of the tube by other sensor devices. Whenever one or another of these devices signals a cosmic ray coming into the equipment from the side, the electronics of the proportional counter can be suppressed so as *not* to count any ionization of atoms that occurs in the detector box.

The detecting device for the cosmic rays works somewhat like the proportional counter itself, except that it can be constructed in the cruder fashion of a

Geiger counter. In a Geiger counter, the first electron is accelerated strongly and immediately by an electric field so that it almost instantly knocks out a second electron from an atom of gas. This second electron almost instantly knocks out a third electron, and so on, so that a vast cascade of electrons is quickly produced throughout the counter, a cascade that takes some time to disappear. In the proportional counter, on the other hand, the burst of electrons is produced only close to the anode wire, and the burst is sudden and soon over. Because the proportional counter works this way, it can distinguish several x-ray quanta separately since the electron pulses from the several quanta usually arrive at different parts of the anode wire, thereby permitting a spatial resolution to be made. Additionally, the sharpness of the pulses themselves often permits a time resolution to be made. This ability to make both time and space resolutions is exactly why counters like the one in Figure 7-3 are called proportional counters. In a Geiger counter, the cascade of electrons is enormous and widespread, and time resolution is poor. Consequently, a Geiger counter cannot distinguish between several x rays or several cosmic rays arriving more or less at the same time.

Surrounding the x-ray detecting equipment by simple counters of the Geiger type is thus the first defense of the x-ray astronomer against unwanted cosmic rays. This defense does not work, however, against the minority of cosmic rays that come directly down the open tube itself—that is, against cosmic rays that happen to come from the same direction as the x rays. Fortunately, the pulses of electrons that arrive at the anode wire for these cosmic rays are more diffuse—in particular, they have fewer sudden onsets—than the pulses due to the x rays. Thus, diffuse pulses are once again rejected by the electronic equipment to which the counter of Figure 7-3 is connected.

There are other problems besides cosmic rays. Electrons are knocked out of atoms, and light is also emitted from atoms. This emission of light is occasioned by the colliding electrons serving to change the states of the atoms, which become excited and then radiate light in the manner discussed in Chapter 2. This radiation of light happens most strongly as the pulse of electrons approaches the anode wire. If no precautions were taken, the light would travel to the metal walls of the box, and the metal walls would act as a cathode from which further electrons would be knocked out. Such secondary electrons entering the box would themselves drift toward the anode wire where they would be accelerated to give rise to further pulses of electrons, and an on-going sequence of similar events would take place. There would thus be a long, unwanted trail of pulses instead of the simple sharp pulse we have assumed to this point. To obviate this difficulty, an *impurity gas* (or *quench gas*) is deliberately included in the detector box—a quench gas chosed for its ability to absorb light and thus prevent it from reaching the walls of the box. Methane, carbon dioxide, and alcohol are used as quench gases.

As well as coping with all the problems that inevitably arise, physicists have to deal with yet another insidious difficulty, the problem of noise. Sequences of improbable events take place from time to time. A cosmic ray may get into the detecting box without triggering a protecting counter, or an electron may acquire a particular energy that enables it to mock an electron produced by a wanted x ray. If an experiment is performed for a short interval of time, improbable events may occur during the particular interval of the experiment causing a false conclusion to be reached, with the improbable events being misinterpreted as a systematic effect. The defense against a mistake of this kind is obviously to run the experiment for a long interval because improbable events will not continue to happen indefinitely. Unfortunately, it is not always possible for x-ray astronomers to run long experiments. Rocket-borne equipment is viable only for minutes. The limit to the length of time available for equipment in a satellite is imposed by the fact that the bit of the sky under observation is not the only region of the sky that the x-ray astronomer wishes to investigate. Therefore, a number of similar observing tubes, each with its own proportional counter, are constructed to operate simultaneously. Having many tubes operate for a short time, with their results added, is just as good as having a single tube continue for a very long time. Figure 7-4 shows the first x-ray satellite, drawn in schematic form, the two banks of observing tubes on opposite sides of the satellite being marked "Collimator."

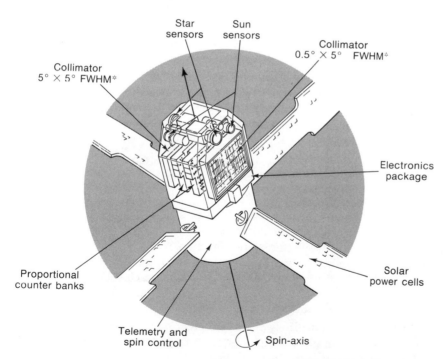

Star sensors

Sun sensors

Collimator
0.5° × 5° FWHM*

Collimator
5° × 5° FWHM*

Electronics package

Proportional counter banks

Solar power cells

Telemetry and spin control

Spin-axis

*Full width half maximum.

FIGURE 7-4

Exploded view of the *Uhuru* (SAS-A) instrumentation. Each collimator defines approximately 840 sq cm effective area for x-ray detection by a proportional counter bank. The nominal energy bandwidth is 2–20 keV, and the limiting time resolution is 0.096 second. The spin control section contains a constant speed momentum wheel and nutation damper for stability. (From Giacconi, R., and H. Gursky, eds., *X-Ray Astronomy.* Hingham, MA: D. Reidel, 1974.)

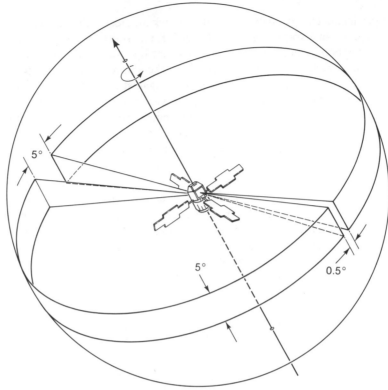

FIGURE 7-5
Observation of the sky by the Uhuru detectors. The spin axis can be commanded anywhere
in the sky, with a stability of about 1° in 1 day. Each collimator sweeps an 5°-wide swath
that is almost a great circle, except that the circles swept by the 5° and 0.5° collimators are
offset ±1.2° from the equatorial plane of the spin axis. (From R. Giacconi and H. Gursky,
eds., *X-Ray Astronomy*. Hingham, Mass.: D. Reidel, 1974.)

Another reason that a single tube cannot be maintained pointing at a particular area of the sky can be seen in Figure 7-4; it can be seen still more clearly in
Figure 7-5. Satellites spin, causing the direction of observation to change all the
time. For the satellite *Uhuru,* there are two sets of tubes, each tube in one set
having a 5° × 5° field of view at any instant and each tube in the other set
having a 0.5° × 5° field. The tubes are all oriented so that the rotation of the
satellite causes them to sweep a 5° strip on the sky, as illustrated in Figure 7-5.

The reader will immediately perceive another problem. With the direction in
which the tubes point thus changing all the time, how can we know to which bit
of the sky the x-ray quanta, counted in a particular small time interval, really
belong? The solution to the problem is clear in principle. There must be a clock
in the satellite, and the x rays coming down the massed banks of tubes must be
counted against the time and the results automatically recorded on tape. The
direction in which the tubes point must also be recorded on tape at each
moment. For this latter recording to be possible, the satellite must "know" how
it is oriented with respect to the sky. For the satellite to know its orientation, the

direction of the axis of spin with respect to the stars and the instantaneous angle of the spin must be determined. There are two ways in which these very necessary data are obtained: (1) by using auxilliary visual telescopes set to sweep across particular bright stars (star sensors) and (2) by employing gyroscopic devices.

So far, nothing has been said of the way x-ray astronomers obtain information about the frequencies of the quanta they observe. For frequencies below 3×10^{18} cycles per second (that is, wavelengths greater than 10^{-8} cm), the energies of the quanta are judged from the strengths of the pulses of electrons that arrive at the anode wires of the proportional counters. For still higher frequencies, the proportional counters can be replaced by counters of a different type, scintillation counters. The essential component of a scintillation counter is a crystal, often a crystal of sodium iodide (NaI), containing a deliberately induced impurity, often thallium (Tl), that acts as a so-called *activator*. Such crystals emit pulses of light whose intensities can be measured when electrons of sufficient energy impinge upon them.

Another method now used in sophisticated equipment to determine the frequencies of x rays goes back to a discovery made long ago by William and Lawrence Bragg. They discovered that crystals scatter x rays through angles that are different for radiation of different wavelengths, as for instance the two wavelengths λ_1 and λ_2 shown in Figure 7-6. The wavelength-separation device

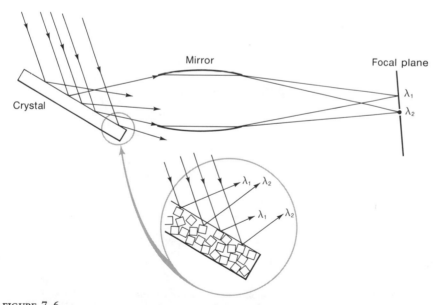

FIGURE 7-6

Functional diagram of a mosaic crystal spectrometer. The incident beam encounters micro-crystals with a range of orientations as shown in the blowup. The reflected beam incident upon the two-dimensional imaging mirror thus contains a range of angles, each angle corresponding to a unique wavelength. The focused image for each angular increment gives the spectral density at the wavelength satisfying the Bragg criterion for that angle. (From R. Giacconi and H. Gursky, eds., *X-Ray Astronomy*. Hingham, Mass.: D. Reidel, 1974.)

of Figure 7-6 is analogous in its effect to the simple glass prism used for visual light (Figure 6-2). The situation is much more difficult to execute for x rays, although the method is similar in principle.

Nothing we have discussed so far will permit x-ray astronomers to distinguish one bit of the field of view of the metal tubes from another bit (at the same instant of time). Nor will they be able to distinguish between an x-ray source that is essentially a point (for example, a star) and one that is a cloudlike object that fills an appreciable fraction of the field of view. Yet, by deliberately including obtacles such as grids of wires in the metal tubes, they can gather information about the detailed sizes of objects. What they must do is introduce one grid close to the mouth of the tube and another grid close to the inner end of the tube in such a way that, for a point source of radiation, one grid shadows the other for a given direction(s) of the source. Then, as the field of view sweeps across a *point source,* the grids sometimes shadow each other. The result is a sharply oscillating signal from the source. For an *extended source,* on the

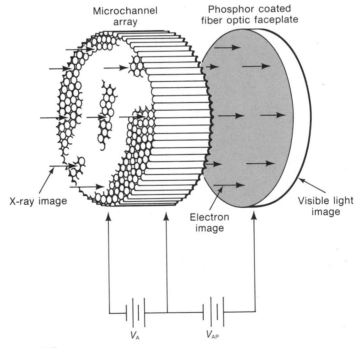

FIGURE 7-7

Two-dimensional channel multiplier array, used as an x-ray image intensifier. The x-ray image is converted to an electron image inside the entrance walls of the microchannel array. The bias V_A along the array multiplies these electrons, while the channel structure preserves the image. The resulting electron image is accelerated by V_{AP} to produce a visible image on a phosphor screen. (From R. Giacconi and H. Gursky, eds., *X-Ray Astronomy.* Hingham, Mass.: D. Reidel, 1974.)

other hand, the oscillations are either much weaker or absent altogether. When such a device is used, the bank of tubes is said to constitute a *modulating collimator*.

To end this section on techniques, Figure 7-7 shows an advanced proposal for an x-ray detection device, one that is easily comprehended—perhaps more easily comprehended than some of the earlier devices.

Once again we have a bank of parallel tubes, in this case sealed circular tubes, with windows at the left to admit the x rays. The tubes contain a gas, say, xenon, and photoelectrons are directed along the tubes by an electric field that also serves to accelerate the electrons so they knock more electrons from the atoms of gas. The result is that pulses of electrons emerge from the inner ends of the tubes. These pulses are then accelerated further, down the voltage V_{AP} of Figure 7-7, until they hit a screen. The result is an optical picture of the original x rays that were incident at the left of the apparatus—a picture, analogous to that of television equipment, that astronomers can then photograph or record digitally for future reference.

§ 7-2. X Rays from the Sun

SOLAR X RAYS CAUSE IONIZATION LAYERS IN THE EARTH'S UPPER ATMOSPHERE

One of the authors remembers how, in the middle 1940s, the question of whether the Sun might emit x rays was considered by astronomers to be highly speculative. Together with D. R. Bates, he suggested that the lower regions of the Earth's ionosphere might be produced by solar x rays. The ionized layers of atoms in the terrestrial atmosphere, usually beginning with the so-called E layer at a height of about 115 km above ground level and extending upward to a height of more than 300 km in the F layer, were then very important for long-range radio communication. During the daytime, radio waves were reflected downward by the E layer. At night, however, the free electrons that constituted the E layer dried up by attaching themselves to atoms, and the radio waves then reached the F layer where they were reflected back to Earth. This phenomenon made longer ranges of communication possible at night than during the day. Occasionally, the whole worldwide network of radio communication would break down due to what was called a *fade-out*. Fade-outs were caused, it was discovered, by the appearance of a temporary layer of free electrons—the D layer—at a lower height of about 80 km. Fade-outs were often found to be contemporaneous with the appearance of large flares on the Sun.

The coincidence in time between the visual appearance of a flare, such as the one in Figure 7-8, and the appearance of free electrons in the D layer showed that whatever had produced the electrons must have traveled from the Sun at essentially the speed of light. This speed made it exceedingly probable that the electon-producing agency was some form of radiation. The radiation had to be capable of ionizing molecules of nitrogen and oxygen, the common gases of the

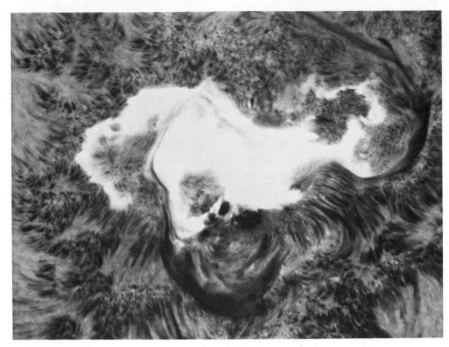

FIGURE 7–8
A flare on the Sun. Flares cause particle streams to be ejected from the Sun. Such jets of particles move rapidly outward and sometimes impinge upon the Earth. (Courtesy of Hale Observatories.)

Earth's atmosphere (visual light will not ionize them), and the radiation had to penetrate down through the atmosphere to a height of about 80 km above ground level. These requirements, when put together, clearly pointed to x rays as the cause of the radio fade-outs.

This issue was of considerable commercial importance, and, therefore, the impetus toward experimental investigation was strong. Small rockets could conveniently be used, and, following World War II with its development of rocket techniques, rocket research on radio fade-outs and their connection with solar activity was quickly underway at the U.S. Naval Research Laboratory (NRL). In 1948, only six years after radio emission from the Sun was first detected, x rays from the Sun were found.

Just as radio emission could be greatly enhanced by activity connected with the solar cycle (see Chapter 4), so could x-ray emission also be greatly enhanced. As it turns out, the Sun emits x rays about 100 million times more strongly at the time of a large solar flare than it does when it is quiescent (when there are few, if any, spots on the Sun).

Decisive proof that radio fade-outs were caused by the emission of x rays from solar flares, over a time scale of about an hour, was obtained in 1956 by T. Chubb and H. Friedman. Equipped with ten small rockets aboard the U.S.S.

Colonial some 500 miles out to sea from San Diego, they were in the habit of firing one rocket each day. On the fourth day of this routine, they received news from R. T. Hansen at the Climax Observatory of the University of Colorado that a large flare was just then breaking out on the Sun. They launched a rocket immediately, and with good luck it reached its full altitude at just the time when the flare activity was at its maximum; they detected an intense burst of x rays. This expedition of Chubb and Friedman had another important outcome that is discussed in the next section.

THE OUTER ATMOSPHERE OF THE SUN IS BOTH HOT AND DYNAMICALLY ACTIVE

The Sun possesses a hot diffuse outer atmosphere with a temperature of from 1 to 3 million degrees K. This very hot atmosphere causes a thin gas to extend, with steadily decreasing density, to great distances from the Sun, as we can see in Figure 7-9. Figure 7-9 is a photograph of the solar corona taken in visible white light during a total eclipse, that is, when the Moon comes directly between

FIGURE 7-9
The solar corona at the total eclipse of June 1973. The form of the corona is indicative of the presence of magnetic forces. (Courtesy of High Altitude Observatory, Boulder, Colorado.)

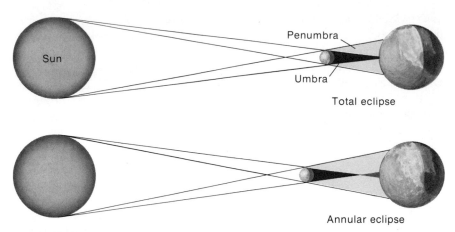

FIGURE 7-10

An eclipse of the Sun occurs when the Moon comes between the Earth and Sun. For the eclipse to be total, the Moon's direct shadow must reach the Earth. Since the Moon's distance from the Earth varies with time—the Moon's orbit around the Earth not being exactly circular—both the situations shown in this figure occur. Annular eclipses are more frequent than total ones. (From J. C. Brandt and S. P. Maran, *New Horizons in Astronomy,* Second Edition. W. H. Freeman and Company. Copyright © 1979.)

the Earth and the Sun, as in Figure 7-10, thereby cutting off the light from the Sun. The structure of the corona with its many streamers is indicative of the presence of magnetic forces. Indeed, Figure 7-9 plainly shows that the Sun behaves like a great magnet.

A hot, dense gas soon cools itself by radiation, as we know from practical experience. A hot, *diffuse* gas like the one in the corona also cools itself, but, because of its low density, it takes much longer to do so than a dense gas. Consequently, it is possible for the coronal gases to be very hot without the emission of radiation being very intense. By very hot, we mean that the particles move at much higher speeds than particles do in the cooler but denser gas at the photosphere of the Sun (see Figure 4-2). Although the coronal particles move fast, they do not go very far in any one direction because they collide repeatedly with other particles, as shown schematically in Figure 7-11. Radiation is emitted during such collisions, the radiation from the very hot gas in the corona being quite different from ordinary visible light because of the high speeds of the collisions (Figure 7-11). The radiation from the corona consists of ultraviolet light and x rays. Once again, we emphasize that, although the corona thus emits radiation of an unusual kind, the radiation is much weaker in its power output than the emission from the denser gases at the photosphere.

The gas in the solar corona is hot enough so that there is no suitably defined outer boundary to the Sun. A *wind* of particles is constantly streaming outward from the corona in the manner of Figure 7-12. The density within the wind as it reaches the Earth ranges from about 1 atom per cubic centimeter on the low

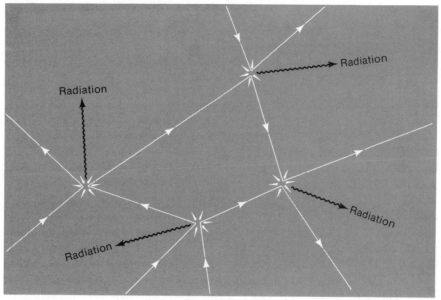

FIGURE 7-11
Schematic representation of particle motions in a hot gas. When one particle collides with another, the directions of the motions tend to be randomized. In a sufficiently hot gas, the collisions may be sufficiently violent for x rays to be emitted.

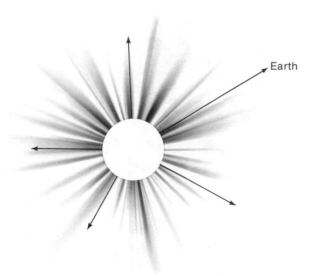

FIGURE 7-12
The Sun emits a steady wind as well as more intense jets of particles. A jet incident on the Earth produces a terrestrial magnetic storm.

side to hundreds or even thousands of atoms per cubic centimeter at maximum intensity. The wind at its strongest can appreciably disturb the magnetic field of the Earth, producing what are known as *magnetic storms*. This wind has other observable effects on planets, on comets, and on the Moon.

Why are the diffuse gases of the corona so hot? The answer to this question appears to be rooted in convective motions below the photosphere, as shown schematically in Figure 7-13, motions whose cause we discuss in Chapter 8. Their effect is to generate sound waves that travel upward into the solar atmosphere, the sound waves becoming more and more violent as the density of the gas decreases with increasing height. In other words, the particles are constantly being shaken backward and forward by these sound waves, and the shaking motion becomes more and more intense as the density of the gas decreases. Collisions of particles eventually cause the wave motions to be dissipated into heat, the heating effect being strongest where the motions are most violent—that is, in the corona. The process is very likely a good deal more complicated than it may seem from this simple description. Magnetic fields undoubtedly also play a role, especially in dictating variations of temperature from one part of the corona to another, probably through energy passing from such fields into the coronal gases. Thus, magnetic fields are almost surely

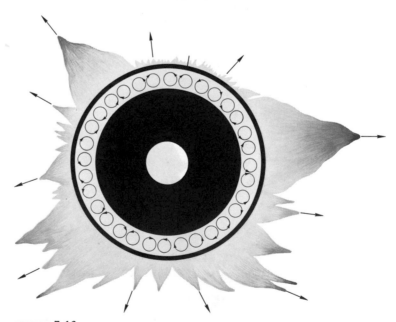

FIGURE 7-13
Schematic representation of the structure of the Sun. Energy is generated in a central core. The energy is then carried by radiation to a subsurface zone where it produces convective motions that may well be responsible for the solar cycle.
(From J. C. Brandt and S. P. Maran, *New Horizons in Astronomy,* Second Edition. W. H. Freeman and Company. Copyright © 1979.)

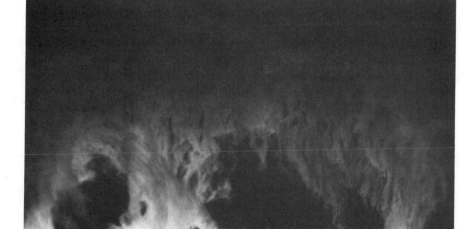

FIGURE 7-14

A giant prominence on the Sun. Gases in the corona are cooling and moving under the influence of both gravity and magnetic forces. (Courtesy of Hale Observatories.)

responsible for particular hot spots in the corona from which exceptionally strong x-ray emission takes place. Hot regions with looped structures such as those shown in Figure 7-14 are undoubtedly associated with magnetic fields, as indeed is the whole range of phenomena (including sunspots) that go by the name *solar activity*.

The level of solar activity determines the intensity of the x-ray emission from the Sun. When the Sun is active with sunspots, with arched structures like the one in Figure 7-14, and with the corona showing a complex of streamers, the x-ray emission is about a hundred thousand to a million times stronger than when the Sun is at its quietest. The x-ray emission from the active Sun, with the regions of strong emission represented by bright zones, is shown dramatically in Plate V.

THE SUN EMITS X RAYS MOST INTENSELY DURING FLARE ACTIVITY

Flares occur much lower down in the solar atmosphere than the hot regions of the corona that give rise to the bright zones of Plate V. Indeed, flares occur in regions not much higher than the photosphere, the level that gives rise to most of the ordinary visible light of the Sun. The process of x-ray emission from flares is the same as that illustrated schematically in Figure 7-11, namely, collisions of fast-moving electrons with other particles. The fast-moving electrons in flares are generated by intense electrical discharges, however, not simply by heat. They are said to be *nonthermal,* and they are analogous in their origin and properties to the fast-moving electrons discussed in Chapter 4 in connection with the emission of synchrotron radiation. Because the electrons in flares have more energy, particle for particle, than the electrons in the corona, the x rays they emit have higher frequencies, frequencies above 10^{19} cycles per second. It is precisely these higher frequencies that permit x rays from flares to penetrate more deeply into the Earth's atmosphere than the x rays from the corona. It is this deeper penetration that produces the temporary D layer that causes radio fade-outs, the fade-outs whose commerical significance led to the first x-ray investigations outside the confinement of the Earth's atmosphere. Indeed, the golden age of x-ray astronomy may be said to have been initiated by these investigations that were spurred on because of their commercial importance.

§7-3. *ScoX-1, the First X-Ray Source Discovered Outside the Solar System*

In 1956, Chubb and Friedman, in addition to fulfilling their avowed purpose of identifying solar flares as the cause of radio fade-outs, made a major discovery, one analogous to Karl Jansky's discovery of radio waves. Just as radio waves from the galaxy were discovered as a general hiss along the Milky Way, Chubb and Friedman found x rays coming diffusely from the many directions to which their x-ray detecting equipment was pointed. As there was a problem in 1933 in understanding how radio waves could be diffusely generated (cosmic synchrotron radiation by high-speed electrons had not yet been thought of in 1933), so there was a mystery in 1956 in understanding how x rays could be diffusely generated. Attempts were, of course, made to apply the same ideas as those we have already described for the Sun. However, in the interstellar gas, the density was too low for the picture of Figure 7-11 to be relevant. There were too few collisions of an appropriate high-energy kind. Nor was it possible for synchrotron radiation to produce x rays diffusely in space because the intensity of the general galactic magnetic field was far too low to do much more than generate waves in the radio band (Jansky's hiss).

As we shall see, there is a process, first suggested by one of the authors, whereby x rays can be produced diffusely in space, but this process could not be thought of in 1956 because it depended on a discovery that was not made until 1965—namely, the microwave background found by Arno A. Penzias and Robert W. Wilson. So, in 1956, the only deduction to be made was that the diffuse x rays found by Chubb and Friedman came from a multitude of comparatively faint sources of small angular size that could not be separately distinguished by the detecting equipment. Unlike the case of radio waves, however, it was clear from the beginning that many of the sources for x rays would be required to lie outside our galaxy because the diffuse x rays were not confined to directions along the Milky Way itself.

About 1960, a second x-ray group began to form, with B. Rossi and G. W. Clark at the Massachusetts Institute of Technology (MIT) and R. Giacconi, F. Paolini, and H. Gursky at American Science and Engineering (ASE). At Lockheed Corporation, a further group under P. Fisher was also beginning to work on x rays. In the light of what was shortly to happen, the first aim of the MIT-ASE team, namely, to detect solar x rays scattered (fluoresced) by the Moon, seems astonishingly minor. At midnight on June 12, 1962, a rocket was launched to an altitude of about 230 km. Two of the three x-ray counters mounted on the rocket functioned correctly over 350 seconds of observation. When the counters were directed toward the south-southwest (geographical direction), a source of soft x rays with a strength of about 5 quanta crossing an area of 1 sq cm each second was found. This was a much stronger signal than had been expected or even hoped for. If it came from a nearby star, then the star had to be emitting x rays with a power output that exceeded the Sun ten-million-fold. This was the source that was later to be named Scorpius X-1 when its position on the sky had been shown to be in the constellation of Scorpius.

It is relevant to note that the rocket used for this crucial astronomical discovery was provided by the U.S. Air Force, not by NASA. Further scientific collaboration between the armed services and institutions such as MIT was soon to be forbidden by the Mansfield Amendment, and, for more than a decade after the discovery of Sco-X1, x-ray astronomy would prove a poor relation of NASA and be relegated to the class of minor projects.

By 1964, the problem for ScoX-1 was clear; it was to determine the position of the source on the sky with sufficient precision to make it worthwhile for a search using a visual telescope such as the 200-inch one on Mount Palomar. The search for an unusual star (or some other unusual object) would have to be made over the area of uncertainty of the x-ray observations, and, if this area were too large, the search would be like looking for a needle in a haystack. The modulation collimator described in Section 7-1 was developed at MIT by M. Oda, H. Bradt, G. Garmire, and G. Spada to restrict the area of uncertainty. The collimator was then built into a rocket payload by H. Gursky at ASE, a payload that was eventually flown on March 8, 1966.

The analysis of the results from this flight yielded two possible positions for the source, namely:

Right Ascension	Declination
$16^h17^m7^s \pm 4^s$	$-15°30'54'' \pm 30''$
$16^h17^m19^s \pm 4^s$	$-15°35'20'' \pm 30''$

The smallness of the errors (indicated by $\pm$) shows the formidable extent to which technologies can improve in only a few years; the errors of the first x-ray measurements had been about $\pm 2°$ or about 250 times larger.

An optical search of the two areas of sky defined by these positions could now profitably be made. In June 1966, an unusual star of the thirteenth magnitude

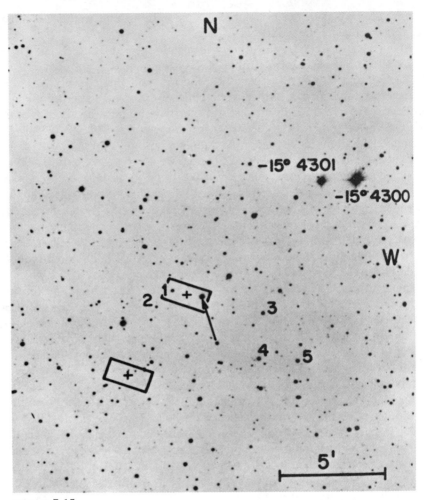

FIGURE 7-15

The x-ray source ScoX-1 was identified at the position shown by the arrow. (Reprinted courtesy of A. R. Sandage, P. Osmer, R. Giacconi, P. Gorenstein, H. Gursky, J. R. Waters, H. Bradt, G. Garmire, B. V. Sreekantan, M. Oda, K. Osawa, and J. Jogaku and *The Astrophysical Journal,* published by The University of Chicago Press; © 1966 The American Astronomical Society.)

was found at the position RA 16ʰ17ᵐ4.3ˢ, Dec. −15°31′13″ by Oda and his colleagues in Tokyo and by A. R. Sandage at Mount Palomar. All the results were pooled and published together by A. R. Sandage, P. Osmer, R. Giacconi, P. Gorenstein, H. Gursky, J. R. Waters, H. Bradt, G. Garmire, B. V. Sreekantan, M. Oda, K. Osawa, and J. Jogaku (1966, *Astrophysical Journal,* Volume 146, p. 316). The Palomar identification is shown by the arrow in Figure 7-15 that points into one of the two areas of uncertainty that had been determined by the x-ray observations.

THE HIGHLY EVOLVED REMNANTS OF STARS ARE OFTEN SOURCES OF X RAYS

The development of x-ray astronomy was to prove uncannily similar to that of radio astronomy, which means that our old friend the Crab nebula had to crop up early in the story. The identification of the Crab nebula by the NRL group under H. Friedman followed in 1963, hard on the heels of ScoX-1. Much more was known, of course, about the physical nature of the Crab nebula than about ScoX-1, the latter being a "star" whose properties were uncertain at this stage. The Crab nebula was known to emit visual light by the synchrotron process, and this emission demanded the presence of both high-energy electrons and a comparatively intense magnetic field within the blue light region shown in the lower part of Figure 2-7. Were there some electrons with energies so high that, instead of emitting ordinary visual light, they emitted x rays? This possibility certainly had to be considered. If it turned out to be correct, the x rays would be coming from an extended region comparable to the visual nebula of Figure 2-7.

There was one other possibility, namely, that there was a highly condensed central object within the nebula. If this second possibility were correct, the x rays would be from a pointlike object, not from the extended visual nebula. The problem was how to distinguish between these two possibilities.

The x-ray equipment of 1964 was not refined enough in its directivity to settle this question in a straightforward way, and a special scenario—one that was analogous to the method used by Hazard in the work that led to the discovery of QSOs in 1963—was employed by the NRL group.

It happened in 1964 that the Moon moved across the position of the Crab nebula. If the x rays were coming from the extended nebula, the Moon would cut them off smoothly over a time interval of about 12 minutes. But if the x rays were coming from a point source, they would be cut off all in an instant. The technical problem lay in having a rocket above the atmosphere during the critical moments during which the Moon reached the central part of the Crab, that is, the part where the pulsar NP 0532 is now known to be located.

On July 7, 1964, a rocket was in place at the correct time, and its x-ray counters showed only a slow change as the Moon crossed the position of NP 0532 (the oscillating starlike object of Figure 4-14). The position of the edge of the Moon at the moment it covered NP 0532 is shown in Figure 7-16. The detailed analysis of the results, obtained in the first place by S. Bowyer, E. Byram, T. Chubb, and H. Friedman in 1964 and later in 1967 by M. Oda and his colleagues (*Astrophysical Journal*, Volume 184, L5), showed that at least 90 percent of the x-ray emission of the Crab nebula came from an extended source that was spread across the area of the broken circle of Figure 7-16. While this conclusion still admitted the possibility that 10 percent of the emission might come from a central source (the now-known pulsar NP 0532), this possibility did not at first seem attractive.

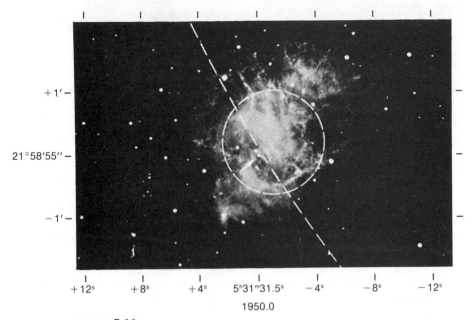

FIGURE 7-16

X-ray source in the Crab nebula (dashed curves) compared to visible light photograph (Hale Observatories). The curve of large radius represents the limb of the Moon as it occulted the center of the x-ray source as seen from a rocket. The center of the x-ray source (after a post-publication correction for the rocket motion) is close to the position of the pulsar. This observation also showed that the x-ray source had a finite size of about two arc-minutes along the direction of the Moon's apparent motion. The dashed circle represents the two-dimensional size and position of the x-ray source as determined by a modulation collimator observation from a sounding rocket. Both observations lacked the resolution needed to detect the discrete contribution of about 10 percent from the pulsar NP 0532 (arrow). (From R. Giacconi and H. Gursky, eds., *X-Ray Astronomy*. Hingham, Mass.: Reidel, 1974, by permission.)

As it turns out, the ghost of NP 0532 would not lie down. Balloon flights of x-ray detection equipment, notably from MIT, soon showed that the Crab was emitting higher and higher x-ray frequencies, much higher than those measured in 1964. It began to seem implausible that such very high frequencies could all come from the extended circle of Figure 7-16. Even so, the problem of how to distinguish between a point source and the extended circle seemed insurmountable (since the Moon would not again sweep over the position of the Crab until about 1973) until NP 0532 was discovered from its radio and optical properties.

As discussed in Chapter 4, NP 0532 emits radio and optical pulses (Figure 4-14) at a rate of 33/second. It became clear, therefore, that the way to distinguish NP 0532 from the dotted circle of Figure 7-16 was to look for flashes of x rays at exactly the same rate of 33/second. The race to detect the x-ray flashes was on, and, in 1969, within a short time interval, the NRL and ASE-MIT groups, the Goddard Space Research Center, and the Research

Center Saclay in France all found them. There seems to be no reason to doubt that the physical processes giving rise to optical and x-ray flashes, shown together for comparison in Figure 7-17, are the same. The common process may well be the one described in Chapter 4 and illustrated there in Figure 4-19.

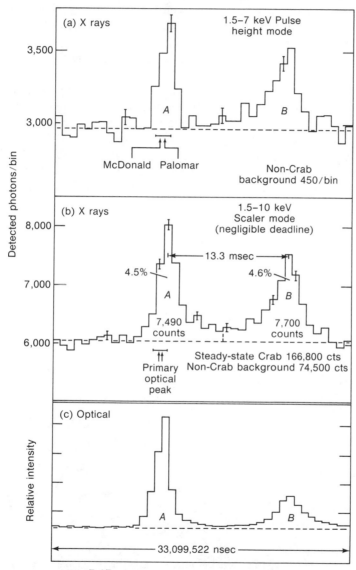

FIGURE 7-17

Comparison of simultaneous visible light and x-ray observations of the time structure of the Crab pulsar. The data are from an observation lasting several minutes. (From R. Giacconi and H. Gursky, eds., *X-Ray Astronomy*. Hingham, Mass.: Reidel, 1974, by permission.)

It was now possible to put together the radio, optical, and x-ray emissions of the Crab. In Chapter 4, we explained the idea of a bandwidth extending from frequency v to a neighboring frequency $v + dv$ and of writing the rate at which energy is received in this bandwidth (crossing unit area at the astronomer's telescope) as $F(v)\,dv$. Choosing the unit area as 1 sq m, taking the time interval over which the energy is received as 1 second, and measuring energy in terms of the watt-second (1 watt-second $= \frac{1}{1000}$ kilowatt-second $= 10^7$ ergs), $F(v)$ has the meaning of the quantity plotted logarithmically on the left-hand scale of Figure 7-18. The horizontal scale of this figure is simply the radiation frequency in cycles per second also plotted logarithmically.

Figure 7-18 shows two other objects as well as the Crab nebula. They are Cassiopeia A, the first radio source to be identified (1948), and "Tycho," and they have an important evolutionary property in common with the Crab. All are residues from the explosions of supernovae (Chapters 4 and 8).

In 1572, Tycho Brahe observed a supernova bright enough to be visible in broad daylight. In 1952, R. Hanbury-Brown and C. Hazard detected a radio source at the position where Tycho had recorded the apparition in the sky, but

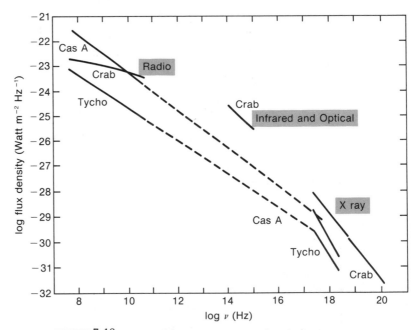

FIGURE 7-18

The electromagnetic spectra of three supernova remnants, the Crab nebula, the supernova observed by Tycho Brahe, and a third case known as Cassiopeia A. The logarithm of the frequency in cycles per second (Hertz) forms the absissa. The ordinate is the logarithm of the flux received, measured in watts per square meter for a bandwidth of 1 cycle per second (Hz). (From R. Giacconi and H. Gursky, eds., *X-Ray Astronomy.* Hingham, Mass.: Reidel, 1974, by permission.)

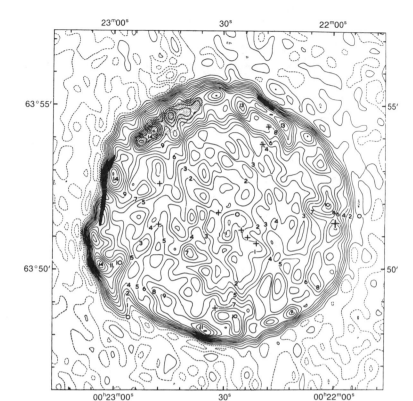

FIGURE 7-19
Radio map of Tycho Brahe's supernova at a wavelength of 21.3 cm. The
scale at left is Declination and at bottom Right Ascension. (After J. Baldwin, 1967, IAU Symp. No. 31,
p. 352.)

now there is little to be seen visually in the direction of Tycho's supernova
except a few faint wisps. In 1967, J. E. Baldwin published a beautiful 21-cm
radio map of this source shown in Figure 7-19.

As far as is known, no supernova was ever recorded at the position of
Cassiopeia A. Cassiopeia A lies in a region of the Milky Way that is subject to
heavy extinction of starlight, and it is thought that this supernova, which
probably occurred only a few centuries ago, was hidden behind obscuring clouds
of interstellar dust, as indeed the majority of supernovas in our galaxy may be.

Returning to Figure 7-18, we should be careful not to misread the relationships of the radio, optical, and x-ray regions. They do not imply that the Crab
emits more energy in the radio region than it does in the optical region and more
energy in the optical than it does in the x-ray region. The situation is exactly the
opposite. The Crab emits about five times more energy as x rays than it does as
visual light, and it emits many thousands of times energy as x rays than it
does as radio waves. The reason for this inversion is that the left-hand scale of
Figure 7-19 has to be multiplied by the bandwidth to obtain energy, and, of
course, the bandwidth is enormously wider (in cycles per second) for x rays than
it is for the lower frequencies.

All in all, the rate of energy emission of the Crab nebula, adding all frequencies and making allowance for the unobserved parts of the spectrum, is about 2.10^{37} ergs/second, which is some 5,000 times the energy output of the Sun. (The power output of the Crab is about 2.10^{27} kW compared to 3.8×10^{23} kW for the Sun.) Where does this power come from? Along the lines of the discussion of the latter part of Section 4-4, the answer is: from the spin of the neutron star that constitutes NP 0532. The power is derived from a slowing down of this spin. The existence of precisely observed pulses from the Crab (Figures 4-14 and 7-18) enabled this idea to be tested through a comparison of the pulse rate with accurate terrestial clocks. If the neutron star is slowing down, the measured pulse rate should decrease steadily with time. And so it does, by about 1 part in 2,000 each year. Making a reasonable estimate for the mass of the neutron star, say, 2 $\odot$, and knowing by calculation what the size of a neutron star of this mass must be, we can easily translate the measured slowing down of the star into a rate of energy loss. When this is done, we find that the neutron star in the Crab nebula is losing its energy of rotation at a rate of about 3.10^{38} ergs/second, which is more than ten times the rate needed to explain all the radio waves, light, and x rays emitted by this remarkable object. So, is the calculation wrong somewhere, or is still more energy emerging from the Crab? Surely there is more energy than we actually observe. At the end of Section 4-4, we saw that, when particles in the atmosphere of a neutron star are spun up to speeds very close to that of light, the lines of force of a magnetically closed region are broken open, and the particles are then able to escape with high energies into the outside world. Not all of the energy of these particles is likely to be restricted to the energy carried by electrons. The energy carried by protons, and perhaps the energy carried by complex atoms, cannot be readily detected. Protons and atoms are fed, as high-energy particles, first into the extended part of the Crab nebula, from which they eventually escape into interstellar space. It is likely that this process is a main source of origin of *cosmic rays*. Thus, the main output of energy from the slowing rotation of NP 0532 probably goes into cosmic rays.

§7-5. *The First X-Ray Galaxy*

Treading once more along an historic path similar to that of radio astronomy, the NRL group under Friedman unequivocally detected the first strong x-ray galaxy in 1970. This was the galaxy M87, the galaxy with the peculiar inner jet shown in Figure 4-24. The jet is a strong source of x rays, but there is also an important extended outer halo of very large volume. The results revealed later by the *Uhuru* survey (discussed in Section 7-6) led to the situation shown schematically in Figure 7-20.

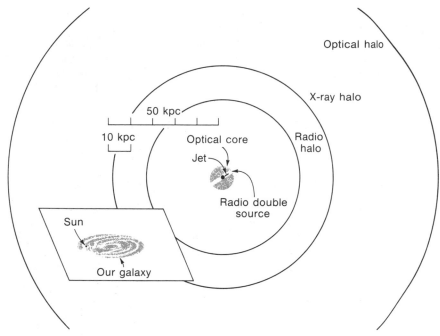

FIGURE 7-20

Schematic view of M87. This gives an idea of the general structure of this active, giant elliptical galaxy. Our own galaxy is shown in the inset for comparison. We know very little about what causes this appearance. It is generally believed that the optical core is stars that condensed out of the primordial medium and that the radio double source and the jet were ejected from the inner region. The origin of the halos is not clear. They may be stimulated by violent activity in the inner regions, or they may be due to matter falling into the gravitational potential well due to the central mass. (From R. Giacconi and H. Gursky, eds., *X-Ray Astronomy*. Hingham, Mass.: Reidel, 1974.)

DIFFUSE X RAYS ARE GENERATED BY THE INVERSE COMPTON PROCESS

The emission from the halo of M87 is on such a huge scale ($10 \text{ kpc} = 3.26 \times 10^4$ light-years) that we must ask what process could produce radiation of such high frequency over such a vast volume. The process cannot be the simple collisions of Figure 7-11 because the gas density could not be nearly high enough for collisions between electrons and protons to be frequent enough to produce such high frequencies. Nor can the process be synchrotron radiation because it would require implausibly high electron energies (or an implausibly high magnetic intensity). However, when radiation quanta are scattered by fast-moving electrons, it often happens that the quanta gain energy; if the electrons move fast enough, this jump of energy by the radiation can be very large. Thus, if visual light (or even radio radiation) is bounced off fast-moving electrons, x rays are produced. This process, known as the *inverse Compton process,* is a strong contender for the emission mechanism for most strong, diffuse sources of x rays, as, for instance, the x rays from the circular patch within the Crab nebula shown in Figure 7-16.

The inverse Compton process requires electrons to be present and moving with speeds approaching that of light, just as the synchrotron process requires them to be present and moving at speeds close to light for the production of radio waves. But, whereas the synchrotron process also requires the presence of a magnetic field, the inverse Compton process requires the presence of *previously existing* lower-frequency radiation. The previously existing radiation can be derived from three distinct kinds of sources. In interstellar space, it can be ordinary starlight, and in this form, the application of the process to the production of x rays was first discussed by J. E. Felton and P. Morrison Or the radiation can be supplied within a particular x-ray source by the source itself, as, for instance, a supply of radiation by the synchrotron process. Thus, the synchrotron process can act as an input for the inverse Compton process, a linkage of the two processes that must be going on within the broken circle of Figure 7-17, where we have fast-moving electrons generating visual light by the synchrotron process. This self-produced light then bounces against the same fast-moving electrons to produce x rays. And the x rays sometimes bounce for a *second time* to produce still higher radiation frequencies, γ rays, which have also been detected from the Crab nebula (by ground-based observations of the effects of the γ rays as they traverse the terrestrial atmosphere).

Yet, the x rays from the enormous halo of Figure 7-20 were still quite a mystery. If they were produced by the inverse Compton process, where did the radiation that was being bounced on fast electrons come from? Not from the synchrotron process. Although the synchrotron process is thought to be responsible for the radio halo of M87 (also marked in Figure 7-20), the intensity of these radio waves is far too small to explain the big output of x rays. And because the x-ray halo is so much larger than the optical size of M87, starlight is also too feeble a supply of radiation for the inverse Compton process. The answer to this difficulty, suggested by one of the authors, lies in the microwave radiation discovered in 1965 by Penzias and Wilson. This radiation has the same energy per unit volume everywhere. Inside galaxies, the energy is about the same as local starlight. Outside galaxies, as in the halo of Figure 7-20, the microwave background has a greater energy per unit volume than the light from the far-away stars of M87 or the stars of all other galaxies. This makes the microwave background radiation the most effective source of the halo x rays of Figure 7-20.

It is interesting to notice that, although M87 is also a radio galaxy, it is far from being the strongest radio galaxy. Indeed, Cygnus A (Figure 4-4), although very much farther away from us than M87, gives a far stronger radio signal than M87. In the *Uhuru* survey described in the next section, Cygnus A was found to be an x-ray source, but a much weaker one than M87. Thus, the radio and x-ray roles for these two galaxies are inverse to each other, with M87 the powerful x-ray source and Cygnus A the powerful radio source. Why should there be this inversion? The answer to this question appears to lie in the length of time that has elapsed since the outbursts in the central zones of the galaxies gave rise to the emission of high-speed electrons. The outburst in M87 appears to have occurred more recently, with the x-ray emission an early manifestation and with the strong radio emission continuing over a much longer time period than the x-ray emission. Although the outburst in Cygnus A was almost certainly the larger, the one in M87 was the more recent.

§7-6. The Uhuru Survey

The *Uhuru* satellite (*Uhuru* is the Swahili word for freedom) was one of NASA's minor projects, but it achieved more that was of lasting scientific worth than most of NASA's major projects. The satellite itself was launched on December 12, 1970, but the planning of its contents by ASE began as early as 1964. It was not until 1966, however, that NASA found a suitable partner for ASE in the Goddard Space Flight Center, which was then assigned to manage the project.

The equipment carried by *Uhuru* was designed to map the whole sky for x-ray sources that had energy fluxes above 2×10^{-10} erg cm^{-2} sec^{-1}. Results were published in what became known as the 3U Catalogue (*Astrophysical Journal Supplement Series 2*, Volume 27). These results, plotted in *galactic coordinates*, are shown in Figure 7-21. The sizes of the points given to the sources are an indication of their measured intensities, with the most intense sources being represented by the largest dots. The plane of the Milky Way follows the central horizontal line of Figure 7-21, the direction to the center of

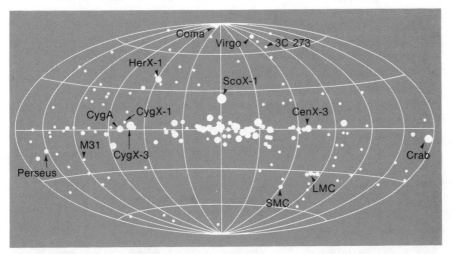

FIGURE 7-21

A map of the x-ray sky in galactic coordinates derived from the 3U catalogue. The location of each x-ray source is approximately shown. The size of the dots is proportionate to the logarithm of the intensity. Several of the sources of outstanding astrophysical interest are shown. (From R. Giacconi and H. Gursky, eds., *X-Ray Astronomy*. Hingham, Mass.: Reidel, 1974.)

our galaxy being at the center of the figure. Just as an ordinary geographical map of the whole world is plotted on a flat map, the whole sky (celestial sphere) is plotted here on a flat map. Points with the same latitude on the outer curve of the figure refer to the same point on the sky. (See the Glossary for a technical specification of galactic coordinates.)

Table 7-1 gives those 3U sources that can be identified with optically visible

TABLE 7-1
Identified x-ray sources

3U0021 + 42	M31
3U0022 + 63	TYCHO SNR
3U0115 − 73	SANDULEAK 160 SMC
3U0254 + 14	Abell 401 (Cluster)
3U0316 + 41	Abell 426 (Perseus cluster)
3U0521 − 72	Large Magellanic cloud
3U0527 − 05	M42
3U0531 + 21	Crab nebula
3U0532 − 66	
3U0539 − 64	Large Megellanic cloud
3U0540 − 69	
3U0820 − 42	Pup A
3U0833 − 45	Vela X
3U0900 − 40	HD 77581
3U0901 − 09	Abell 754 (Cluster)
3U1044 − 30	Abell 1060 (Cluster)
3U1118 − 60	Cen X-3
3U1144 + 19	Abell 1367
3U1207 + 30	NGC 4151
3U1224 + 02	3C 273
3U1228 + 12	Virgo Cluster (M87)
3U1231 + 07	IC 3576
3U1247 − 41	NGC 4696 (Cluster)
3U1257 + 28	Abell 1656 (Coma cluster)
3U1322 − 42	NGC 5128 (Cen A)
3U1551 + 15	Hercules cluster
3U1617 − 15	ScoX-1
3U1653 + 35	HZ HER
3U1700 − 37	HD 153919
3U1706 + 78	Abell 2256 (Cluster)
3U1956 + 35	HDE 226868 (CygX-1)
3U1957 + 40	CygA = 3C 405
3U2142 + 38	CygX-2
3U2321 + 58	CasA

From R. Giacconi and H. Gursky, eds., *X-Ray Astronomy*. Hingham, Mass.: Reidel, 1974.

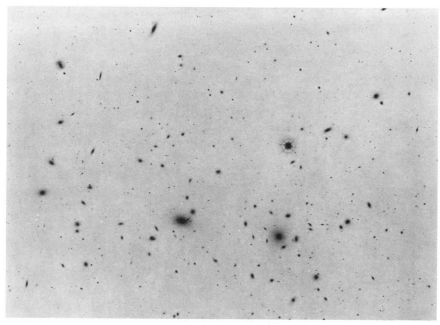

FIGURE 7-22
Part of the rich cluster of elliptical and spiral galaxies in Coma Berenices, about 400 million light-years distant from us. Clusters of this kind play an important role in establishing very large distances. (Courtesy of Kitt Peak National Observatory.)

objects, one of which is shown in Figure 7-22. We have encountered a number of these sources before. Going down Table 7-1 and picking out the previously discussed objects, we have: M31, the Andromeda nebula; Tycho's supernova; the Crab nebula; 3C 273, the first QSO to be discovered and now the first QSO shown to be an x-ray source; M87, the first x-ray galaxy; NGC 5128, the first radio galaxy to be optically identified; ScoX-1, the first x-ray source; Cygnus A, the first radio galaxy to be observed; and Cassiopeia A, the first radio source to be identifed optically.

We now want to mention some other members of this list. The Large and Small Magellanic Clouds (LMC and SMC) are discussed in Chapter 9 (see Figures 9-11 and 9-12), and a number of star systems are considered in more detail in Section 7-8. These latter are HD 77581; Cen X-3; HZ Her (HerX-1); HD 153919; and HDE 226868 (CygX-1), the strong black-hole candidate.

The numbers that follow the designation 3U in the first column of Table 7-1 specify the position of the object on the sky in terms of right ascension and declination. Thus, 3U1956 + 35 (CygX-1) was at R.A. 19^h56^m and Dec. + 35° on 00.00 hour, January 1, 1950. Plotting the sources with respect to galactic coordinates (rather than R.A. and Dec.) as in Figure 7-21 immediately gives us a rough-and-ready separation of those sources that belong to our own galaxy from those that do not. Notice the heavy concentration of the brighter sources toward the center of the galaxy. Clearly the great majority, if not all, of these must be sources within our galaxy. Most, if not all, of the brighter sources along the plane of the galaxy also lie within the galaxy. Off the central horizontal line of Figure 7-21, however, most (but *not all*) of the sources are distant objects outside our galaxy. HerX-1, and ScoX-1 itself, are examples of sources that belong to our galaxy but nevertheless lie outside the plane of the galaxy.

It is easy to predict from Figure 7-21 what would happen for further x-ray surveys carried through to fainter sensitivity limits. Just as in the surveys of radio sources, most of the fainter x-ray sources would be distributed uniformly over the sky because they would be sources outside our galaxy. More and more, the fainter sources would contribute to the diffuse x-ray background that was discovered in 1956.

CLUSTERS OF GALAXIES TEND TO BE X-RAY SOURCES

The source marked "Coma" in Figure 7-21 refers to the Coma cluster of galaxies shown in Figure 7-22, whereas 3U1551 + 15 is the Hercules cluster shown in Figure 7-23. There are also seven other sources in Table 7-1 that are associated with clusters of galaxies in the catalogue compiled by G. Abell. Evidently then, clusters of galaxies must be prolific sources of x rays. Figure 7-24 is centered on the local group of galaxies to which our own galaxy belongs and whose members are set out in detail in Table 7-2. Figure 7-24 is a sche-

FIGURE 7-23
A rich cluster in Hercules, about 600 million light-years distant from us, containing many spiral forms. (Courtesy of Hale Observatories.)

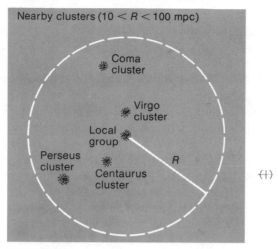

Nearby clusters (10 < R < 100 mpc)

FIGURE 7-24
Clusters of galaxies within a range of 100 mpc. At
greater distances, the cluster distribution would
become more uniform. (From R. Giacconi and
H. Gursky, eds., *X-Ray Astronomy*. Hingham,
Mass.: Reidel, 1974.)

TABLE 7-2
The local group

Description	Approximate distance (light-years)	Right ascension		Declination		Class
Our galaxy						Sb
NGC 224 (M31)	2×10^6	00h	40m	41°	00′	Sb
NGC 221 (M32)	2×10^6	00	40	40	35	Ell (Dwarf)
NGC 205	2×10^6	00	38	41	25	Ell (Dwarf)
NGC 598 (M33)	2×10^6	01	31	30	25	Sc
NGC 147	2×10^6	00	30	48	15	Ell (Dwarf)
NGC 185	2×10^6	00	36	48	05	Ell (Dwarf)
IC 1613	2×10^6	01	03	01	50	Irr (Dwarf)
NGC 6822	1.5×10^6	19	42	− 14	55	Irr (Dwarf)
Leo I	9×10^5	10	06	12	35	Ell (Dwarf)
Fornax	8×10^5	02	38	− 34	45	Ell (Dwarf)
Leo II	7×10^5	11	11	22	25	Ell (Dwarf)
Draco	3×10^5	17	19	58	00	Ell (Dwarf)
Sculptor	3×10^5	00	58	− 34	00	Ell (Dwarf)
Ursa Minor	2×10^5	15	08	67	20	Ell (Dwarf)
Large Magellanic cloud	1.7×10^5	05		− 69°		Irr (Dwarf)
Small Magellanic cloud	1.7×10^5	01		− 73°		Irr (Dwarf)
Possible members						
Maffei 1	3×10^6	02	33	59	25	Ell
Maffei 2		02	38	59	25	Ell

matic representation of a spherical volume with a radius of about 10^8 parsecs (pc) (1 pc = 3.26 light-years) that contains NGC 5128 (Centaurus cluster) and the Coma and Perseus clusters. The other clusters in Table 7-1 lie outside this spherical volume, most of them far outside it.

Two theories have been suggested to explain the emission of x rays from clusters of galaxies. One is the inverse Compton process that arises from the bouncing of the microwave background radiation off high-speed electrons escaping from galaxies within the clusters. The other theory, proposed in 1971 by J. Gunn and J. Gott, goes back to the particle collision idea illustrated in Figure 7-11. This return to a collision model is at first surprising because the density of the gas lying between the galaxies of a cluster must be small. The point of reviving the collision idea was to suggest that the amount of intergalactic gas within clusters might be much larger than astronomers had been in the habit of thinking. Although in each volume element the rate at which collisions occur must be small, the total for the whole volume might nevertheless be considerable since the total volume of a cluster of galaxies is immense (on the order of 10^{75} cm^3).

§7-7. *Close Binaries and Eclipsing Binaries*

This section is introductory to Section 7.8. Systems with two stars, usually called binaries, revolve around a common center, the *barycenter,* in the manner of Figure 7-25. The orbits may be ellipses rather than the circles shown in Figure 7-25, but for binaries in which the two components are close to each

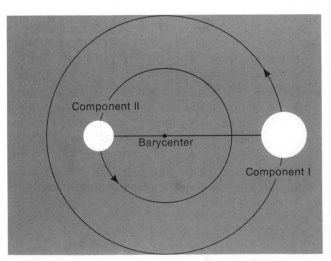

FIGURE 7-25
Two stars, forming a binary system, moving in orbits in the same period about a common barycenter.

other, the orbits usually approximate circles. Here and in the following section, we are largely concerned with such close binaries. It is worth noting that the gravitational field of each component distorts the shape of the other component, but this refinement is omitted from the present discussion.

A simple relation exists between the masses of the stars and their distances from the barycenter. Writing the masses as m_1 and m_2 and the distances as a_1 and a_2, we have the relation $m_1 a_1 = m_2 a_2$. We do not have to explain this relation because the barycenter is *defined* so it is true. Because the line joining the two stars always passes through the barycenter, the stars necessarily move around their orbits in the same period of time, say, P. Writing v_1, v_2 for their speeds of motion, we also have

$$P = \frac{2\pi a_1}{v_1} = \frac{2\pi a_2}{v_2},$$

and from $m_1 a_1 = m_2 a_2$, together with these equations for P, we can see that $m_1/m_2 = v_2/v_1$. It follows that an observational determination of v_2/v_1 would immediately give the mass ratio of the two stars.

Let us see next what we can determine from observation. Because we are considering the binary pair to be not too far apart, we may not be able to see the two stars individually. So how can we separate v_1 and v_2? When we analyze the visual light from such a system with the aid of a spectroscope, we hope to find the spectrum lines from both stars. If we do find both spectrum lines, it will not be difficult to tell which comes from which star. How will we do this? When one star moves toward us, the other moves away, as in Figure 7-26. And as the stars move in their orbits, their roles in this respect alternate, as can be seen from Figure 7-16. This oscillation of motion, with the two stars out-of-phase with

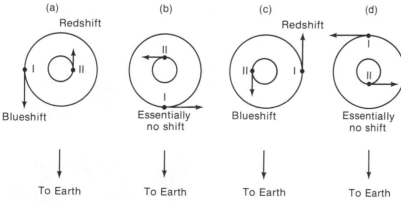

FIGURE 7-26

In position (a), the light from component I is blueshifted for a terrestrial observer, while the light from component II is redshifted. The opposite situation occurs in position (c). In positions (b) and (d), the frequency of radiation from both stellar components is essentially unaffected by the binary motion.

each other, causes corresponding oscillations in the frequencies of the spectrum lines from each star. This oscillation happens because of the so-called Doppler shift, which is discussed in Chapter 10 and in more detail in Appendix C. There is thus an oscillatory out-of-step effect that permits the spectrum lines from one star to be separated from the spectrum lines of the other star.

Using the ideas of the Doppler shift, simply by observing the frequency of one set of spectrum lines relative to the other (assuming both sets to be present), we can find the ratio v_1/v_2. Therefore, we can determine the ratio of the masses of the two stars from $m_2/m_1 = v_1/v_2$. On further investigation (using Kepler's third law, which appears when we consider the *gravitational interaction* holding the stars in their orbits), it turns out that m_1 and m_2 could be determined separately *if v_1 and v_2* could be obtained separately. Unfortunately, this cannot (in general) be done because the Doppler shift determines the speed of motion of a star only when the motion happens to be either directly toward us or directly away from us. In general, the line of sight to the Earth would be inclined to the orbital plane as in Figure 7-27, not in the orbital plane as was assumed in Figure 7-26.

In the situation in Figure 7-27, the stars have components of motion transverse to the line of sight, and the Doppler shift is unable to give us anything more than the ratio v_1/v_2. There is a special case, however, where we can determine that the line of sight must be essentially in the orbital plane, the case of *eclipsing binaries* illustrated in Figure 7-28.

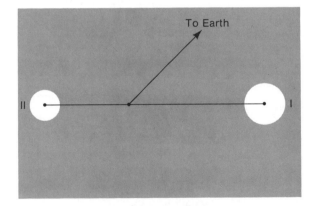

FIGURE 7-27
In general, the line of sight from a binary system to the Earth is inclined at an appreciable angle to the plane of the binary motion. In such a system, the Doppler-shift method yields the ratio of the speeds of the component stars in their orbits, but not their actual speeds.

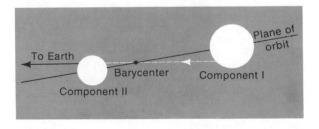

FIGURE 7-28
When the line of sight from a binary system to the Earth happens to make a small angle with the orbital plane of such a system, the two stars eclipse each other.

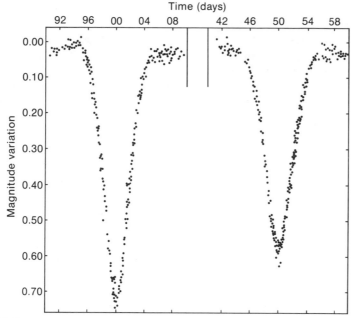

FIGURE 7-29

In exceptional cases where the line of sight happens to make a small angle with the plane of the orbital motion, the Doppler method does indeed determine the speeds. Such exceptional cases can be detected from the periodic eclipses that occur in the manner of Figure 7-29. Here we have the example of the binary WW Aurigae, with variations of magnitude plotted upward and with time in days plotted horizontally. What is the period P in this case? (Reprinted courtesy of C. M. Huffer and Z. Kopal and *The Astrophysical Journal*, published by The University of Chicago Press; © 1951 The American Astronomical Society.)

Because the stars have a finite size and because we are considering them to be not too far apart, the binary motion in such a system must cause one star to pass across our line of sight of the other. Star I will eclipse Star II, either completely or partially, during some part of its orbit. The total light we observe from the binary system must, therefore, be reduced. There will also be a stage in the motion when Star II will similarly eclipse Star I, and there will again be a reduction in the light we receive from the system. Consequently, in every complete period, there will be two reductions of light. These reductions will be from two eclipses that will usually not be equal because, in general, the two stars will not be equally bright nor will they have just the same radius. A faint star passing in front of a bright one cuts off more of the light than the bright star passing in front of the faint one. An example of such a double light reduction is shown in Figure 7-29 for the star WW Aurigae.

The conditions required for the determination of both v_1 and v_2 can thus be taken to exist for eclipsing binaries. Hence, separate masses m_1 and m_2 can be obtained for their component stars. A great deal of what we know concerning the masses of stars comes from this method of investigation. These particular data have played a critical role in the development of astronomy since stars are found to behave in markedly different ways depending on their masses. It would be difficult to separate the various categories of behavior, which we study in Chapter 8, if mass values were not available for individual stars.

Figure 7-30 is an example of an eclipsing system in which the fainter component II happens to be a red giant of large size, as is the case for the star Algol. The figure shows four positions for the smaller, brighter component I in its orbital motion relative to II. Over the broken portion of the orbit of I relative to II, the brighter component is eclipsed. This leads to the deeper of the two dips of the light curve shown schematically in the figure.

Cases of this last kind are of interest to x-ray astronomy because material from the larger, fainter star is often captured by the gravitational attraction of the smaller star. This slopping of matter from a big giant star over to a more compact companion is not so relevant when component I of Figure 7-30 is a highly luminous visual star as it is in the case of Algol. But if the compact companion is a neutron star (or a black hole), the material falls into an extremely strong gravitational field that can lead to the emission of x rays. Figure 7-31 is a schematic representation of this case, with the compact star now emitting little or no visual light (and, therefore, being designated II in the figure).

If component II of Figure 7-31 emits x rays, then, by observing the x rays, we can know whether the star is an eclipsing binary. If it is an eclipsing binary, the x rays will be shut off over the broken part of the orbit of II. This situation is an indication that the line of sight from the Earth does not lie very much out of the orbital plane of the binary. Consequently, the velocity v_1 of component I can be determined with reasonable accuracy from the Doppler shift of the spectrum lines in the visual light of I. Also, the period P can be found from the

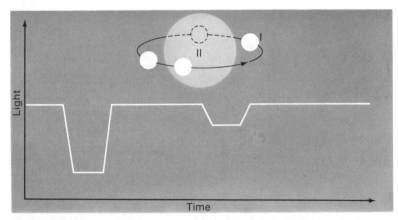

FIGURE 7-30

A schematic illustration of the fact that, when the two components are of unequal luminosities, the eclipse of the brighter component by the fainter one produces a deeper eclipse than does the reverse situation. Often the component of greater luminosity is a hot blue star of smaller size than its giant companion. This is the situation for the star Algol.

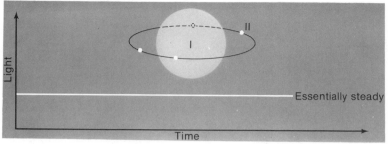

FIGURE 7-31

Sometimes one of the components is a neutron star, too small in size to produce an effective eclipse of the brighter component. In such systems, the eclipses have no significant effect on the light, but the binary motion still gives a Doppler effect.

Doppler oscillation of the spectrum lines of I. Knowing v_1, P, and using Kepler's third law, we can find the mass m_2 for the unseen component *if* the mass m_1 of the visible component is otherwise known.

If m_1 and m_2 are measured in terms of the Sun's mass, if P is measured in years, and if $a_1 + a_2$ is measured in terms of the Earth's average distance from the Sun, Kepler's law requires $m_1 + m_2 = (a_1 + a_2)^3/P^2$. Using $m_1 a_1 = m_2 a_2$, we get $m_2^3 = (m_1 + m_2)^2 a_1^3/P^2$, and by $P = 2\pi a_1/v_1$, we get $(2\pi m_2)^3 = (m_1 + m_2)^2 v_1^3 P$. If the inclination of the line of sight to the plane of the binary orbit is now small enough for v_1 to be given to sufficient accuracy by the Doppler shift of the spectrum lines of the visible star, and if the visible star belongs to a special type of otherwise known mass (so that m_1 is known), the preceding equation determines m_2 since P is also known from the oscillation of the spectrum lines.

§ 7-8. *X-Ray Binaries and Black Holes*

Table 7-3 gives a description of starlike x-ray sources discovered in the *Uhuru* survey. The second and third columns of the table describe the variability of the x-ray emission of the various sources. The fourth column gives the lowest energy for which x-ray quanta were counted, and, where this amount changed, the range is given. (Note that x rays with an energy of 1 keV have a frequency of 2.42×10^{17} cycles per second, and each quantum has an energy of 1.6×10^{-9} ergs.) The fifth column lists the optically visible star found at the position of the x-ray source, and the sixth column is the estimated distance (or distance range) of the optically visible star. The designation BoIb refers to the spectrum type (see Section 3.3) of the optically visible star, that is, of component I in Section 7.7. The seventh column gives the x-ray power output (for the range from 2 to 10 keV in the energies of the x-ray quanta), using as distance either the value specified for D or the distance specified in the sixth column of the table.

TABLE 7-3
Characteristics of selected x-ray stars

Source	Short-term variability	Long-term variability	Cutoff (keV)	Optical candidate	Distance (kpc)	Luminosity 2–10 keV (10^{36} erg/second)	Radio emission
HerX-1 (3U1653 + 35)	1.24-s pulsations	X-ray eclipses; 1.7-day period; 35 ± 2-day modulation	1.5 ⟶ 3.2	Hz Her; 15–13 mag. 1.7-day binary; extended lows	2–6	≲0.6 ⟶ 10 for D = 6 kpc	No
CenX-3 (3U1118 − 60)	4.84-s pulsations	X-ray eclipses; 2.1-day period; extended lows	1.5 ⟶ 4.2	Cen-3; 13 mag. Oor B; 2.1-day binary	5–10	~0 ⟶ 30 for D = 10 kpc	No
CygX-1 (3U1956 + 35)	Irreg. variability in times of 1 ms	One slow (~ days) transition Mar. 71	≲1.5	HDE 226868; 9 mag. B0Ib; 5.6-day binary	2.5	2 ⟶ 10 before Mar. 71 transition 0.6 ⟶ 3 after	No before transition yes after
3U0900 − 40 (GX263 + 3, Vela XR 1)	Irreg. variability in times of 0.4 s	X-ray eclipses; 9-day period; slow (~h) flares	2.5 ⟶ 4.4	HD 77581; 6 mag. B0Ib; binary	1.3	≲0.4 ⟶ 40	No
3U1700 − 37	Irreg. variability in times of 0.1 s	X-ray eclipses; 3.4-day period	2.1 ⟶ 5.5	HD 153919; 6 mag. binary 07f	1.7	≲0.03 ⟶ 3	No
ScoX-1 (3U1617 − 15)	Irreg. variability in times of min	Slow (~10-min-h) flares	~0.5 (variable)	ScoX-1; blue; 12–13 mag.	?	1 ⟶ 3 for D = 300 pc	Yes
CygX-2 (3U2142 + 38)	Irreg. variability in times of min	?	~0.5	CygX-2; 14 mag. sdG	0.6	0.2 ⟶ 0.4	Yes
CygX-3 (3U2030 + 40)	Irreg. variability in times of min	X-ray variations with 4.8-h period; slow (~days) variations	2.9 ⟶ 4.0	None. IR source has 4.8-h period	10	≲20 ⟶ 60	Yes
SMCX-1 (3U0115 − 73)	Irreg. variability in times of min	X-ray eclipses; 3.9-day period; extended lows	1.5 ⟶ 3.0	Sk 160; 13 mag. B0Ib	60	≲30 ⟶ 300	No

From R. Giacconi and H. Gursky, eds., *X-Ray Astronomy.* Hingham, Mass.: Reidel, 1974.

The sources can be divided into three groups:

1. HerX-1 and CenX-3, which show stable x-ray pulsations in a time scale of seconds.

2. Irregularly fluctuating sources like CygX-1, which undergo large changes of intensity in a time scale of seconds.

3. More slowly varying irregular sources like ScoX-1, which are comparatively steady on a time scale of seconds, but which can change by 100 percent or more in a time scale of hours.

In the previous section we mentioned that the favored idea for the generation of x rays in these systems is the one first suggested by Burbidge, in which material from the optically visible component slops over onto the highly compact object. This highly compact object may be a magnetized neutron star, as shown in the two schematic representations of Figure 7-32. On the left, the visible component is assumed to fill completely its so-called *Roche lobe*, through the cross-over constriction of which material circulates between the stars. On

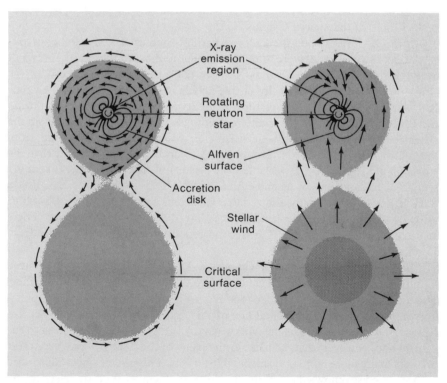

FIGURE 7-32

Schematic representation of the rotating neutron star model for pulsating x-ray stars. Both accretion disk and stellar wind cases are shown. (From R. Giacconi and H. Gursky, eds., *X-Ray Astronomy.* Hingham, Mass.: Reidel, 1974.)

the right, material reaches the neutron star through a stellar wind from the visible component. This stellar wind is analogous to the solar wind discussed in Section 7.2. It is important to realize that the scale of the neutron star has been enormously exaggerated in Figure 7-32 for illustrative reasons. The diameter of the neutron star is exceedingly small compared to the scale of the Roche lobes (the figure-eight curve).

These ideas have been applied with considerable success to HerX-1, perhaps the best understood compact x-ray source. From Table 7-3, we can see that HerX-1 pulsates on a time scale of 1.24 seconds, undergoes x-ray eclipses every 1.7 days, and becomes unobservable every ~35 days for some 20 days or more. Historically, the determination of the 1.7-day eclipses of HerX-1 was important in establishing the connection with binary systems—in the case of HerX-1, with the 1.7-day periodicity of the optically variable star HZ Herculis.

With the *Uhuru* survey completed, x-ray astronomy emerged from its pioneering stages into its golden age. The golden age is still very much with us as the number of research papers appearing in the astronomical literature clearly testifies. From here on, we must leave the interested reader to an independent study of further developments, with one exception. We end this chapter with perhaps the most dramatic discovery of x-ray astronomy—the likely presence of a black hole in the source Cygnus X-1.

Cygnus X-1 was among the ten or so earliest sources of x rays to be discovered. Its x-ray properties were somewhat similar to those of the Crab nebula, but here the similarity ended. Cygnus X-1 was not a strong radio source like the Crab, nor was there any optically visible nebula in its vicinity. Because Cygnus X-1 was different in its x-ray features from ScoX-1, the suspicion grew during the 1960s that Cygnus X-1 was different not just in degree but in kind from the other known sources. By 1971, this suspicion was amply confirmed. Data from the *Uhuru* survey, and from groups at MIT and NRL, all showed that the emission of Cygnus X-1 is variable over periods of time that ranged from 0.1 second up to some tens of seconds. And in 1974, R. E. Rothschild, E. A. Boldt, S. Holt, and P. J. Serlemitos showed that variations occurred over times as short as 0.001 second. No single period of variation could be found, however. There was nothing like the regular 1.24-second repetition of the x-ray pulses from HerX-1.

The next step was to identify Cygnus X-1 with an optically visible object. Compared to the optical identification of ScoX-1 discussed in Section 7.3, there was a difficulty that can be understood from Figure 7-21. ScoX-1 lies appreciably above the central horizontal line of this figure, which means that ScoX-1 lies well out of the plane of the Milky Way. Cygnus X-1 lies very close to the central line, however; that is, it lies in the plane of the Milky Way where there are very many more faint, optically visible objects. This circumstance demanded a more precise knowledge of the position of Cygnus X-1 (if a successful identification were to be made) than had been necessary for ScoX-1.

This problem was solved in an ingenious way. Radio sources are much less densely crowded on the sky than are optically visible objects, especially when the latter are in the plane of the Milky Way. Consequently, when a faint radio source was discovered in 1971 in the vicinity of Cygnus X-1 by L. Braes and G. Miley, and by C. Wade and R. Hjellming, it was argued that the x-ray source and the radio source were one and the same object. But now the position of the radio source could be determined with much more accuracy than the position of the x-ray source. Indeed, the position of the radio source could be determined to within a margin of only about 1″, which was amply good enough for optical identification to be made. The optically visible object found in this way turned out to be a binary star with a period, $P = 5.6$ days, that had long been known to be variable in its light emission. It had long been in astronomical catalogues under the designation HDE 226868. The identification was made in 1972 by B. Warner and P. Murdin, and by C. Bolton.

Only one component of this binary is directly visible. The unseen component is the one identified as the x-ray source. From its spectrum lines, the visible component was evidently a very massive star. It was of spectrum type B (see Section 3-3), of especially high luminosity for its type, and with fine details that astronomers denote as BoIb. The best judgment to be made for the mass of the visible star was that it exceeded 20 times the mass of the Sun, a judgment made from the known masses of other stars of spectrum class BoIb.

We turn now to the formula given at the end of Section 7-7. By putting $m_1 = 20 \odot, P = 5.6$ days $= 0.0153$ year, and by obtaining the velocity v_1 from the Doppler-shift method described in Section 7-7, we can calculate the mass of the x-ray star to be at least $5 \odot$, and this is significantly too large for a neutron star. The extreme rapidity of the x-ray fluctuations, only 0.001 second, shows that the unseen object must be very small in size. If it is not a neutron star, what can it be? The answer is a black hole, the properties of which we study in some detail in Chapter 11.

ARGUMENTS AGAINST CYGNUS X-1 BEING A
BLACK HOLE ARE UNLIKELY TO BE CORRECT

Can there be a weakness in the argument we have just outlined? First, we note that it is *not* a weakness that the formula given at the end of Section 7-7 assumes the line of sight to the Earth to lie in the plane of the binary orbit. Calculation shows that if this were not so, the mass m_2 of the x-ray source would turn out to be *even larger* than five times the Sun. A possible weakness does lie in assigning the value $m_1 = 20$ times the Sun for the optically visible BoIb component. This step in the argument was debated by astronomers, but in the end it was felt that $m_1 = 20$ was more likely to be an underestimate than an overestimate, with the likelihood that $m_2 = 5 \odot$ was also an underestimate for the mass of the x-ray source. Could the association of Cygnus X-1 with the radio source of Braes and Miley, and of Wade and Hjellming, be in error? Might the positioning of the radio source close to the x-ray source be only a coincidence?

The subsequent identification of the radio source with HDE 226868 made the possibility of coincidence remote. A second coincidence, that HDE 226868 just happened to possess an optically invisible component, would in that case also have to be postulated. Looking back over the *Uhuru* data put the matter beyond doubt, however. It was found that Cygnus X-1 had undergone an unusual x-ray fluctuation in March and April 1971; radio observations in Holland showed that a new component of radio emission had appeared at the Cygnus X-1 location at just this time. It would be remarkable indeed if these simultaneous changes in two very different wavebands had occurred in two distinct, physically unrelated objects. If, however, the different wavebands were part of the radiative emission of the same object, the simultaneity of drastic change could be readily understood. Hence, it would seem as if Cygnus X-1 is the first reasonably well-attested case of a black hole, a discovery of profound scientific importance as we shall see in Chapter 11.

General Problems and Questions

1. Compare x-ray telescopes with optical telescopes.
2. In what respects are methods of x-ray detection similar to, and dissimilar from, the methods used for the detection of visual light?
3. Write an essay on the emission of x rays from the sun. How does this emission affect human society?
4. What are (i) the solar corona, (ii) the solar wind, and (iii) a terrestrial magnetic storm?
5. Relate the history of the discovery of the first x-ray source outside the solar system.
6. Discuss the emission of x rays from highly evolved stellar remnants, with special reference to the case of the Crab nebula.
7. Describe the generation of x rays by the inverse Compton process.
8. Discuss the *Uhuru* survey. What kinds of x-ray sources were discovered in this survey?
9. What are (i) binary stars, (ii) eclipsing binaries, and (iii) x-ray binaries?
10. Describe the argument for regarding x-ray source Cygnus X-1 as a black hole.

Part II:
The Strong and Weak Interactions

Chapter 8
Atoms, Nuclei, and
the Evolution of Stars

§8-1. The Need for a Stellar Energy Source

In Chapter 3, we noticed that stars form by condensation in gas clouds such as the Orion nebula. Gravitation pulls both gas clouds and the protostars within them together. However, internal pressure arising from the kinetic energy of particle motions resists the gravitational forces to such a degree that, at any moment during the condensation, the pressure forces and the gravitational forces are nearly in balance. The balance cannot be exact because radiation from the outer surface into space slowly weakens the ability of the pressure forces to resist further condensation. Yet, we saw that a stage represented by the zero-age main sequence in the H–R diagram is eventually reached when the leakage of heat from the surface is at last compensated for by a process of energy production by nuclear reactions. At this stage we take the star to be formed. What are these reactions? This question will occupy us in this chapter.

Historically, it was a hard question to answer because during the nineteenth century a point of view which held that matter consists of atoms that are *indestructible* developed. There are many different kinds of atoms that were thought to be all separate from each other, each with its own distinctive properties. If this had been true, there could be no conversion of one form of atom into another form of atom. Yet, this conversion is the essential characteristic of a *nuclear* reaction. Many reactions were studied, but they were all ones that we now call *chemical*, reactions in which certain elementary substances never changed. These substances could be combined into associations called molecules, but it was always possible to recover the "elements," as they became called, from such molecules by a process of dissociation, that is, of breakup. Each element was taken to consist of a very large number of identical units, the atoms. The fact that the elements were never found to change in any chemical reaction was then explained by the indestructibility of the basic atoms themselves.

Techniques that enabled chemists to measure standard amounts of the elements (samples containing the same number of atoms) eventually became available. Consequently, it became possible, simply by weighing the samples, to arrange the elements in an order 1, 2, 3, . . . , with element 1 the lightest sample (hydrogen), element 2 the next lightest sample (helium), and so on. This method of ordering* gives a present-day list of 103 elements, set out in Table 8-1. Also given in this table are the relative abundances of the elements as they are believed to have existed in the gas cloud from which the Sun was formed. These relative abundances are scaled to give the element silicon (Si) a standard value of 10^6. That is to say, relative to a million silicon atoms, the other atoms existed in the numbers given in Table 8-1. Unlike the similar table given in Chapter 1, which was classified by elements, Table 8-1 is classified by the *isotopes* of the elements. The meaning of this additional detail will become clearer as we proceed.

ATOMS CAN CHANGE FROM ONE KIND TO ANOTHER

Although the multiplicity of reactions studied by chemists seemed to support the indestructible nature of atoms, there was cause for disquiet even before the end of the nineteenth century. Some form of energy production had to be making good the energy that the Sun was constantly radiating into space. From fossils found in rocks, it was known that the Sun must have been radiating for a long period of time pretty much as it does now. Geologists had estimated ages on the order of 100 million years for these rocks. The problem was to understand how the Sun could have maintained itself this long.

In 1854, H. von Helmholtz (1821–1894) had already considered the possibility that the Sun might still be condensing—that is, might still be a protostar—in which case the radiated energy could be coming from the compression caused by the gravitational forces. When Lord Kelvin (1824–1907) later examined this idea in detail, he found that, if it were true, the Sun could not be more than about 20 million years old, less than the geologists were claiming for the rocks containing the fossils. So puzzled was Kelvin about this discrepancy that

---

*For a reason that will appear later, this ordering by weight has been inverted for the pairs Co, Ni, and Te, I.

TABLE 8-1
Abundances of nuclei

Element	A	Isotopic abundances (%)	Cosmic abundances (Si = 10^6)
1 H	1	~ 100	3.18×10^{10}
	2		5.2×10^5
2 He	3		$\sim 3.7 \times 10^5$
	4	~ 100	2.21×10^9
3 Li	6	7.42	3.67
	7	92.58	45.8
4 Be	9	100	0.81
5 B	10	19.64	68.7
	11	80.36	281.3
6 C	12	98.89	1.17×10^7
	13	1.11	1.31×10^5
7 N	14	99.634	3.63×10^6
	15	0.366	1.33×10^4
8 O	16	99.759	2.14×10^7
	17	0.0374	8040
	18	0.2039	4.38×10^4
9 F	19	100	2450
10 Ne	20	(88.89)	3.06×10^6
	21	(0.27)	9290
	22	(10.84)	3.73×10^5
11 Na	23	100	6.0×10^4
12 Mg	24	78.70	8.35×10^5
	25	10.13	1.07×10^5
	26	11.17	1.19×10^5
13 Al	27	100	8.5×10^5
14 Si	28	92.21	9.22×10^5
	29	4.70	4.70×10^4
	30	3.09	3.09×10^4
15 P	31	100	9600
16 S	32	95.0	4.75×10^5
	33	0.760	3800
	34	4.22	2.11×10^4
	36	0.0136	68
17 Cl	35	75.529	4310
	37	24.471	1390

Element	A	Isotopic abundances (%)	Cosmic abundances (Si = 10^6)
18 Ar	36	84.2	9.87×10^4
	38	15.8	1.85×10^4
	40		~ 20 ?
19 K	39	93.10	3910
	40		5.76
	41	6.88	289
20 Ca	40	96.97	6.99×10^4
	42	0.64	461
	43	0.145	105
	44	2.06	1490
	46	0.0033	2.38
	48	0.185	133
21 Sc	45	100	35
22 Ti	46	7.93	220
	47	7.28	202
	48	73.94	2050
	49	5.51	153
	50	5.34	148
23 V	50	0.24	0.63
	51	99.76	261
24 Cr	50	4.31	547
	52	83.7	1.06×10^4
	53	9.55	1210
	54	2.38	302
25 Mn	55	100	9300
26 Fe	54	5.82	4.83×10^4
	56	91.66	7.61×10^5
	57	2.19	1.82×10^4
	58	0.33	2740
27 Co	59	100	2210
28 Ni	58	67.88	3.26×10^4
	60	26.23	1.26×10^4
	61	1.19	571
	62	3.66	1760
	64	1.08	518
29 Cu	63	69.09	373
	65	30.91	167

(*continued*)

TABLE 8-1 (*continued*)

Element	A	Isotopic abundances (%)	Cosmic abundances (Si = 10⁶)	Element	A	Isotopic abundances (%)	Cosmic abundances (Si = 10⁶)
30 Zn	64	48.89	608	41 Nb	93	100	1.4
	66	27.81	346	42 Mo	92	15.84	0.634
	67	4.11	51.1		94	9.04	0.362
	68	18.57	231		95	15.72	0.629
	70	0.62	7.71		96	16.53	0.661
31 Ga	69	60.4	29.0		97	9.46	0.378
	71	39.6	19.0		98	23.78	0.951
32 Ge	70	20.52	23.6		100	9.63	0.385
	72	27.43	31.5	44 Ru	96	5.51	0.105
	73	7.76	8.92		98	1.87	0.0355
	74	36.54	42.0		99	12.72	0.242
	76	7.76	8.92		100	12.62	0.240
33 As	75	100	6.6		101	17.07	0.324
34 Se	74	0.87	0.58		102	31.61	0.601
	76	9.02	6.06		104	18.58	0.353
	77	7.58	5.09	45 Rh	103	100	0.4
	78	23.52	15.8	46 Pd	102	0.96	0.0125
	80	49.82	33.5		104	10.97	0.143
	82	9.19	6.18		105	22.23	0.289
35 Br	79	50.537	6.82		106	27.33	0.355
	81	49.463	6.68		108	26.71	0.347
36 Kr	78	0.354	0.166		110	11.81	0.154
	80	2.27	1.06	47 Ag	107	51.35	0.231
	82	11.56	5.41		109	48.65	0.219
	83	11.55	5.41	48 Cd	106	1.215	0.0180
	84	56.90	26.6		108	0.875	0.0130
	86	17.37	8.13		110	12.39	0.124
37 Rb	85	72.15	4.16		111	12.75	0.189
	87		1.72		112	24.07	0.356
38 Sr	84	0.56	0.151		113	12.26	0.181
	86	9.86	2.65		114	28.86	0.427
	87		1.77		116	7.58	0.112
	88	82.56	22.2	49 In	113	4.28	0.008
39 Y	89	100	4.8		115	95.72	0.181
40 Zr	90	51.46	14.4	50 Sn	112	0.96	0.0346
	91	11.23	3.14		114	0.66	0.0238
	92	17.11	4.79		115	0.35	0.0126
	94	17.40	4.87		116	14.30	0.515
	96	2.80	0.784		117	7.61	0.274
					118	24.03	0.865

Element	A	Isotopic abundances (%)	Cosmic abundances (Si = 10^6)	Element	A	Isotopic abundances (%)	Cosmic abundances (Si = 10^6)
50 Sn	119	8.58	0.309	60 Nd	142	27.11	0.211
	120	32.85	1.18		143	12.17	0.0949
	122	4.72	0.170		144	23.85	0.186
	124	5.94	0.214		145	8.30	0.0647
					146	17.22	0.134
51 Sb	121	57.25	0.181		148	5.73	0.0447
	123	42.75	0.135		150	5.62	0.0438
52 Te	120	0.089	0.0057	62 Sm	144	3.09	0.00698
	122	2.46	0.158		147		0.0349
	123	0.87	0.056		148	11.24	0.0254
	124	4.61	0.296		149	13.83	0.0313
	125	6.99	0.449		150	7.44	0.0168
	126	18.71	1.20		152	26.72	0.0604
	128	31.79	2.04		154	22.71	0.0513
	130	34.48	2.21				
53 I	127	100	1.09	63 Eu	151	47.82	0.0406
					153	52.18	0.0444
54 Xe	124	0.126	0.00678	64 Gd	152	0.200	0.000594
	126	0.115	0.00619		154	2.15	0.00639
	128	2.17	0.117		155	14.73	0.0437
	129	27.5	1.48		156	20.47	0.0608
	130	4.26	0.229		157	15.68	0.0466
	131	21.4	1.15		158	24.87	0.0739
	132	26.0	1.40		160	21.90	0.0650
	134	10.17	0.547				
	136	8.39	0.451	65 Tb	159	100	0.055
55 Cs	133	100	0.387	66 Dy	156	0.0524	0.000189
					158	0.0902	0.000325
56 Ba	130	0.101	0.00485		160	2.294	0.00826
	132	0.097	0.00466		161	18.88	0.0680
	134	2.42	0.116		162	25.53	0.0919
	135	6.59	0.316		163	24.97	0.08099
	136	7.81	0.375		164	28.18	0.101
	137	11.32	0.543				
	138	71.66	3.44	67 Ho	165	100	0.079
57 La	138		0.00041	68 Er	162	0.136	0.000306
	139	99.911	0.445		164	1.56	0.00351
					166	33.41	0.0752
58 Ce	136	0.193	0.00228		167	22.94	0.516
	138	0.250	0.00295		168	27.07	0.0609
	140	88.48	1.04		170	14.88	0.0335
	142	11.07	0.131				
59 Pr	141	100	0.149	69 Tm	169	100	0.034

(continued)

TABLE 8-1 (*continued*)

Element	A	Isotopic abundances (%)	Cosmic abundances (Si = 10^6)
70 Yb	168	0.135	0.000292
	170	3.03	0.00654
	171	14.31	0.0309
	172	21.82	0.0471
	173	16.13	0.0348
	174	31.84	0.0688
	176	12.73	0.0275
71 Lu	175	97.41	0.0351
	176		0.00108
72 Hf	174	0.18	0.00038
	176	5.20	0.0109
	177	18.50	0.0389
	178	27.14	0.0570
	179	13.75	0.0289
	180	35.24	0.0740
73 Ta	180	0.0123	0.00000258
	181	99.9877	0.0210
74 W	180	0.135	0.000216
	182	26.41	0.0422
	183	14.40	0.0230
	184	30.64	0.0490
	186	28.41	0.0454
75 Re	185	37.07	0.0185
	187		0.0341
76 Os	184	0.018	0.000135
	186	1.29	0.00968
	187		0.0088
	188	13.3	0.0998
	189	16.1	0.121
	190	26.4	0.198
	192	41.0	0.308

Element	A	Isotopic abundances (%)	Cosmic abundances (Si = 10^6)
77 Ir	191	37.3	0.267
	193	62.7	0.450
78 Pt	190	0.0127	0.000178
	192	0.78	0.0109
	194	32.9	0.461
	195	33.8	0.473
	196	25.3	0.354
	198	7.21	0.101
79 Au	197	100	0.202
80 Hg	196	0.146	0.000584
	198	10.2	0.0408
	199	16.84	0.0674
	200	23.13	0.0925
	201	13.22	0.0529
	202	29.80	0.119
	204	6.85	0.0274
81 Tl	203	29.50	0.0567
	205	70.50	0.135
82 Pb	204	1.97	0.0788
	206	18.83	0.753
	207	20.60	0.824
	208	58.55	2.34
83 Bi	209	100	0.143
90 Th	232	100	0.058
92 U	235		0.0063
	238		0.0199

SOURCE: A. G. W. Cameron, *Space Science Reviews* 15 (1970), 121.

he suggested the geologists must be in error. Indeed they were, but not in the sense argued by Kelvin. The oldest rocks containing fossils of living creatures are very much older than 100 million years; 3,000 million years is the present-day estimate! Thus, the discrepancy is real—much worse than Kelvin thought it to be—the clear-cut implication being that Helmholtz's idea could not be correct.

The first step toward a resolution of this problem of the solar energy supply came in 1896 with the quite unexpected discovery of radioactivity by Henri Bequerel (1852–1908). The discovery arose from an investigation into the properties of x rays, which had been produced for the first time by Wilhelm Röntgen (1845–1923) only a few months earlier. As we have already noted, x rays are a form of radiation like ordinary light or like ultraviolet light. But in 1896, the nature of x rays was unclear—as the designation by the letter x implies! An erroneous idea that x rays were somehow produced when ordinary light was shone onto atoms came to be generally held, and scientists began to try out various kinds of atoms from this point of view. Becquerel was lucky in trying the element uranium. Light was shone on a sample of material containing uranium that was then placed below a photographic plate wrapped in black paper with a thin sheet of silver between, the idea being that only x rays would be able to penetrate the silver. After storing this simple arrangement for a while, he found the photographic plate fogged. Triumph! The idea worked; x rays really were generated by shining light onto uranium. But notice now the enormous power of experiment to correct wrong ideas. Becquerel repeated the experiment without shining light on his uranium material, and still the photographic plate became fogged. Consequently, it followed that the uranium was doing something all by itself. But what?

THE HEAVIEST ATOMS CHANGE BY DISINTEGRATION

The next step was to see if other kinds of atoms also produced the effect that had been discovered by Becquerel. Mme. Curie (1867–1934) found that, of the elements then known, only uranium and thorium had this property of spontaneously fogging a photographic plate, a property that became known as *radioactivity*. Together with her husband, she went on to concentrate, from large quantities of uranium ore, a small quantity of a substance that became known as *radium,* a name suggested by the intense radioactivity it was capable of producing. The suspicion that atoms of uranium were changing into different atoms—atoms of radium—soon became a certainty. Indeed, radium is atom number 88 in Table 8-1, whereas uranium is atom number 92.

Where did this process of spontaneous transformation of one form of atom into another end? From the abnormal quantities of lead found in association with uranium ores, the answer appeared to be that radioactivity ended at lead: lead was the ultimate product. This answer turned out to be correct. The scheme of transformations eventually worked out by Ernest Rutherford (1871–1937) and Frederick Soddy (1877–1956) has the principal form shown in Table 8-2. Rutherford had discovered that at each transformation something was emitted, and that there were two kinds of "something," which he called α rays and β rays. It was just these "somethings," particularly the β rays, that produced the

TABLE 8-2
Radioactive decay of uranium

Parent element	Characteristic decay time	Ray emitted	Daughter element
Uranium I[a]	4.5×10^9 years	α	Thorium I
Thorium I	24.1 days	β	Protoactinium
Protoactinium	1.2 minutes	β	Uranium II
Uranium II	2.5×10^5 years	α	Thorium II
Thorium II	8×10^4 years	α	Radium
Radium	1620 years	α	Radon
Radon	3.8 days	α	Polonium I
Polonium I	3 minutes	α	Lead I
Lead I	27 minutes	β	Bismuth I
Bismuth I	20 minutes	β	Polonium II
Polonium II	1.64×10^{-4} seconds	α	Lead II
Lead II	22 years	β	Bismuth II
Bismuth II	5 days	β	Polonium III
Polonium III	138 days	α	Lead III (stable)

[a] The Roman numerals refer to different forms of the same element. Thus Uranium I and Uranium II are different forms of uranium.

fogging of a photographic plate found by Becquerel. In 1909, Rutherford and Royds obtained the result that the α rays consisted of the nuclei of helium atoms, whereas the β rays were shown to be fast-moving electrons; the "rays" were particles. The surprise in Table 8-2 is to find that starting from uranium (element 92), after it emits an α particle and two β particles, we nevertheless arrive back again at element 92. A similar situation occurs for element 84, polonium (named for Poland, the country of Mme. Curie's birth). These cases provide a strong indication that it is possible to have *more than one form of the same chemical atom.*

LIGHTER ATOMS CAN ALSO CHANGE BY ELECTRON EMISSION

At first, it was thought that radioactivity was only associated with atoms heavier than lead. It therefore came as a further surprise when N. R. Campbell found that samples of both potassium and rubidium also emitted β rays, but very weakly, so that they had not been noticed in Mme. Curie's experiment. As a consequence of these emissions, potassium atoms were able to change into calcium atoms, and rubidium atoms into strontium atoms. The nineteenth-century concept of indestructible atoms was dead. Atoms *could* change into one another. When they did so, either an α ray (helium) or a β ray (electron) was emitted, *and energy was released in the motion of these emitted particles.* Consequently, there was now a new form of energy production that might be important in the Sun and stars.

The concept of energy derived from the radioactivity of naturally occurring elements was important, not because it turned out to be the right idea but because it freed physicists and astronomers from the straightjacket of nineteenth-century chemistry. It became possible to think in terms of transformations of all manner of atoms, one into another. The trouble with the idea that naturally occurring radioactivity is the source of stellar energy will be clear from the relative abundance values given in Table 8-1. There is far too little uranium, thorium, or any of the elements heavier than lead in stellar material for appreciable quantities of energy to be generated by it.

It was worry over this situation that caused A. S. Eddington (1882–1944) to turn in 1920 to a suggestion of J. Perrin. Instead of being concerned with the heaviest of the elements, Perrin's suggestion involved the two lightest elements, hydrogen and helium. His proposal was that if four atoms of hydrogen could be converted into one atom of helium, energy would be released in adequate quantity, since, unlike uranium, hydrogen is present in the Sun in great quantity as can be seen from the abundance values of Table 8-1. From 1905, the time of Einstein's special theory of relativity, scientists had been familiar with the concept that mass and energy are quantities of a similar type. Hence, because four atoms of hydrogen have a mass greater than the mass of one atom of helium by about one part in 125, it followed that Perrin's proposed transformation would indeed yield energy:

$$4H \longrightarrow He + \text{energy.}$$

Perrin's proposal turned out to be correct. Indeed, it *had* to be correct, for there is no other way, consistent with the composition given in Table 8-1, to explain the energy supply that the Sun clearly has. Yet, in 1920, the details of how four atoms of hydrogen could be converted to one atom of helium were quite obscure. Next we shall examine these details, and for this we must concern ourselves more closely with the structure of atoms, specifically, with the properties of atomic nuclei.

§8-4. *Nuclei and Particles*

FOR NEUTRAL ATOMS, THE NUMBER OF ELECTRONS EQUALS THE NUMBER OF PROTONS

Electrons all have the same electric charge, and consequently they repel each other electrically. Like charges repel; unlike charges attract. Why, then, do the electrons in an atom not simply fly apart? Why do they stay in the atom? The answer is that the electrons are bound to the atom by a charge of opposite sign carried in a small but massive nucleus. Nuclei contain protons, and the proton has been found to carry an electric charge equal in magnitude but opposite in sign to that of the electron.

Sometimes atoms contain fewer electrons than protons, in which case the atom is said to be *ionized*. The ionized atom then has a strong tendency to acquire electrons, until the number of electrons comes into balance with the number of protons, when the atom becomes *neutral*. There are a few atoms, hydrogen and oxygen being two of them, that occasionally carry their electron-acquiring tendency too far, so that they end up by having one electron too many—that is, one more than the number of protons. When in this state, hydrogen and oxygen are said to have formed *negative ions,* an odd property that turns out to have important effects in the atmospheres of stars and in our own terrestrial atmosphere. This oddity apart, atoms have a marked tendency to become neutral under cool conditions, and it is with the combination of neutral atoms into molecules that the science of chemistry is mainly concerned. What now becomes clear is that, whereas the chemical properties of an atom arise directly from the electron shell structure, the number of electrons, and hence the electron shell structure, is itself controlled by the nucleus. The chemistry of an atom is therefore determined by the number of protons in the nucleus, *and this is the deeper meaning of the number by which we ordered the elements in Table 8-1.* It is usual to refer to this number by Z.

The electrons of an atom form an extensive lightweight cloud surrounding a small massive nucleus. The mass of the nucleus exceeds that of the electron cloud several thousandfold, and, if we think of the nucleus of a typical atom, say, iron, as being a centimeter in diameter, then the electron cloud is about the size of a baseball park. In actual size, however, both the nucleus and the electron cloud are small compared to practical dimensions. In Figures 8-1 and 8-2, where some familiar objects and quantities, and some not so familiar, have been marked, we show lengths logarithmically going from the nuclei of atoms to the greatest observed distances in the universe.

THERE ARE NUCLEAR FORCES AS WELL AS ELECTRICAL FORCES IN ATOMS

We have seen that the electrons of an atom do not fly apart because they are held to the atom by the electrical attraction of the protons in the nucleus. But since charges of the same type repel each other, and since the protons all have the same charge, why do the protons not immediately burst apart? To answer this

FIGURE 8-1

The range of lengths in meters, from the nuclei of atoms up to stars. The relative sizes of planets, and of the smallest stars—white dwarfs and neutron stars—compared to the Sun, are also shown here. (After L. Pauling, *College Chemistry,* 3rd ed. W. H. Freeman and Company, Copyright © 1964.)

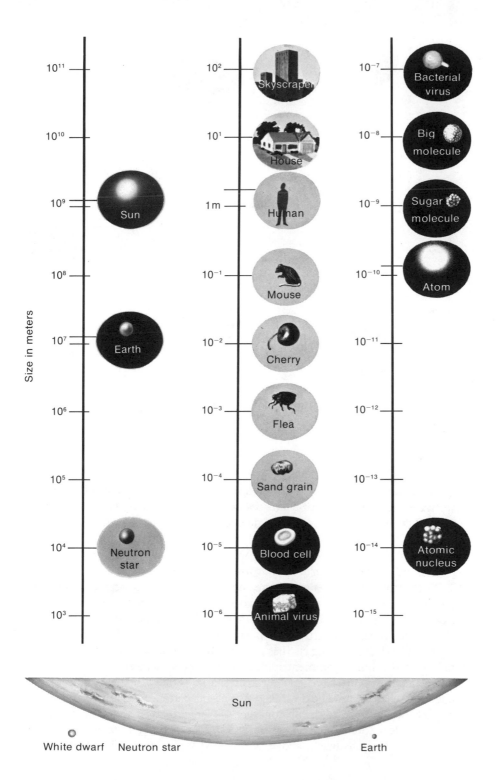

Size in meters

10^{11}
10^{10}
10^{9} — Sun
10^{8}
10^{7} — Earth
10^{6}
10^{5}
10^{4} — Neutron star
10^{3}

10^{2} — Skyscraper
10^{1} — House
1 m — Human
10^{-1} — Mouse
10^{-2} — Cherry
10^{-3} — Flea
10^{-4} — Sand grain
10^{-5} — Blood cell
10^{-6} — Animal virus

10^{-7} — Bacterial virus
10^{-8} — Big molecule
10^{-9} — Sugar molecule
10^{-10} — Atom
10^{-11}
10^{-12}
10^{-13}
10^{-14} — Atomic nucleus
10^{-15}

Sun

White dwarf   Neutron star

Earth

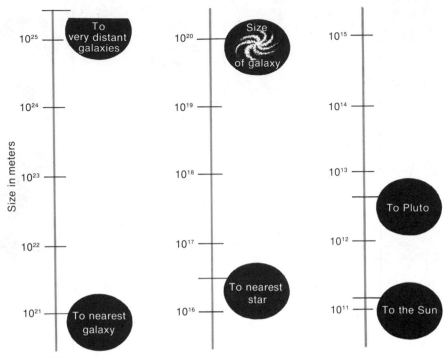

FIGURE 8-2
The sizes of objects, and the length scales, ranging from stars to the most remote galaxies.
(After L. Pauling, *College Chemistry*, 3rd ed. W. H. Freeman and Company. Copyright ©
1964.)

question, we must postulate the existence of an entirely new force quite different
from electrical and gravitational forces. We call it the *nuclear force*. It must work
to bind the protons together. Since no such force is observed when protons are
far apart from each other, we have to suppose that, unlike electrical and
gravitational forces that do operate for particles that are far apart, *the nuclear
force is active only for particles that are close together.*

The picture is not yet complete, however, for an important reason that
emerges as soon as we consider in detail the actual masses of nuclei. For
example, the nucleus of the helium atom, which contains two protons, has a mass
almost four times greater than the normal hydrogen atom, which contains one
proton. Plainly, something else besides protons must be present in the nucleus of
helium, and a similar argument applies to all elements heavier than helium. This
"something" cannot have any electric charge, otherwise the whole of our
chemical scheme would be destroyed. We need a particle without electric charge
but otherwise similar to the proton, that is, similar in mass and in being subject
to the nuclear force. Such a particle was in fact discovered experimentally in
1932 by James Chadwick, and it was named the *neutron*. The nuclear force

operates equally to hold protons to protons, neutrons to neutrons, and protons to neutrons, so that all the particles of the nucleus have the property of being able to attract each other. Notice, however, that if for some reason a particle were to manage to move appreciably outside the nucleus, it would not be attracted back again because the nuclear force between the particle and the nucleus must cease to operate when the two become separate. And if such a particle contained one or more protons, it would actually be repelled by the electrical effect of the other protons remaining in the nucleus, and so it would gain speed as it separated. This is just the situation for the α rays observed to emerge from the nuclei of the uranium atoms, the α rays that Rutherford and his colleagues identified with nuclei of helium atoms. Thus, the phenomenon of radioactivity, first observed by Becquerel, is now subject to the beginnings of an explanation: when a heavy nucleus of an atom of large Z emits α rays, it is simply losing bits of itself.

The nuclear force that we have considered is said to arise from a *strong interaction*, strong because the energies involved are much greater than the electrical interaction of Part I.

THERE IS A WEAK INTERACTION IN ATOMS AS WELL AS THE ELECTRICAL AND STRONG INTERACTIONS

What, then, were the transformations observed in 1907 by N. R. Campbell, in which the emission of β rays (electrons) was found to change rubidium to strontium? For rubidium, we have $Z = 37$; for strontium, $Z = 38$; that is, we have 37 and 38 protons, respectively. Thus, the emission of an electron from rubidium is accompanied by the appearance of an extra proton in the nucleus of the atom, the implication being that a neutron n changes into a proton p with the emission of an electron e, $n \longrightarrow p + e$. This transformation is balanced in terms of electric charge. The neutron has zero charge, and although the proton has a charge, the proton charge is just balanced by the opposite charge of the electron.* Electric charge is said to be *conserved* by the transformation. But energy was not conserved in this way of writing the transition from neutron to proton. This lack of balance of energy eventually led Wolfgang Pauli (1900–1958) to suggest that a fourth particle, say, $\bar{\nu}$, had to be involved, $n \longrightarrow p + e + \bar{\nu}$. This further particle could not have electric charge nor could it have appreciable mass, otherwise it would already have been detected experimentally, but it could have energy, and it could serve to maintain the conservation of energy.

Astronomer Walter Baade told one of the authors that, when he was dining with Pauli one day, Pauli exclaimed, "Today I have done the worst thing for a theoretical physicist. I have invented something which can never be detected experimentally." Baade immediately offered to bet a crate of champagne that the elusive $\bar{\nu}$ would one day prove amenable to experimental discovery. Pauli

---

*Historically, the proton charge was taken to be positive and the electron charge negative. A reversed choice, with the electron taken as positive and the proton negative, would be equally possible.

accepted, unwisely failing to specify a time limit, which made it impossible for him ever to win the bet. Baade collected his crate of champagne (as one of us can testify, having helped Baade consume a bottle of it) when, just over 20 years later, in 1953 C. L. Cowan and F. Reines did indeed succeed in detecting Pauli's particle.

Shortly after the dinner with Baade, Pauli's arguments for the existence of $\bar{\nu}$ were being discussed at a seminar in Rome. Confusion arose between $\bar{\nu}$ and the newly discovered neutron. In some exasperation, Enrico Fermi explained to the assembled company that Pauli's particle was not the massive neutron at all. It was only a "little neutron," and the Italian diminutive, neutrino, has been used for Pauli's particle ever since.

If $n \longrightarrow p + e + \bar{\nu}$, why not $p + e + \bar{\nu} \longrightarrow n$ since all other detailed transformations in physics are reversible? It would be hard to observe this reverse form, however, because three particles—one of them the elusive $\bar{\nu}$—would have to come together in order to produce the neutron. But then why always associate two particles with the proton and none with the neutron? Why not symmetrize the situation, writing $p + e \longrightarrow n + \nu$, or $n + \nu \longrightarrow p + e$? Written in this form, ν is called the *neutrino*. Our previous $\bar{\nu}$, the particle detected by Cowan and Reines, is nowadays called the *antineutrino*.

Are there actual data to support such a way of writing an inverse transformation from proton to neutron? N. R. Campbell in 1907 also found β rays from the element potassium. A close examination of the β-decay process in potassium shows that, although most decays result in the production of calcium, something different happens to about 10 percent of the decaying atoms. Argon is produced instead of calcium. This involves a change from $Z = 19$ to $Z = 18$, the reverse of potassium going to calcium, which requires Z to change from 19 to 20. Evidently then, potassium can go both ways. A neutron can go to a proton, or a proton can go to a neutron. The proton goes to a neutron by joining with an electron from the innermost shell of the surrounding electron cloud; in short, we have $p + e \longrightarrow n + \nu$.

With clear-cut data thus available for both $n \longrightarrow p + e + \bar{\nu}$ and $p + e \longrightarrow n + \nu$, why not a transition like $p \longrightarrow n + \nu + ?$, where ? is some new particle? Since neither $n \longrightarrow p + e + \bar{\nu}$ nor $p + e \longrightarrow n + \nu$ involves any change of the total electric charge, there should be no change for $p \longrightarrow n + \nu + ?$, which means that ? cannot be an ordinary electron—otherwise we should have a positively charged proton changing into three particles having a total *negative* charge. Thus, we write $p \longrightarrow n + \nu + e^+$, with e^+ a particle like an electron but having positive charge. Such e^+ particles were in fact detected in 1932 by C. D. Anderson, and by P. M. S. Blackett and G. P. S. Occhialini, at just about the time Pauli was postulating the existence of the neutrino. These positively charged electronlike particles are called *positrons*. The various ways in which these decays can occur are shown in Figure 8-3.

The decays shown in this figure are crucial to astronomy, and they are said to arise from the *weak interaction,* so-called because these transitions happen slowly even though quite high energies are often involved.

$$n \longrightarrow p + e + \bar{\nu}$$
(neutron) (proton) (electron) (antineutrino)

$$n + \nu \longrightarrow p + e$$
(neutron) (neutrino) (proton) (electron)

$$p + e \longrightarrow n + \nu$$
(proton) (electron) (neutron) (neutrino)

$$p \longrightarrow n + \nu + e^{+}$$
(proton) (neutron) (neutrino) (positron)

FIGURE 8-3
The various ways in which the process of β decay can occur.

By now we have encountered a number of different particles, and to keep order it is useful to classify them into families. In the family containing the electron there is a corresponding particle known as the muon (μ). Like the electron, the muon has negative charge. Indeed, apart from having a mass rather more than 200 times that of the electron, the muon seems to be exactly like the electron. Recent experiments have shown there to be two kinds of neutrino, written as ν_e and ν_μ, with ν_e the same as ν, $p + e \longrightarrow n + \nu_e$, and ν_μ a neutrino similarly associated with muons, $p + \mu \longrightarrow n + \nu_\mu$. The four particles e, μ, ν_e, ν_μ, form a family known as the *leptons,* the lightweights. Some physicists think that this family should have six members and that one of the two remaining members should be much more massive, in which case the name lepton would become a misnomer. What is certain, however, is that the proton and neutron belong to a family of eight particles known as *baryons,* the heavyweights. The two families of leptons and baryons appear to be distinct in the sense that a member of one family has not been observed (yet!) to change into a member of the other family although connections between them can be established through the agency of a third family of particles called *mesons.* Table 8-3 gives the main properties of the members of these three families.

Corresponding to the four leptons e, μ, ν_e, ν_μ, there is a family of *antileptons,* written as e^{+}, μ^{+}, $\bar{\nu}_e$, $\bar{\nu}_\mu$, of which we have already encountered e^{+} and ν_e, and corresponding to the family of baryons, there is a family of eight *antibaryons.* There is no separate family of antimesons, however. Certain other families have been experimentally detected in recent years, but they do not concern us in this book. In most astronomical problems, only two baryons, p and n, are involved, and only two leptons, e and ν_e, so that the situation in astronomy is usually simpler than it is in physics generally.

ISOTOPES HAVE THE SAME NUMBER OF PROTONS

Write N for the number of neutrons contained in a nucleus and Z for the number of protons. The sum $A = N + Z$ is called the atomic number or sometimes the

TABLE 8-3
Families of particles

Family	Electric charge	Mass[a]	Mean lifetime (seconds)	Common decay products	Antiparticle
LEPTONS					
μ (mu minus)	$-e$	106	2.2×10^{-6}	$e \, \nu_\mu \, \bar{\nu}_e$	μ^+ (mu plus)
e (electron)	$-e$	0.511	stable		e^+ (positron)
ν_e (neutrino)	0	0	stable		$\bar{\nu}_e$ (antineutrino)
ν_μ (mu neutrino)	0	0	stable		$\bar{\nu}_\mu$ (mu antineutrino)
BARYONS					
p (proton)	$+e$	938.26	stable		$\bar{p}$ (antiproton)
n (neutron)	0	939.55	930	$p \, e \, \bar{\nu}_e$	$\bar{n}$ (antineutron)
λ (lambda)	0	1115.6	2.5×10^{-10}	$p \, \pi^-$ $n \, \pi^0$	$\bar{\lambda}$ (antilambda)
Σ^+ (sigma plus)	$+e$	1189.4	8.0×10^{-11}	$p \, \pi^0$ $n \, \pi^+$	$\bar{\Sigma}^-$ (anti sigma minus)
Σ^0 (sigma zero)	0	1192.5	less than 10^{-14}	λ + radiation	$\bar{\Sigma}^0$ (anti sigma zero)
Σ^- (sigma minus)	$-e$	1197.3	1.5×10^{-10}	$n \, \pi^-$	$\bar{\Sigma}^+$ (anti sigma plus)
Ξ^- (xi minus)	$-e$	1321.2	1.7×10^{-10}	$\lambda \, \pi^-$	$\bar{\Xi}^+$ (anti xi plus)
Ξ^0 (xi zero)	0	1314.7	3.0×10^{-10}	$\lambda \, \pi^0$	$\bar{\Xi}^0$ (anti xi zero)
MESONS					
π^+ (pi plus)	$+e$	139.6	2.6×10^{-8}	$\mu^+ \, \nu_\mu$	π^- (pi minus)
π^- (pi minus)	$-e$	139.6	2.6×10^{-8}	$\mu \, \bar{\nu}_\mu$	π^+ (pi plus)
π^0 (pi zero)	0	135.0	10^{-16}	radiation	π^0 (pi zero)
K^+ (K plus)	$+e$	493.8	1.2×10^{-8}	$\mu^+ \, \nu_\mu, \, \pi^+ \, \pi^0$	K^- (K minus)
K^- (K minus)	$-e$	493.8	1.2×10^{-8}	$\mu \, \bar{\nu}_\mu, \, \pi^- \pi^0$	K^+ (K plus)
K^0 (K zero)	0	497.8	8.6×10^{-11} (Fast-decay mode) 5.4×10^{-8} (Slow-decay mode)	$\pi^+ \, \pi^-, \, 2\pi^0$ $3\pi^0, \, \pi^+ \, \pi^- \, \pi^0,$ $\pi^+ \, \mu \bar{\nu}_\mu, \, \pi^+ \, e \, \bar{\nu}_e,$ $\pi^- \, \mu^+ \, \nu_\mu, \, \pi^- \, e^+ \, \nu_e$	$\bar{K}^0$ (anti-K zero)
$\bar{K}^0$ (anti-K zero)	0	497.8	Same decays as K^0		K^0 (K zero)
η (eta)	0	548.8		$3\pi^0, \, \pi^0 \, \pi^+ \, \pi^-$ $\pi^+ \, \pi^-$ + radiation radiation only	η (eta)

[a] The particle masses are given here in terms of 1 MeV (million electron volts) as the unit of energy. To relate this quantity to everyday units, note that 1 MeV is the same as a power of 1 kW operating for 1.6×10^{-16} seconds. Note also that, in all decays, the products always have the same total electric charge as the parent particle. Can you find particle-antiparticle regularities among the decay products? The particles K^0, $\bar{K}^0$ are peculiar in having two decay modes. As a project, seek to discover a further peculiarity about these two particles.

mass number. If we imagine Z to stay fixed but N to vary, we have a situation in which the basic chemical properties of the neutral atom do not change since the chemical properties are determined by Z alone. Atoms with the same Z, but different N, are called *isotopes* of each other. *Isotopes all belong to the same element.*

UNSTABLE NUCLEI CAN CHANGE INTO STABLE NUCLEI THROUGH THE EFFECT OF THE WEAK INTERACTION

The values of Z and N for a nucleus cannot be chosen arbitrarily. If *for a specified Z* we were to choose N too large, the nucleus would simply disgorge neutrons until a suitably smaller value of N was attained. And if N were chosen too small, protons or helium nuclei would similarly be disgorged. Nevertheless, for a specified Z, many values of N are in general permitted; that is, many isotopes of the element of specified Z can exist. Yet, this does not mean that all such permitted isotopes are permanently *stable*. For any one isotope, we may have $n \longrightarrow p + e + \bar{\nu}_e$, increasing Z by one unit and thereby changing the element, or we may have one or the other of $p \longrightarrow n + \nu_e + e^+$, $p + e \longrightarrow n + \nu_e$, decreasing Z by one unit. Only if these so-called β processes do *not* occur is the isotope stable.

Table 8-1 lists all the stable isotopes, and it also contains a few unstable ones that are of special importance in astronomy. It is of interest that every stable nucleus is actually found among naturally occurring materials.

§8-5. Nuclear Energy and the Energy of the Stars

If we imagine protons and neutrons added together from a large initial separation to form a nucleus, energy will be yielded during the aggregation process because the nuclear force pulls the particles together once they are within close enough range of each other. Write B for this energy of formation, and consider the quantity B/A. This average binding energy per neutron or proton will in general be different from one nucleus to another. Figure 8-4 shows B/A plotted against A for all the naturally occurring nuclei.*

THE ABUNDANCES OF THE ELEMENTS IN NATURE ARE RELATED TO THE ENERGIES OF BINDING OF THEIR NUCLEI

For comparison with Figure 8-4, the relative abundances of the naturally occurring nuclei are plotted in Figure 8-5, a schematic representation of Figure

---

*The unit of energy used in this figure is not of great importance in our discussion, but it is one that is frequently used in physics: the energy acquired by an electron in moving through a static electric field with a potential of 1 V is called the *electron volt* (eV). An electron with an energy of 1 eV has a speed of about 600 km/second. A million electron volts (MeV) is the unit in Figure 8-4.

226

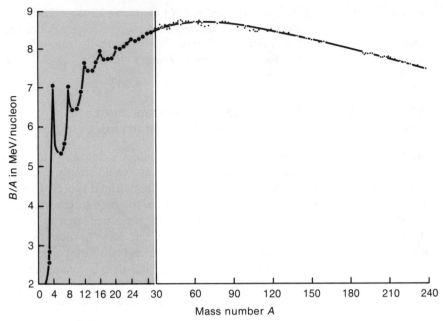

FIGURE 8-4

The binding energy per unit mass number plotted for all the naturally occurring nuclei.

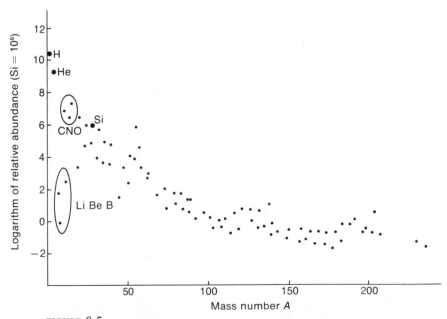

FIGURE 8-5

The relative abundances of the naturally occurring nuclei. The abundances are highest for hydrogen and helium. Notice from Figure 8-4 that it is just these light nuclei that can yield the greatest energy.

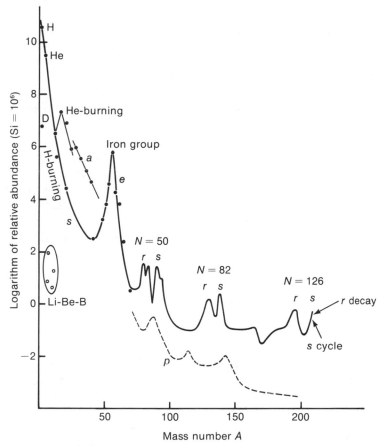

FIGURE 8-6
The relative abundances of Figure 8-5 plotted in a schematic form.

8-5 being given in Figure 8-6. The curve marked α in Figure 8-6 is drawn through the abundance values for $A = 28, 32, 36, 40$. The nuclei at these values of A can be thought of as being made up of helium nuclei, that is, made up of α particles, and hence the designation α. The α curve of Figure 8-6 is seen to lie well above the curve joining the abundances of the non-α nuclei. Comparison with Figure 8-4 shows that the curve marked "He-burning" passes through the nuclei at the sharply peaked maxima of the B/A curve, a connection that can hardly be accidental. Then again, Figure 8-6 has a peak, often called the *iron peak* since the element iron sits at the top of it, that is close to the general broad maximum of the B/A curve. These connections show a relation between the physical properties of the nuclei and their naturally occurring abundances. This relation indicates an origin of a physical kind. Just as plants and animals have been produced by evolution, not by special creation, so the nuclei seem to have been produced by some form of physical evolution, not by special creation.

We shall return to Figure 8-5 later, but before leaving it here let us notice once again how very small the abundances, *logarithmically plotted,* are for nuclei of large A. This is the reason that the radioactive decay of elements of large A, like uranium and thorium, cannot be responsible for the energy produced by the stars; there is far too little of these materials.

The curve of Figure 8-4 shows that the most strongly bound nuclei occur at a broad, flat maximum with A about 60. Energy can therefore be obtained either by adding nuclei of small A, so that the mass number of the product nucleus increases toward A about 60, or by removing pieces from nuclei of very large mass number, as by the emission of α particles from uranium. Energy could only be lost, however, by attempting either to add or to divide nuclei with mass numbers near 60. The atoms corresponding to such nuclei are the well-known metals—iron, nickel, chromium, manganese, cobalt, copper, zinc, titanium, and vanadium. Although these metals play a dominant role in the economy of modern society, they are nothing in terms of energy but the "ashes" into which lighter nuclei can be "burned" or into which heavier nuclei can be broken down. The iron peak of Figure 8-6 is indeed just a pile of cosmic ashes that has resulted from the burning of lighter nuclei. In the remainder of this chapter, we trace the details of the way this happened.

The curve of Figure 8-4 rises particularly sharply from hydrogen at $A = 1$ to helium at $A = 4$. Thus, the largest energy yield comes from adding the simplest *and the most abundant* atoms in just the process suggested by Perrin, namely $4\ ^{1}\text{H} \longrightarrow\ ^{4}\text{He}$. (We are now adopting the convention of writing the value of A as a superscript in front of the chemical symbol. Since the chemical symbol defines Z, the neutron number N is given immediately by the subtraction $A - Z$.)

MAIN-SEQUENCE STARS OF SMALL MASS DERIVE MOST OF THEIR ENERGY FROM THE PROTON–PROTON CHAIN, WHEREAS MAIN-SEQUENCE STARS OF LARGER MASS DERIVE MOST OF THEIR ENERGY FROM THE CARBON–NITROGEN CYCLE

Unlike the situation in 1920, when Eddington accepted Perrin's suggestion, we know today many ways to convert hydrogen to helium. Two of these ways are of special importance inside stars. A complex of nuclear reactions known as *the proton–proton chain,* shown in detail in Figure 8-7, is mainly responsible for the conversion of hydrogen to helium inside stars of comparatively small mass like the Sun. Another complex of reactions, given in Figure 8-8 and known as the *carbon–nitrogen cycle,* is of most importance inside main-sequence stars of larger mass. The proton–proton chain was first outlined in 1938 by H. A. Bethe and C. L. Critchfield, and the carbon–nitrogen cycle was proposed by Bethe in 1939. These first steps were considerably extended and developed by later workers, notably by C. C. Lauritsen and William A. Fowler at the California Institute of Technology.

Energy release (MeV)

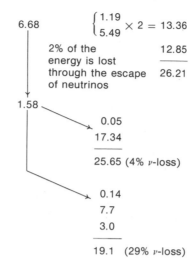

$^1H + {}^1H \longrightarrow {}^2D + e^+ + \nu_e$

$^2D + {}^1H \longrightarrow {}^3He$

$^3He + {}^3He \longrightarrow {}^4He + 2\,{}^1H$

or

$^3He + {}^4He \longrightarrow {}^7Be$

$^7Be + e \longrightarrow {}^7Li + \nu_e$

$^7Li + {}^1H \longrightarrow 2\,{}^4He$

or

$^7Be + {}^1H \longrightarrow {}^8B$

$^8B \longrightarrow {}^8Be + e^+ + \nu_e$

$^8Be \longrightarrow 2\,{}^4He$

$4\,{}^1H \longrightarrow {}^4He$

6.68

$\left\{ \begin{array}{l} 1.19 \\ 5.49 \end{array} \times 2 = 13.36 \right.$

2% of the energy is lost through the escape of neutrinos

12.85

26.21

1.58

0.05

17.34

25.65 (4% ν-loss)

0.14

7.7

3.0

19.1 (29% ν-loss)

FIGURE 8-7
The proton-proton chain.
The MeV unit is equal
to 1.6×10^{-16} kWs.

Energy release

$^{12}C + {}^1H \longrightarrow {}^{13}N$ 1.95

$^{13}N \longrightarrow {}^{13}C + e^+ + \nu_e$ 1.50

$^{13}C + {}^1H \longrightarrow {}^{14}N$ 7.54

$^{14}N + {}^1H \longrightarrow {}^{15}O$ 7.35

$^{15}O \longrightarrow {}^{15}N + e^+ + \nu_e$ 1.73

$^{15}N + {}^1H \longrightarrow {}^{12}C + {}^4He$ 4.96

25.03 MeV

(6% of the energy
is lost through the
escape of neutrinos)

In 1 case in 1,000, there
are the alternative reactions:

$^{15}N + {}^1H \longrightarrow {}^{16}O$

$^{16}O + {}^1H \longrightarrow {}^{17}F$

$^{17}F \longrightarrow {}^{17}O + e^+ + \nu_e$

$^{17}O + {}^1H \longrightarrow {}^{14}N + {}^4He$ or

$^{17}O + {}^1H \longrightarrow {}^{18}F$

$^{18}F \longrightarrow {}^{18}O + e^+ + \nu_e$

$^{18}O + {}^1H \longrightarrow {}^{15}N + {}^4He$

FIGURE 8-8
The carbon-nitrogen cycle.
The MeV unit is equal
to 1.6×10^{-16} kWs.

There is a general rule to the effect that reactions involving nuclei of larger positive electric charge—that is, with more protons—go slower than reactions involving nuclei of smaller positive charge. This is due to the fact that charges of the same sign repel each other, and the repulsion is greater when the nuclear charge is larger. The particles must therefore approach each other at higher

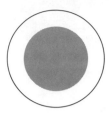

FIGURE 8-9
Left: Star with mass similar to that of the Sun. Right: More massive star, on upper part of main sequence. In each case, the shading denotes the region where energy is transported mainly by convection. In the unshaded region, transport is by radiation.

speeds and at higher temperatures when the charge is large than when the charge is small. According to the system of reactions in Figure 8-7, it is possible to generate helium from hydrogen without involving any nucleus that has a charge of more than two units, $Z = 2$ for ^{3}He, whereas in the system of Figure 8-8, helium production involves ^{15}N with $Z = 7$. Consequently, we expect the system of Figure 8-7 to operate at a lower temperature than that of Figure 8-8, and this is correct. Thus, the reactions of Figure 8-7 control the energy production in stars with central temperatures up to about 2×10^7 K, whereas the reactions of Figure 8-8 control the situations at temperatures higher than this. It is indeed just this difference of behavior with respect to temperature that causes the processes in Figure 8-7 to be operative in main-sequence stars of small mass, and those in Figure 8-8 to be operative in main-sequence stars of large mass.

The difference here leads to an interesting difference of structure between stars like the Sun and main-sequence stars of larger mass. This is illustrated in Figure 8-9. In the shaded parts of each star, energy is transported outward mainly by convection, that is, by a boiling motion of the stellar material, whereas in the unshaded parts, transport of energy is by radiation. The structures of the two kinds of stars are completely opposite in character. In massive main-sequence stars, we have convection near the center and radiation outside, whereas in solar-type stars we have radiation carrying the energy throughout the inner portions and convection on the outside. It is this transport of energy by convection in the outer regions of the Sun that probably accounts for the highly complicated behavior that is observed for the gases of the solar atmosphere. Figures 8-10, 8-11, and 8-12 show examples of this behavior.

The system of reactions of Figure 8-7 would operate even faster at comparatively low stellar temperatures if it were not for a peculiar feature of the first reaction. Two protons approach each other, which they can do with ease, because only $Z = 1$ is involved. However, the two protons simply come apart again, $p + p \longrightarrow p + p$, unless one of them switches into a neutron through $p \longrightarrow n + e^+ + \nu_e$. The neutron and proton then hold together as a *deuteron*, denoted in Figure 8-7 by the symbol D.

FIGURE 8-10

A giant prominence with a size of ∼ 200,000 km on the Sun. Gases in the corona are cooling and moving under the influence of both gravity and magnetic forces. (Courtesy of Hale Observatories.)

FIGURE 8-11

The solar corona at the total eclipse of June 1973. The form of the corona is indicative of the presence of magnetic forces. (Courtesy of High Altitude Observatory, Boulder, Colorado.)

FIGURE 8-12
A large flare, about 100,000 km across. Flares cause particle streams to be ejected from the
Sun. Such jets of particles move rapidly outward and sometimes impinge upon the Earth.
(Courtesy of Hale Observatories.)

According to Figure 8-7, there are two possible branch points. Our objective
of building ^{4}He can be attained directly from ^{3}He $+$ ^{3}He $\longrightarrow$ ^{4}He $+$ $2p$, or we
can have ^{3}He $+$ ^{4}He $\longrightarrow$ ^{7}Be $+$ energy. The relative probability of these two
routes is well known from laboratory data. Then, in the route via ^{7}Be, there is
a further branch, depending on whether ^{7}Be first decays to ^{7}Li or first acquires a
proton, thereby forming ^{8}B. The probabilities at this second branch are also
known from laboratory data. No matter which branch is followed, the outcome
is always to build ^{4}He. It has taken us some time and effort to arrive at Figures
8-7 and 8-8, but the problem is not inherently simple, as can be seen from the
fact that man-made fusion reactors capable of building ^{4}He are still a dream for
the future. In the preceding work, we have shown how a process that cannot yet
be carried out terrestrially takes place in the stars.

§8-6. *The Evolution of Stars*

THE ENERGY SUPPLY FROM HYDROGEN
TO HELIUM CONVERSION DOES NOT LAST FOREVER

The generation of energy inside stars has an inevitable quality about it. We saw
in Chapter 3 that protostars are drawn together by gravitation until compression
raises their internal temperatures high enough for nuclear reactions to produce
energy on a scale sufficient to compensate over long periods of time for the
radiation that is being lost continuously from their surfaces. At this stage, newly
formed stars are said to populate the zero-age main sequence. In this chapter,
especially in Figures 8-7 and 8-8, we have formulated the detailed nuclear
reactions responsible for this energy generation.

In Chapter 3, we also saw that the amount of time for which hydrogen-to-helium conversion can generate energy depends markedly on the mass of the star. For solar-type stars, energy will be generated by this process for at least 10^{10} years, but for stars of large mass, the time scale is much shorter, only 10^6 years for a mass of 20 $\odot$. The latter interval is very short compared to the terrestrial geological time scale, and it is therefore extremely short compared to the age of our galaxy. It follows that we must consider what happens next.

EXHAUSTION OF HYDROGEN AT THE CENTER
OF A STAR LEADS TO AN INCREASE OF ITS RADIUS

Suppose we return to the luminosity-color diagram—the Hertzsprung–Russell diagram—as in Figure 8-13, with the zero-age main sequence going from lower right to upper left. It is natural to ask how the stars behave in this diagram as

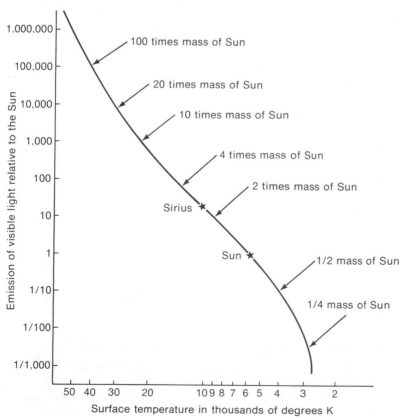

FIGURE 8-13
Newly formed stars lie in the Hertzsprung-Russell diagram on a zero-age main sequence, their position on this sequence being determined by mass. Notice that luminosities with respect to *visual light* are plotted here.

hydrogen-burning proceeds. The answer to this question, determined by extensive calculations, is shown schematically in Figure 8-14. Stars with masses comparable to that of the Sun evolve to the right and markedly upward, whereas stars of large mass evolve mainly toward the right. The effect produces a kind of funneling of stars into a certain region of the diagram. This region is characterized by a red color and by quite high luminosity, a combination that requires such evolving stars to have radii much greater than that of the Sun, usually from 10 to 100 times the Sun's radius (and in some cases even more), as can be seen from Figure 3-10. It is indeed because of their large radii that stars in this part of the diagram are called *giants*.

It is interesting to compare these expectations with the observations of Figure 8-15, which gives a logarithmic plot of visual luminosity versus surface temperature—that is, color—for the nearest stars, denoted by crosses, and for the brightest stars, denoted by circles. The remarkable feature of Figure 8-15 is that, whereas the nearest stars are either comparable to, or fainter than, the Sun, the brightest stars as they appear in the sky are intrinsically more luminous and are undergoing evolution toward the giant region of the diagram, exactly as was predicted in the schematic drawing of Figure 8-14. Many of the brightest stars are evolving away from the main sequence, and they are therefore already approaching hydrogen exhaustion.

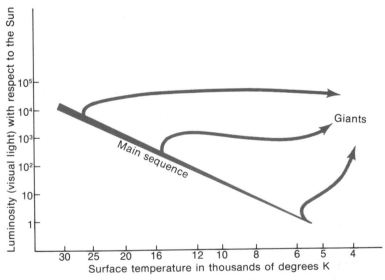

FIGURE 8-14

As more and more hydrogen is converted into helium, stars evolve from the main sequence toward the right of the Hertzsprung-Russell diagram. They do so in such a way that they all arrive at what is known as the giant region of the diagram, so-called because the radii of stars in this region are large, comparable in scale to the radius of the Earth's orbit around the Sun.

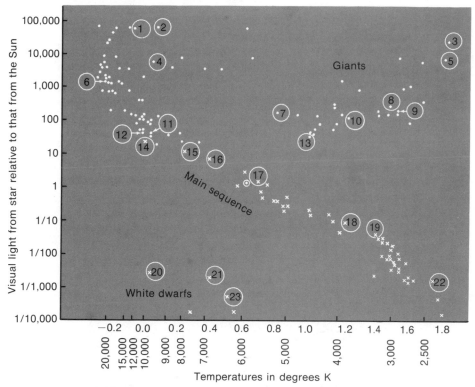

FIGURE 8-15

The positions in the Hertzsprung-Russell diagram of the nearest stars (x), and of the stars which appear brightest to the eye (•). Key: 1, Rigel; 2, Deneb; 3, Betelgeuse; 4, Polaris; 5, Antares; 6, Spica; 7, Capella; 8, Mira; 9, Aldebaran; 10, Arcturus; 11, Castor; 12, Vega; 13, Pollux; 14, Sirius A; 15, Altair; 16, Procyon A; 17, Alpha Centauri; 18, 61 Cygnus A; 19, 61 Cygnus B; 20, 40 Eridani B; 21, Sirius B; 22, Barnard's Star; 23, Procyon B.

Two points concerning Figure 8-15 should be noted. First, in the representations of the luminosity-color diagram given in Chapter 3, total luminosities involving radiation of all frequencies were plotted. In Figure 8-15, however, the luminosities are given for visual frequencies only. The difference is not appreciable for stars with colors in the green-yellow region, but the difference can be very significant for blue and red stars since the ultraviolet and the infrared ranges of frequency are both excluded from the visual luminosities. Contours of equal radius, using the solar radius as unit, are shown in the visual-luminosity plot of Figure 8-16. A comparison with Figure 3-10 makes clear the effect of changing from total luminosities to visual luminosities.

The second point concerns the distances of the stars given in Figure 8-15. Distances must be known in order for observed fluxes to be converted into intrinsic luminosities. The distances were determined by methods discussed in Chapter 9.

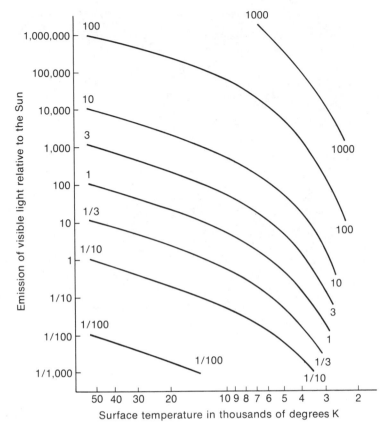

FIGURE 8-16
Contours of equal stellar radii with the solar radius as unit. These contours are different from those of Figure 3-10 because the left-hand scale is for radiation at visual frequencies only, whereas the scale is for radiation of all frequencies in Figure 3-10.

If we anticipate that discussion, even more striking evidence for the evolution of stars away from the main sequence and into the region of the giants can be given. Figure 8-17 shows the result of superimposing the different regions of the H–R diagram occupied by the stars of several associated groups or clusters. The effect of evolutionary funneling into the giant region is clearly shown from this side-by-side comparison. The clusters represented here are not by any means of the same age, as can be seen from the approximate ages marked at the right-hand margin of the diagram.

It should be noted that the luminosities of the giant members of the cluster h and χ Persei are less than those of the members that lie near the main sequence because here we are using visual luminosities; most of the radiation from these giants lies in the infrared and is not included in the visual luminosities.

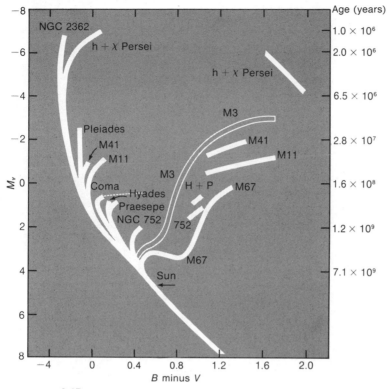
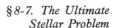

FIGURE 8-17

A composite representation of stars belonging to a number of open clusters (solid lines) and to the globular cluster M3. The ages shown on the left-hand scale refer to clusters departing from the main sequence at the marked luminosity levels. (Reprinted courtesy of A. R. Sandage and *The Astrophysical Journal*, published by The University of Chicago Press; © 1957 The American Astronomical Society.)

§8-7. *The Ultimate Stellar Problem*

EVEN NUCLEAR ENERGY IS NOT ETERNAL

By the time a star evolves away from the main sequence, it is already in its death throes. Although it is true that the general trend of evolution is from the main sequence and toward the giant region of the color-luminosity diagram, the detailed behavior of an individual star can be very complex, with sudden movements both up and down and left and right, as if the star were searching for some resting place. Figure 8-18 shows in schematic form a part of the path followed by a solar-type star. Each twist or turn may generally be related to some distinct physical process beginning or ceasing to operate, as for instance the ignition of a new nuclear fuel or the exhaustion of an old fuel.

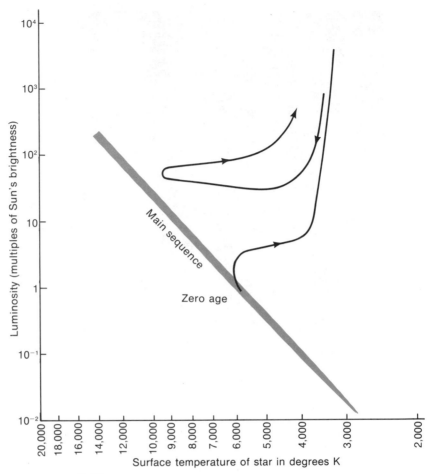

FIGURE 8-18
The schematic track of a solar mass star.

If we return to Figure 8-4, which gave the energy yield per nuclear particle, B/A, plotted against the mass number A, it is again important to notice that much the largest possible yield of nuclear energy comes from the conversion of hydrogen to helium. Once a star begins to exhaust this possibility, most of its energy availability has gone. As long as the star contains nuclei with mass numbers appreciably less than 60, the region of the broad maximum of Figure 8-4, it can generate energy in other ways, but they are more complex and less effective than the simple conversion of hydrogen to helium. What stars are doing in following paths like that of Figure 8-18 or like that of Figure 8-19 for stars of 7 ⊙ and 9 ⊙ is seeking these alternative possibilities. In Table 8-4, these further energy-producing processes and typical temperatures at which they occur in the

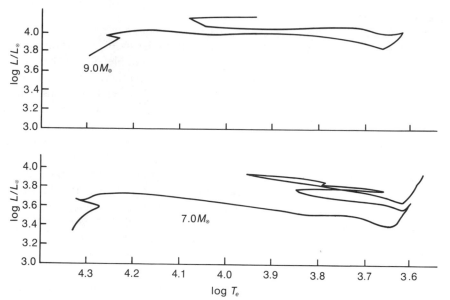

FIGURE 8-19

Schematic tracks of massive stars. (The track for 9 ⊙ is from I. Iben, Jr., *Astrophysical Journal,* vol. 140, 1964, p. 1631. University of Chicago Press. The track for 7 ⊙ is from E. Hofmeister, R. Kippenhahn, and A. Weigert, *Stellar Evolution,* Stein and Cameron, eds., New York, Plenum Press, 1966.)

TABLE 8-4
Burning of light elements

Process	Typical temperature of operation (K)
$3\ {}^4\text{He} \longrightarrow {}^{12}\text{C}$	$1\text{--}2 \times 10^8$
${}^{12}\text{C} + {}^4\text{He} \longrightarrow {}^{16}\text{O}$	2×10^8
$2\ {}^{12}\text{C} \longrightarrow {}^4\text{He},\ {}^{20}\text{Ne},\ {}^{24}\text{Mg}$	8×10^8
$2\ {}^{16}\text{O} \longrightarrow {}^4\text{He},\ {}^{28}\text{Si},\ {}^{32}\text{S}$	1.5×10^9
$2\ {}^{28}\text{Si} \longrightarrow {}^{56}\text{Ni}$	3.5×10^9

stars are outlined. At the end of the table, we have arrived at $A = 56$, which lies close to the maximum of Figure 8-4, and so indeed we have now arrived at the end of the road for nuclear-energy availability. At the end of Table 8-4 the material has been converted into nuclear ashes.

It will also be seen that the temperature values in the right-hand column of Table 8-4 rise as nuclei of larger and larger mass numbers are involved. Since the electrical forces increase with the mass number, higher speeds of motion are required in order that the nuclei can approach close enough to each other for the short-range nuclear forces to produce a general reordering of the protons and neutrons of which the nuclei are composed. For example, in the reaction $^{12}C + ^{12}C$, we have two similar nuclei approaching each other, each containing six protons, whereas for $^{16}O + ^{16}O$, the nuclei each have eight protons. The electrical forces are therefore greater for oxygen-burning than for carbon-burning, and a higher temperature is needed to give higher speeds for the former. The temperatures rise from about 10^8 K for helium-burning, $3 \, ^4He \longrightarrow ^{12}C$, to about 3.5×10^9 K for silicon-burning, $2 \, ^{28}Si \longrightarrow ^{56}Ni$.

Only the main effects of these processes of fusion are shown in Table 8-4. Many other reactions of lesser importance also take place. Many of the elements ranging in mass number from sulfur to nickel are produced, particularly during silicon-burning. We shall return to such processes, called *element synthesis,* later in this chapter. Here, we simply notice that by the two decays, $^{56}Ni \longrightarrow ^{56}Co + e^+ + \nu_e$, $^{56}Co \longrightarrow ^{56}Fe + e^+ + \nu_e$, the nickel arising from silicon-burning is converted into ^{56}Fe, the most common isotope of iron. The iron we find in our everyday world is believed to have been formed in this way in a stellar furnace at a temperature of 3.5×10^9 K.

THE NUCLEAR EVOLUTION OF A STAR CAUSES A COMPLEX LAYERING IN ITS STRUCTURE

Several of the processes set out in Table 8-4 can occur simultaneously in the same star, with the process of highest temperature occurring nearest to the center. A situation involving all the stages of Table 8-4, applicable to the late evolution of a star of large mass, say, 20 ☉, is shown in Figure 8-20. The figure is schematic, the radii of the various shells not being drawn accurately to scale. It is the outer low-density envelope of hydrogen and helium, enormous compared to the inner dense core of silicon, that causes such a star to lie in the giant region of the luminosity-color diagram.

Silicon-burning occurs at the center of the core. Other burning processes occur at the interfaces between one zone and another. Working inward from the surface, hydrogen-burning can occur at the interface between the first and the second shells, that is, between the zones marked H + He, He; helium-burning can occur at the interface between the zones marked He and C + O; carbon-burning at the interface between the C + O and O + Ne + Mg zones; and oxygen-burning at the interface between the latter zone and the core. The different interfaces can vary markedly in their contributions to the nuclear-energy production, and indeed it is just such variations that cause the star to make sudden changes in the path it follows in the luminosity-color diagram.

When material burns at an interface between two shells, there is an evolution in which the products of the nuclear reactions become added to the inner shell. Each shell gains material at its outer surface and loses material at its inner surface. Depending on the relative rates of burning at its inner and outer

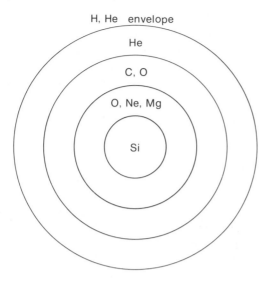

H, He envelope

He

C, O

O, Ne, Mg

Si

FIGURE 8-20
The layered structure and varied chemical composition of a star of mass 20 ⊙ at an advanced evolutionary stage.

surfaces, a shell may either gain or lose mass. The investigation of such a complex nuclear evolution demands careful and extended calculations that can be performed only with the aid of a powerful digital computer. One cannot easily judge by general intuition how the evolution will develop or exactly what move the star will next make in the luminosity-color diagram.

But all this "sound and fury" is of no ultimate avail. However complicated the structure a star may adopt, there is no gainsaying the fact that its energy sources are limited, that it tends to exhaust them at an increasing rate, and that the nuclear-energy production must eventually fail to make good the losses from the star into outer space. This lack of finality in the evolution along complex paths like those of Figures 8-18 and 8-19 leads us to ask if there can be any final resting state for the star, any ultimate graveyard in the luminosity-color diagram.

STELLAR REMNANTS HAVE VERY SMALL RADII WITH THEIR MATERIAL AT VERY HIGH DENSITIES

There can be no final static state as long as a star goes on losing energy into space; thus, to achieve such a state, radiation must stop. The luminosity must die to zero, which means that the temperature of the surface must go to zero. This is an impossible condition as long as the interior of the star is hot, for then there must always be a steady flow of energy from the interior to the surface, where it will be lost to space—contradicting our requirement. It is necessary, therefore, that the temperature of the interior shall also go to zero. But then how is the interior material to develop a pressure adequate to support the weight of the star? Only if we can resolve this dilemma can there be any final resting place for the star.

From the point of view of classical physics, this dilemma was unresolvable, but with the development of modern quantum mechanics (Chapter 2), it was realized that a new form of pressure would arise in cold matter of very high density, on the order of 10^5–10^6 g/cm^3, roughly a ton to a cube the size of a sugar lump. Thus, if a star were to contract to this very high density, a pressure adequate to support the weight of the star might be possible. Such highly compressed stars do in fact exist and are known as *white dwarfs*. Several of them occur among the nearest stars, and they were already plotted in Figure 8-15. Here was the graveyard we were seeking (at any rate, one of the graveyards).

It is tempting now to argue that, in its final stages of evolution, a star consumes its last nuclear-energy resources and then contracts to a white-dwarf state, gradually cooling, emitting less and less radiation into space, and ultimately ending its life as a cold, inert body of exceedingly high density. This line of argument is destroyed, however, by the circumstance that the new form of pressure existing in a white dwarf is not adequate to support the weight if the mass of the star exceeds 1.3 to 1.4 $\odot$ (the exact limit here depends on the precise nuclear composition of the star). Although such a scheme could indeed be applied to a solar-mass star, it cannot be used for a star of large mass like the stars with the evolutionary tracks shown in Figure 8-19.

For many years, astronomers thought the *only* way to meet this difficulty would be for a massive star to eject a large fraction of its material back into space, where the material would eventually join the clouds of gas that exist along the plane of the Milky Way. Two arguments could be advanced to support this idea, one theoretical, the other practical. It is a curious fact that the nuclear reactions yielding the most energy, those that convert hydrogen to helium, never lead to an explosive disintegration of a star. Later stages of the nuclear evolution, like the $^{16}O + ^{16}O$ reaction of Table 8-4, are potentially explosive, however. Although delivering less energy in total than the hydrogen-to-helium conversion, oxygen-burning, if it should become unstable, would be capable of suddenly lifting off the outer shells of a star with the structure of Figure 8-20. This would suddenly reduce the mass of the star, perhaps to less than the range 1.3 to 1.4 $\odot$, in which case the remaining residue could at last settle down to becoming a white-dwarf star.

Practical support for this idea came from observations showing clearly that stars do lose mass by violent ejection processes. This is even true for solar-type stars, the ones that might have reached the white-dwarf state without the need for mass ejection. Solar-type stars go to the white-dwarf state without completing the nuclear evolution of Table 8-4. After a star has zigzagged for a while in the giant region of the luminosity-color diagram, a situation is reached where the central regions consist largely of carbon and oxygen as in Figure 8-21, where the outer envelope of hydrogen and helium is expelled from the star. The hot exposed core cools rapidly, following a track in the luminosity-color diagram of the form shown in Figure 8-22. The expulsion of the hydrogen and helium seems to occur in a number of puffs rather than as a single explosion. Such a puff of expanding gas is susceptible to observation since it shines by absorbing light from the parent star. The *planetary nebulae,* of which two examples are

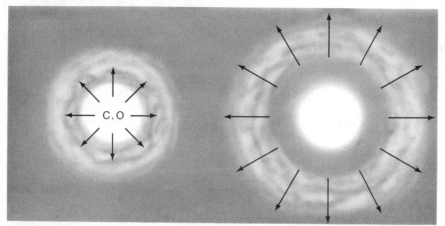

FIGURE 8-21
A schematic representation of the ejection of the outer envelope of a highly evolved star of solar mass.

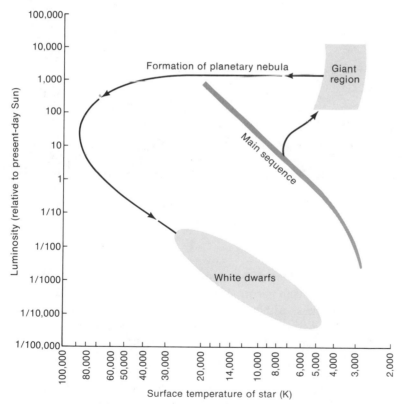

FIGURE 8-22
Evolution of a star of solar mass to the white-dwarf region of the Hertz-sprung-Russell diagram. Planetary nebulae are believed to be formed during the final sweep of the evolutionary track.

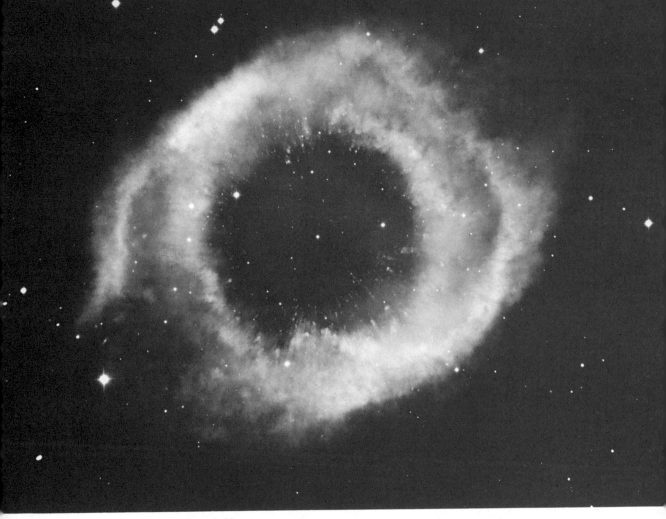

FIGURE 8-23
The planetary nebula NGC 7293, about 2 light-years in diameter.
(Courtesy of Hale Observatories.)

shown in Figure 8-23 and Plate VI, are thought to arise in this way. Thus, the planetary nebulae are a phenomenon associated with the death throes of stars of comparatively small mass. Although spectacular in appearance, they contain rather little gas, a small fraction of $\odot$, and so they are not to be thought of as being in any way similar to nebulae like the one in Orion that contains very much larger masses of gas.

Stars of solar type evolve as in Figure 8-22 to a white-dwarf state before oxygen-burning occurs, and so they avoid the kind of violently explosive situation that can arise from $^{16}O + {}^{16}O$. Stars of much larger mass reach the oxygen-burning phase, however, and are thus exposed to the possibility of a bomblike detonation. Stellar outbursts are in fact observed, as can be seen from the dramatic comparison of Figure 8-24a and b. Such explosions are called *supernovae*. The supernova seen in Figure 8-24b, but not in Figure 8-24a, is comparable in brightness to the whole of the parent galaxy. This is plainly an outburst of very great intensity, requiring a nuclear instability affecting the whole of one, or more, of the shells of Figure 8-20. A mass on the order of $10 \odot$ has to be concerned in such explosions, the scale of which exceeds by 10^{27} the energy released in a man-made hydrogen bomb. Stars in their death throes are capable of a degree of violence quite outside our terrestrial experience.

This degree of violence was the basis on which it was argued that explosions, arising out of nuclear instability, enabled stars of initially large mass to reduce their remaining residues to solar order, which then evolved by cooling into white dwarfs. Yet, there were causes for disquiet. How were unstable stars able to judge the intensity of explosion with sufficient precision to ensure that the residues always had masses below the white-dwarf limit of $1.3–1.4 \odot$? Could there be cases in which the star experienced total disintegration with no residue being left?

Because of these questions, it became imperative to investigate this whole problem by careful calculations using powerful digital computers. Although this program of investigation is still not complete, several conclusions relating to our

(a)   (b)

June 1959   May 1972

FIGURE 8-24
Notice the appearance of the bright supernova, believed to be associated with the galaxy. (Courtesy of Hale Observatories.)

questions have emerged. After some controversy, it now seems agreed that the picture we have just discussed is indeed valid for stars of initial mass up to about 5 ⊙. A white-dwarf residue seems to be left over after an explosion that adjusts itself appropriately so that the white-dwarf limit of 1.3–1.4 ⊙ is not exceeded. However, the situation for stars of much larger initial mass, say, 20 ⊙, has turned out to be quite different from the older picture.

The difference for stars of initially large mass depends on a circumstance that was already suspected by 1960, namely, that the core of such a star—the silicon core of Figure 8-20—is too massive and too dense for it ever to become a white dwarf. Thus, even if the outer shells of the star were entirely lost by an explosion, the remaining core could not end its life in the graveyard of the white dwarfs. What has now been found is that such a core evolves to a quite different graveyard, one of a dramatic and remarkable kind.

Recall that the pressure responsible for supporting the weight of a white dwarf cannot be understood in terms of classical physics because in classical physics there is no motion to supply pressure when the temperature is zero. In quantum physics, however, there is always motion, even though the temperature is zero, because the particles do not follow unique paths, there are always paths having appreciable probability besides the paths corresponding to zero motion. And as the density increases, the paths of one particle correlate with paths of other particles in a way that gives higher and higher probability to those that represent rapid motion. In short, high density forces the particles to move fast, and it is this fast movement that generates the pressure needed to support the weight of the star.

In white dwarfs, at densities of the order 10^6 g/cm^3, the fastest-moving particles are electrons, and it is therefore electrons that are mainly responsible for the pressure in these stars. Suppose now we increase the density substantially above 10^6 g/cm^3. What happens? From what has just been said, we might expect still more rapid electron motions, and consequently a still more marked rise of pressure. This would indeed be correct if the total number of electrons remained always the same, but because of $p + e \longrightarrow n + \nu_e$, this need not be so. In fact, it is not so. For densities rising into the range from 10^9 to 10^{10} g/cm^3, protons and electrons are squeezed into neutrons. It follows that the pressure supplied by the electrons therefore tends to disappear as the density rises. Does the pressure now disappear entirely? No, because exactly the same argument applies to the motions of the neutrons. The neutrons also supply pressure although for them to do so effectively requires a very much higher density than 10^6 g/cm^3. Effective pressure from neutrons, sufficient to withstand the weight of a star, requires densities on the fantastic order of 10^{14} g/cm^3; that is, 10^8 tons to a cube the size of a sugar lump.

This excursion into physics gives a clue to what happens ultimately in the problem of a star of large initial mass. The core becomes too massive and of too high a density for evolution to the white-dwarf state. Instead, after explosion has driven off the outer shells of the star in the manner of Figure 8-25, the core

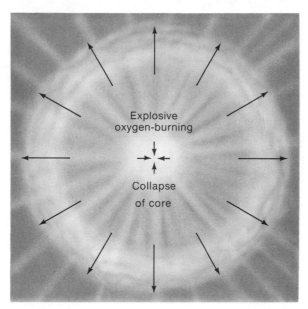

FIGURE 8-25
Schematic representation of the supernova outburst of a
massive star.

shrinks very markedly until the density becomes sufficient for the neutron
pressure to become important, that is, sufficient to support the weight of the
remaining residue. When this stage is reached, the radius has become tiny—only
a few percent of the radius of the Earth, as indeed was shown already in Figure
8-2. The core is said to have then become a *neutron star*.

The concept of a neutron star had already been formulated on theoretical
grounds by 1932.* The idea that such stars might exist remained a speculation,
however, until *pulsars* were discovered in 1967. The pulsars that we studied in
Chapter 4 have turned out to be neutron stars, and they are thought to be the
residues of exploding stars. The pulsar is also a graveyard, but a less placid one
than that of the white dwarfs.

§8-8. The History of Matter

The material of our galaxy is believed by astronomers to have been composed
initially of hydrogen and helium with few, if any, of the heavier elements. The
evidence for this view is that old stars contain a much lower concentration of

---

*Although the neutron had not yet been discovered in 1928, J. Frenkel had already considered
the possibility of stars with central densities of 10^{14}–10^{15} g/cm^3.

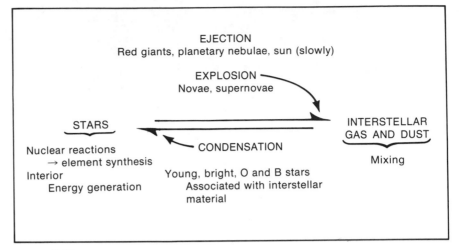

FIGURE 8-26
A cyclical process in which matter is transferred backward and forward between stars and the interstellar gas.

elements having Z greater than 2 than do younger stars. How have elements heavier than helium been produced? By nuclear processes, occurring inside stars, that act to produce fusion of lighter nuclei into heavier ones in accordance with the scheme laid out in Table 8-4.

Subsequent to their production inside stars, the elements are broadcast into space by stellar explosions. We have the cyclical picture of Figure 8-26, in which material is transferred backward and forward between the interstellar gas and the stars. Gas condenses into stars, but material is later returned from the stars to the interstellar gas by the processes we have just been studying.

It is of interest to estimate the amount of material circulated around the loop of Figure 8-26. For this estimate we need to know how often supernovae occur. It has been found that there is about one per 30 years per galaxy, a result coming from the observation of supernovae in many galaxies. Supernovae are very bright indeed, as can be seen from Figure 8-24, so they can be detected even in quite distant galaxies. Hence, their average frequency can be determined from a day-by-day patrol of several hundred galaxies. Since there is no reason to think our galaxy is any different from the others in its supernova rate, we can take one per 30 years to be a satisfactory basis for calculating the total number of supernovae that have occurred during the whole life of our galaxy.

The implication of these considerations is that the common materials of our daily world, the carbon that is the basis of life, the oxygen we breathe, and the metals we use, have all experienced the cycle of Figure 8-26. They were produced in stellar furnaces at temperatures upward of a billion degrees, and they had already been flung violently into space *before the Sun and planets of our system were formed.* The constituents of our daily world have not been fashioned here *in situ* within the solar system itself, but they were fashioned many aeons ago within the galaxy, and they were fashioned through the strong and weak interactions of matter. The parent stars are by now faint white dwarfs or superdense neutron stars that we have no means of identifying.

General Problems and Questions

1. Explain the logical reason why stars must possess an energy source of a nongravitational and nonchemical nature. Why did the views of the scientists of the nineteenth century impede the search for this energy source?
2. What are the elements, and how are they ordered? How many different elements are known?
3. What is radioactivity, and how was it discovered?
4. What is an α particle? What is a β ray? What is a γ ray?
5. What was Perrin's suggestion for the main source of the energy of the stars?
6. Describe the electronic shell structure of atoms, and discuss the relation of shell structure to the periodic table of Mendeleev.
7. How does the idea of shell structure help to explain the properties of molecules?
8. What is the relation of the number of electrons in an atom to the number of protons?
9. Why does electrical repulsion not cause the protons in the nucleus of an atom to burst apart? What is the general size relationship of the nucleus to the scale of the electron shells?
10. What is a neutron?
11. Discuss the interrelations, usually referred to as β processes, between the neutron, proton, electron, and neutrino. Give a general idea of the mass ratios of these particles.
12. What are antiparticles?
13. What are the isotopes of an element?
14. Explain why, for small mass numbers, it is necessary to fuse nuclei to obtain energy, whereas the opposite process of fission is necessary for very large mass numbers.
15. Compare in general terms the energy yield of a nuclear fuel with that of a chemical fuel.
16. Write down in detail the nuclear reactions of the proton–proton chain.
17. Write down in detail the nuclear reactions of the carbon–nitrogen cycle.
18. What criterion in main-sequence stars determines whether the proton–proton chain or the carbon–nitrogen cycle is the more important? How does the overall structure of a main-sequence star depend on this issue?
19. What is meant by the evolution of a star?
20. Draw a sketch of the Hertzsprung–Russell diagram to illustrate how stars evolve away from the main sequence. Why is a giant star so named?

21. Give an account of the nuclear reactions that occur in stars after hydrogen has become exhausted in the central regions. What is the general effect of these further nuclear reactions on the evolution of a star?

22. How high can the temperature rise in an evolving star of large mass?

23. What is a white-dwarf star, and how does the pressure inside it differ from the form of pressure within the Sun?

24. Describe supernovae, and discuss a sequence of events that could lead to their occurrence.

25. How is it thought that planetary nebulae are formed?

26. How does a neutron star differ from a white dwarf?

27. Write an essay on the history of matter.

Chapter 9
The Measurement
of Astronomical Distances

Astronomers would like to be able to determine the zero-age main sequence (discussed in Chapter 3) directly from a group of recently condensed stars. Since no such group is locally available, the lower part of the main sequence is determined for a comoving group of stars known as the Hyades (in the constellation of Taurus) that have a uniform age of about 10^9 years. This is not considered old enough for the lower right-hand part of the Hyades in the H–R diagram of Figure 9-1 to have been much affected by evolution. It will be seen from Figure 9-1 that a lower envelope for the Hyades stars is indeed quite well defined.

§9-1. The Hyades Main Sequence

The first step in establishing a series of links, extending from nearby stars out to vast distances of thousands of millions of light-years, consists in understanding how Figure 9-1 has been obtained for the Hyades stars. In particular, we are concerned with the method used to determine luminosities. Surface tempera-

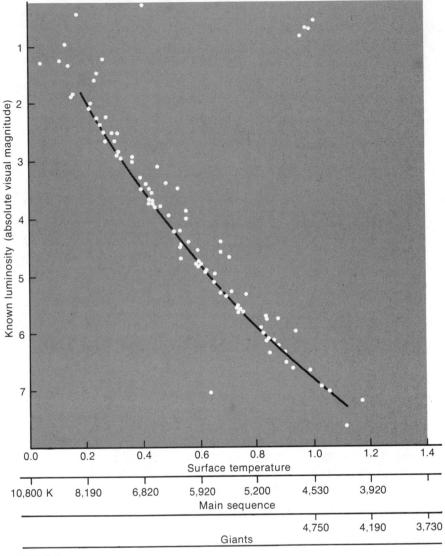

FIGURE 9-1
The Hyades stars, showing the well-defined lower envelope. This envelope is taken to determine the zero-age main sequence. For the positioning of the stars on the vertical luminosity scale, see Table 9-1.

tures can be estimated either from the colors or from the spectra of the stars, in accordance with the ideas already discussed in Chapter 3.

Luminosity determinations, both for the Hyades stars and for other luminous objects, are made from the equation

$$L = (4\pi d^2) \cdot f,$$

in which the distance **d** and the flux **f** are known. (Known quantities are in heavy type.) The flux **f** is directly measurable using the modern electronic methods discussed in Chapter 3. Throughout this chapter, we shall take **f** to be so measured, not only for the Hyades stars but for all the objects that are discussed here. If an object is obscured appreciably by interstellar dust (Chapter 6), we will suppose that we have a suitable way to correct the observed flux and that this correction has been made in the specification of **f**.

The preceding equation leads naturally to values of L expressed in ergs per second or in kilowatts ($1\,\text{kW} = 10^{10}$ ergs/second). Since L for the Sun is 3.8×10^{23} kW, luminosities (in terms of the Sun as the unit) are therefore easily obtained, and these luminosities can be converted into the magnitudes of Figure 9-1 with the aid of Table 9-1.

TABLE 9-1
Relation of luminosity and magnitude

Magnitude	Luminosity in solar units
−0.38	100.00
2.12	10.00
4.62	1.00
7.12	0.10
9.62	0.01

It is evident that the determination of L for any object depends on its distance d. To develop these ideas, let us start with the case of a star for which we know that the line of sight from the solar system is at right angles to the motion of the star, as in Figure 9-2. For the moment, let us also suppose the speed V to be known. Then, the star in question moves through a distance tV in a specified time t, and the line of sight changes, as in Figure 9-3, by the small angle

FIGURE 9-2
The special case in which the direction of motion of a star happens to be at right angles to the line of sight to the Earth.

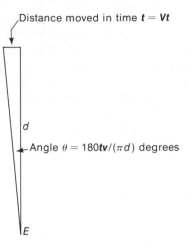

Distance moved in time $t = Vt$

d

Angle $\theta = 180tv/(\pi d)$ degrees

E

FIGURE 9-3

In a time t, such a star, moving with speed V, changes its direction by the small angle 180 $tV/\pi d$ degrees, where d is the distance of the star.

$\theta = 180 \, tV/(\pi d)$ in degrees, where d is the distance of the star. To see that this formula is correct, imagine the distance tV to form a small element along the circumference of a circle of radius d, as in Figure 9-4. The total length of the circumference is $2\pi d$; so the distance tV is a fraction $tV/(2\pi d)$ of the whole circumference. This means that the angle θ we are seeking must be a fraction $tV/(2\pi d)$ of the whole angle around a circle, which is 360°. Hence, $\theta = 360 \, tV/(2\pi d)$ in degrees. Now suppose θ is measured so that we can write $\theta = 180 \, tV/(\pi d)$. This immediately determines d, namely, $d = 180 \, tV/(\pi\theta)$.

The conditions we assumed, motion of the star at a known speed V at right angles to the line of sight, are of course unrealistic since we have no general means of possessing such knowledge. To broaden the discussion slightly, let us dispense with the right angle of Figure 9-2, going to the situation of Figure 9-5 with the speed V and the angle NES supposed known. All we need to do now is replace V in the formula of the previous paragraph by $V(EN/ES)$, and the same result holds good, namely, $D = 180 \, tV(EN/ES)/\pi\theta$. Therefore, the key issues are how the speed V is known and how the angle NES is known.

DOPPLER-SHIFT MEASUREMENTS ARE NEEDED

The first question is subject to the second question being answerable. The first question can then be decided by noting that, if NES is known, the ratio NS/ES is known, and the product $V(NS/ES)$ is the component of the velocity of the star along the line of sight from the Earth. The latter is measurable by the Doppler-shift method mentioned in Section 7-7 and discussed in detail in Chapters 10 and 12.

The situation therefore turns on how the angle NES is known, and it is here that the relevance of having a comoving group of stars emerges. The question could not be answered for a single star, but it can be answered for a separated group of stars with a common speed and a common direction of motion relative to the solar system.

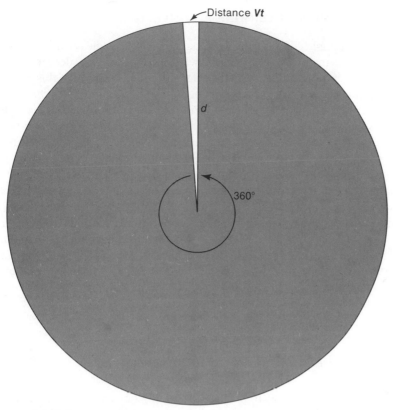

← Distance **Vt**

d

360°

FIGURE 9-4

With the distance **t**V being a fraction **V**t/(2πd) of the circumference of a circle of radius *d*, the angle subtended by the element at the center of the circle must be 360 **t**V/(2πd) degrees.

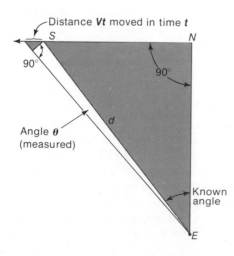

← Distance **Vt** moved in time **t**

S   N

90°

90°

Angle **θ** (measured)

d

Known angle

E

FIGURE 9-5

In general, the line of sight from the Earth is *not* at right angles to the direction of the star's motion. The small heavily marked triangle has the same shape as the large heavily marked triangle.

Because the triangle *NES* is right-angled, knowing the angle *NES* is equivalent to knowing the angle *NSE*, and this knowledge is in turn equivalent to knowing the direction of motion relative to the solar system. If we think of this direction being used as the polar axis of a system of latitude and longitude on the celestial sphere (see Section 3-5), then the projection of the star *S* on the celestial sphere will move slowly along a meridian toward one of the poles. This property will hold for all the stars of a comoving group that have proper motions toward the same pole, known as the *convergent point* of the group.

Conversely, if the observed proper motions of a group of stars are found to determine a convergent point, the group has a common motion with respect to the solar system, and the line from the Earth toward the convergent point is the direction of the common motion. The observed proper motions of the Hyades were found to determine a convergent point. Hence, the common direction of motion of the stars of the Hyades with respect to the solar system was determined, and the angle *NSE* for each star could be found. The distances d of the Hyades stars were also obtained this way. Measurements of f and $L = (4\pi d^2) \cdot f$ then gave the luminosity values of Figure 9-1.

§9-2. The Use of the Hyades Main Sequence

With the Hyades thus used to delinate the main sequence, suppose we observe the color and the flux f of a star whose distance is unknown. From the color, and also from the discrete absorption lines revealed by a spectroscopic frequency analysis, we may be able to decide that the star belongs to the main sequence. If we can, then we can easily use the observed color to decide where on the main sequence the star must be placed, as indicated in Figure 9-6. The luminosity L can now be read off on the left-hand scale, and the distance of the star is then immediately given by $d = \sqrt{L/(4\pi f)}$. The distance in such a case is said to be *spectroscopically* determined. With some care, we can extend this method to stars not of the main-sequence type as was done for some of the brighter stars of Figure 8-15.

Figure 8-17 showed the luminosity-color distribution for the stars of some open clusters. The colors were again obtained by analyzing the frequency distribution of the light from each individual star. The L values were obtained with the aid of the main sequence, but in a somewhat different way from the one just discussed. The stars of a cluster all have effectively the same distance d, so that their observed flux values immediately give their relative luminosities. This equality of distance fixes the shape of the star distribution in the luminosity-color diagram, but an ambiguity in the absolute luminosity still remains. The ambiguity can be removed, however, in the manner of Figure 9-7, if we adjust the absolute luminosity level so that the main-sequence part of the cluster

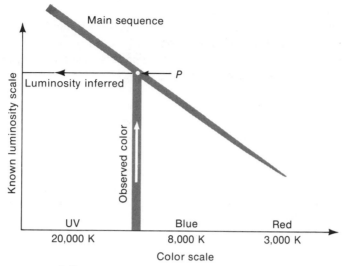

FIGURE 9-6

The observed color of a main-sequence star is used to determine its luminosity. A star at point P on the main sequence has the observed color.

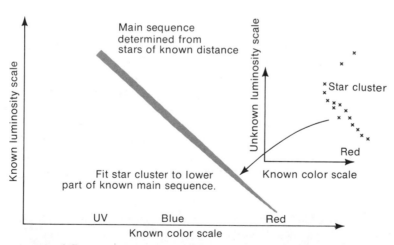

FIGURE 9-7

The scale of the luminosity distribution of the stars of a cluster is determined by fitting the lower part of the distribution to the zero-age main sequence in such a way that the color scales coincide.

distribution falls on the known Hyades main sequence of Figure 9-1 in such a way that the color scales fit together concordantly. With L thus known for every star in the cluster, the distance follows from $d = \sqrt{L/(4\pi f)}$, which is, of course, equivalent to $L = (4\pi d^2) \cdot f$. A similar procedure has been followed for each of the so-called open star clusters of Figure 8-17.

Errors in such determinations of distance arise for two reasons. The stars of the Hyades and of the cluster in question will not usually be of just the same age or of exactly the same chemical composition. Provided that the fitting is done for the lower right-hand part of the main sequence, neither of these sources of error is serious, however.

We consider next how Figure 8-17 can be used—that is, how clusters can be used—in connection with a distance indicator of a very different and remarkable kind. The ideas, which will take a while to develop, can perhaps best be introduced by referring to the discussion of stellar evolution in Chapter 8.

§9-3. *The Cepheid Variables*

Evolving stars of appreciable mass, say from 3 ⊙ to 20 ⊙, follow evolutionary paths that zigzag back and forth in the giant region of the luminosity-color diagram, as we saw in Figure 8-19. During these wanderings, stars occasionally develop pulsational instability. This is a dynamic condition in which the radius of the star goes through a long series of periodic oscillations. Such oscillations are illustrated in Figure 9-8.

The outer regions swell in size for a while and then fall back to a condition of minimum radius, after which the outward motion occurs again. Although for some stars the cycles of expansion and contraction are irregular, for other stars the cycles are regular, the time and amount of expansion and contraction being exactly repeated. Although the physical reasons for these pulsations have been understood for about 30 years, it has been only within the last few years, with the aid of computers, that astronomers have been able to examine the situation at all thoroughly. The subtleties involved reflect the unusual nature of the whole phenomenon. Dramatic phenomena of infrequent occurrence are usually complex in their nature and therefore awkward to analyze in detail.

FIGURE 9-8
Oscillation in size of a Cepheid.

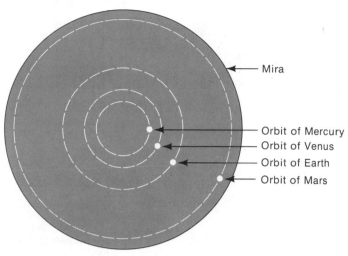

FIGURE 9-9
The oscillating star Mira at maximum size would contain the orbits of the
four inner planets of our solar system, shown here schematically.

An important class of stars that show regular pulsations is known as the
Cepheid variables. These stars play a critical role in astronomical distance
measurements since the Cepheids bridge the gap between measurements of
comparatively small distances of a few thousand light-years and the first stage of
measuring a really large distance—that to the galaxy M31 shown in Figure 4-7.
If the Cepheids had not existed, an alternative way to bridge this gap could
certainly have been found, but it would not have been as accurate as the method
of the Cepheids.

During a pulsation, the radius of a Cepheid varies by about 10 percent, as
shown in Figure 9-8. This is considerably less than the range of pulsation of
certain other kinds of variable stars. The irregularly pulsating star Mira, for
example, varies in radius by about 20 percent. If we drew a diagram like Figure
9-8 for Mira, it would contain the first four planetary orbits of the solar system,
as in Figure 9-9. The visual light emitted by Mira changes enormously during
its cycle. At its brightest, Mira appears as a red star of second magnitude; at its
faintest, it cannot be seen at all by the naked eye. To ancient astronomers, Mira
was an astonishing phenomenon, a star appearing regularly in the sky every
eleven months and then disappearing again! The name itself, Mira, means "Mi-
raculous." The fact that Mira was so well known to early astronomers shows
the care with which they watched the sky.

Figure 9-10 shows the way in which the light of the Cepheids varies with the
oscillation of Figure 9-8. It is from the light variation that Cepheid variables are
discovered. Astronomers do not, however, examine stars one by one until a star
that shows variation is discovered. Two photographic plates are taken of the
same region of the sky at different times. Stars with light variation will, in
general, have changed from one time to the other, and this change will show up

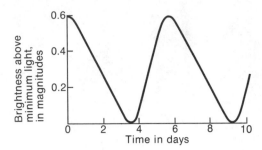

FIGURE 9-10

The light curve of δ Cephei, with period 5.37 days, measured in yellow light. (From J. C. Brandt and S. P. Maran, *New Horizons in Astronomy,* Second Edition. W. H. Freeman and Company. Copyright © 1979.)

immediately if the two plates are viewed in rapid alternation with each other. Not all the variable stars discovered in this way will turn out to be Cepheids; others will also be found by this simple technique. The Cepheids are those that show regular light variations of the form of Figure 9-10. The time required for a complete oscillation of the light can be measured with considerable accuracy, and it is called the *period* of the Cepheid. The period of the star shown in Figure 9-10 is 5.37 days. This is the star δ Cephei, the first to be found, by John Goodricke in 1784. The periods of Cepheid variables range from 1 to 2 days at the short end to more than 100 days at the long end.

Early in this century, an epoch-making discovery relating the periods of the Cepheid variables to their intrinsic luminosities was made. It was found that the longer was the period P, the greater was the luminosity L. The discovery was made in 1912 by Henrietta Leavitt, an astronomer of the Harvard College Observatory working in South Africa. Two great clouds of stars can be seen in the southern skies, known as the Magellanic Clouds, the Small Magellanic Cloud (SMC) and the Large Magellanic Cloud (LMC), shown in Figures 9-11 and 9-12. These objects are not ordinary clusters of stars but small galaxies in their own right. We know today that both Clouds are situated outside our galaxy although they are associated with it.

The distances of the Magellanic Clouds were not known to Ms. Leavitt, however. What was relevant to her observations was that both Clouds contain many Cepheid variables. In each Cloud, the variables could all be considered as essentially the same distance away from us. And since there was very little dust to produce obscuration in the Small Cloud (but not in the Large Cloud, which has a good deal of dust), the fluxes of the Small Cloud variables therefore gave the *relative values* of their luminosities. Thus, the fact that Ms. Leavitt found a relation between the periods P and the fluxes f meant there had to be a relation between P and the intrinsic luminosities L.

Harlow Shapley then realized that, if L could be determined *in an independent way* for any one Cepheid, the whole of the relation discovered by Ms. Leavitt could be represented as a relation between P and L. This would mean that an observational determination of P for any Cepheid, whether in the Magellanic

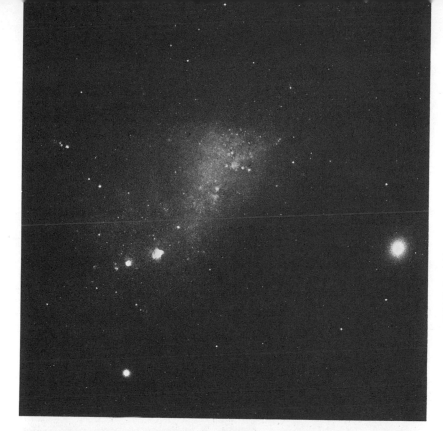

FIGURE 9-11
The Small Magellanic Cloud, about 10,000 light-years in diameter. (Courtesy of Dr. V. C. Reddish, Science Research Council.)

FIGURE 9-12
The Large Magellanic Cloud, about 30,000 light-years in diameter. (Courtesy of Dr. V. M. Blanco, Cerro Tololo Inter-American Observatory.)

Clouds or not, would give L. Hence, the distance d of the Cepheid would follow immediately from $d = \sqrt{L/(4\pi f)}$. Therefore, the critical problem was to make an independent determination of L for at least one Cepheid, and preferably, of course, for several Cepheids.

THE PERIOD–LUMINOSITY RELATION OF THE CEPHEIDS IS NOW CALIBRATED FROM THE MAIN SEQUENCE

Currently, several star clusters are known to contain a Cepheid, and the distances of these open clusters can be determined by using the main sequence—the method shown schematically in Figure 9-7. With the cluster distances d thus known, the L values of the Cepheids follow from $L = (4\pi d^2) \cdot f$, and the calibration problem is solved. The resulting relation between periods and intrinsic luminosities is shown in Figure 9-13. We can therefore calculate the distance of any other stellar group that happens to contain a Cepheid variable provided the group is not so distant that the flux f is too small to permit accurate measurement. The procedure is very simple in principle. Measure f and P for the Cepheid in question. Read off L from Figure 9-13, and work out $d = \sqrt{L/(4\pi f)}$. The logical steps used in establishing this method are:

Lower main sequence (Hyades)
⟶ Open clusters
⟶ Cepheids

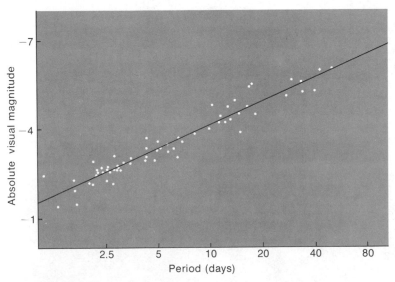

FIGURE 9-13
The observed relation between the periods and average visual magnitudes of Cepheids. (Adapted from H. Arp, *Astronomical Journal*, vol. 65, 1960, p. 426.)

The Cepheids now used for establishing the calibration of the *P-L* relation, those in the so-called open clusters, were not known until the 1950s, when the first of them was discovered by J. B. Irwin. Shapley had to use a method different from the ones described in this chapter to calculate the distances of the nearest few Cepheids in our galaxy. Unfortunately, even the nearest Cepheids are outside the range of the 200 or 300 light-years within which this other method can be used with saisfactory accuracy. Nevertheless, in spite of this difficulty, the use of Cepheid variables proved so powerful that Shapley was able to establish a result of far-reaching importance. He was able to show that our galaxy is very much larger than it had been thought to be. Most astronomers at that time, 1917, thought that our galaxy was only a few thousand light-years in size, and that the solar system lay quite near the center, instead of being about 30,000 light-years out from the center as we now know it to be. This old erroneous picture came from the refusal of orthodox astronomical opinion to recognize that obscuration by dust occurs along the plane of the galaxy, that is, from supposing that our local "swimming hole" was the entire galaxy.

Within a few years of Shapley's work, another and even greater astronomical revolution occurred. Hubble discovered Cepheids in the galaxies M31, M33, and NGC 6822, and thus he was able to calculate distances to galaxies outside our own. This calculation settled a controversy that had developed in the nineteenth century and had raged throughout the first two decades of this century. When Messier compiled his catalogue more than 150 years ago, all prominent "diffuse objects" were listed together as they were when the New General Catalogue (NGC) was compiled. Thus, M42 (NGC 1976) was the Orion nebula, shown in Figure 3-14, an object within our own galaxy, whereas M31 (NGC 224) was the Andromeda nebula, a giant galaxy outside our own. This meant that gas clouds illuminated by stars inside our own galaxy were listed alongside external galaxies. All were classified as "nebulae." Although it had been clear from observations made in the latter part of the nineteenth century that a class of nebulae with remarkable spiral forms could be distinguished among the Messier catalogue of nebulae, most astronomers continued to think of all such diffuse objects as clouds within our own galaxy. There were a few strong protagonists for the opposite view, that the spiral nebulae were really very distant systems like our own Milky Way—the Englishman R. A. Proctor in the last century, R. F. Sanford, then of the Lick Observatory, and K. Lundmark of Sweden in this century—but many distinguished men were ranged against them, including Shapley himself. After his pioneering work on the scale of our galaxy, it is curious to find Shapley taking a pedestrian point of view in this other controversy. Perhaps, having shown our galaxy to be very large, he had the subconscious hope that everything belonged to it!*

---

*Dr. J. Faulkner has pointed out that Shapley's handling of the calibration of the Cepheids involved two errors, one producing an exaggerated scale for our galaxy and the other leading to underestimates for the distances of the Magellanic Clouds. These two errors taken together brought the Clouds within the confines of the galaxy, an erroneous result that influenced Shapley toward believing that there was nothing at all outside our galaxy.

Hubble first announced his proof of the existence of galaxies outside our own in the Halley Lecture for 1925 delivered at Oxford University. An account of the lecture appeared shortly thereafter in the *Observatory* magazine. It is an interesting indication of the leisurely pace of science in those days that Hubble did not trouble to publish the details of his work on M31 until 1929.

§9-4. *Extending the Distance Range*

Hubble found that both the galaxies M31 (Figure 4-7) and M33 (Figure 9-14) contained stars that were significantly brighter than the brightest Cepheids. With the distance d of either galaxy known from the Cepheids, he was able to

FIGURE 9–14
The galaxy M33 (NGC 598), about 30,000 light-years in diameter.
(Courtesy of Hale Observatories.)

calculate the L value for the brightest stars, using $L = (4\pi d^2) \cdot f$. Observing what seemed to be stars in more distant galaxies, and assuming that the brightest stars in these galaxies would have the same intrinsic L values as the brightest stars in M31 and M33, he was able to calculate their distances from $d = \sqrt{L/(4\pi f)}$. Starting from the Cepheids, the logical chain was therefore:

> Cepheids
> $\longrightarrow$ Nearest galaxies
> $\longrightarrow$ Brightest stars
> $\longrightarrow$ More distant galaxies.

The class of "more distant galaxy" reached in this way actually contained about a thousand galaxies, including a cluster of galaxies in the constellation of Virgo. This sample of "more distant galaxy" was then assumed to be large enough that one could pick out a typical "brightest galaxy." The brightest galaxies turned out to be of smooth globular or elliptical forms, like M87, shown in Figure 9-15. The brightest galaxies in much more distant rich clusters were then taken to be intrinsically similar to M87. Once again, there was an inversion in the use of the flux equation. First, with the distance d obtained to the Virgo cluster of galaxies, and with M87 a member of this cluster, the intrinsic luminosity of M87 was calculated from $L = (4\pi d^2) \cdot f$. Then, with the intrinsic luminosity of a still more remote galaxy taken to be the same as M87, the distance was calculated from $d = \sqrt{L/(4\pi f)}$. Starting from the Cepheids, the chain of logic had become:

> Cepheids
> $\longrightarrow$ Nearest galaxies
> $\longrightarrow$ Brightest stars
> $\longrightarrow$ More distant galaxies
> $\longrightarrow$ Giant globular or elliptical galaxies
> $\longrightarrow$ Still more distant clusters of galaxies.

There has turned out to be a snag in this otherwise extremely imaginative and effective scheme. The objects that Hubble distinguished in the Virgo cluster of galaxies, and which he took to be stars like the brightest stars of M31 and M33, have turned out instead to be bright nebulae, objects rather like the whole of the Orion nebula (Figure 3-14). Such gas clouds emit light by absorbing the radiation, not from a single bright star, but from a whole group of bright stars contained within them. Hence, the objects distinguished by Hubble had intrinsic luminosities that were more characteristic of a group of stars than of a single star. As a result, Hubble had underestimated the distance of the Virgo cluster, and hence, all the distances in the final links of the chain of measurement.

To overcome this difficulty, A. R. Sandage decided to work with the brightest gaseous nebulae instead of with the brightest stars. He found that M31 and particularly M33 are rich in such nebulae, and he calculated their intrinsic

FIGURE 9-15
The brightest of the galaxies have globular or elliptic forms, like the well-known galaxy M87 (NGC 4486) in Virgo. Note the many clusters in the outer regions of M87. (Courtesy of Hale Observatories.)

luminosities, again from $L = (4\pi d^2) \cdot f$, with d known from the Cepheids. Assuming bright nebulae in the more distant galaxies of the Virgo cluster to be similar, Sandage in his early work modified Hubble's scheme to the form:

Cepheids
$\longrightarrow$ Nearest galaxies
$\longrightarrow$ Brightest nebulae
$\longrightarrow$ Virgo cluster
$\longrightarrow$ Giant galaxies of Virgo cluster
$\longrightarrow$ More distant clusters containing similar giant galaxies.

More recently, Sandage adopted what he felt to be a still better procedure. It turned out that the brightest members of a class of spiral galaxy are remarkably similar to each other. An example of this class of galaxy is shown in Figure 9-16. With the help of these giant spiral galaxies, one can reach a wider sample of galaxies than the Virgo cluster before determining L values for the brightest

FIGURE 9-16
The galaxy M101 (NGC 5457), with a diameter of ∼150,000 light-years, a giant Sc. (Courtesy of Kitt Peak National Observatory.)

giant galaxies. With this modification included, we now write down the whole logical chain whereby astronomical distances are measured:

Lower main sequence (Hyades)
⟶ Open clusters
⟶ Cepheids
⟶ Nearby galaxies
⟶ Brightest nebulae
⟶ Bright spiral galaxies
⟶ Moderately distant galaxies
⟶ Giant galaxies
⟶ Very distant clusters containing giant galaxies.

When astronomers assert that such and such a galaxy is at a distance of several thousand million light-years, their statement rests on the strength of this chain.

One is compelled to ask: How strong is it? There is no generally agreed answer to this question. Our impression is that each of the links is quite accurate, by which we mean that we would expect some error but perhaps not more than 10 percent at any one link. Yet, with so many links in the whole chain, a cumulative error of 30 percent might still be present in the final outcome. If this assessment of the situation is not overoptimistic, the achievement of establishing the great distances in the universe to such a degree of accuracy must be judged a very major one indeed.

General Problems and Questions

1. The observed flux f from an astronomical object is related to its distance d and to its luminosity L by the equation $f = L/4\pi d^2$. If L were also known, how would you use this equation to calculate d? Alternatively, if d were known, how would you use the equation to calculate L?

2. Satisfy yourself that you understand why a sphere of radius r, situated at a distance d that is large compared to r, has an apparent angular size θ given in degrees by the formula $\theta = 360 \, r/(\pi d)$.

3. Use the result of Problem 2 to explain how the Earth's annual motion around the Sun can be used to calculate the distances of nearby stars. At what order of distance does this trigonometric method of distance determination become seriously inaccurate?

4. How can the zero-age main sequence be used to calculate distances that go well beyond the range of the method of Problem 3?

Part III:
The Gravitational
Interaction

Chapter 10
The Laws of Motion
and Gravitation

§10-1. Introduction

The purpose of this book has been to help the reader understand various observed astronomical phenomena with the help of the tools provided by basic physics. To this end, we have seen in previous chapters how the electrical interaction and the strong and the weak interactions help explain such diverse phenomena as radiation in all its different forms—the emission of x rays, microwaves, radio waves, and visible light—and the structure of stars and the genesis of elements in stellar nuclear furnaces. We have also seen that the inputs have not all been one way, that is, knowledge passing from physics to astronomy. In most cases, astronomy has also helped our understanding of the basic laws of physics. This is due to the fact that the so-called basic laws of physics have been arrived at only from laboratory experiments, and these experiments are necessarily limited by our terrestrial environment. The laws, however, should apply to conditions and situations far beyond our terrestrial limitations; at least, that is the principle that guides theoretical physicists. Astronomy provides a cosmic laboratory with a far wider range of physical conditions than that ever possible in any terrestrial environment.

Let us review some situations where astronomy has provided a cosmic laboratory for the test of physical laws. The largest man-made particle-accelerator, the *Fermilab,* shown in Plates VII and VII, produces particles with energies up to $\sim 10^{12}$ eV. Compare this with the energies in cosmic rays, upward of 10^{20} eV, that is, more than 100 million times the energy achieved in the terrestrial laboratory. Although controlled thermal fusion of hydrogen to helium is still beyond present-day technology, this process takes place in stars' interiors. Indeed, the stars have taken the idea of thermonuclear fusion as far as it is theoretically possible. The energies involved in the explosions of strong radio sources are $\sim 10^{36}$ times the energy released in the explosion of a megaton H-bomb.

Clearly, in such cases physicists have the opportunity of testing their theories well beyond the laboratory range of the Earth. Indeed, the applications to astronomical phenomena represent the only way in which physicists can really test the universality or limitations of their basic laws.

GRAVITATIONAL INTERACTION IS SMALL IN THE LABORATORY BUT IMPORTANT FOR LARGE MASSES OVER BIG DISTANCES

This discussion brings us to gravitational interaction, the last of the four basic interactions of matter to be discussed in this book. Historically, however, gravitation was not the last but the first of these four interactions to be discussed. It was in 1665, the year of the great plague, that Isaac Newton (Figure 10-1), sitting in his home garden in Woolsthorpe, England, saw an apple drop and started to think about the reason for its drop. This speculation is said to have led him to gravitation. Newton finally came up with his famous inverse square law of gravitation, which he describes in his book, *Philosophiae Naturalis Principia Mathematica*, published in 1687.* As we shall see in this chapter, this law was immensely successful, and it provided explanations for a wide range of phenomena. Yet, it made very little impact on the growth of modern physics as we know it today. This is due to the fact that, by its very nature, gravitation is more amenable to astronomical applications than to laboratory ones.

To see this fact, look at the inverse square law of gravitation,

$$F = G \frac{m_1 m_2}{r^2}.$$

This law states that the force F between two particles of masses m_1 and m_2 increases in proportion to m_1 and m_2 and decreases inversely as the square of their distance r apart. The constant G of gravitation is so small that, with one exception, the force F is very small on the Earth. For example, to atomic physicists looking at the hydrogen atom, the *electrostatic force* between the electron and the proton is of the order of

10,000,000,000,000,000,000,000,000,000,000,000,000,000

---

*It is somewhat mysterious that Newton should have taken 22 years (from 1665 to 1687) to publish such an important law. Did he really arrive at it in 1665? Correspondence between Hooke and Newton around 1679 suggests that, in the beginning, Robert Hooke (1635–1703) had a clearer understanding of the significance of the law than Newton.

FIGURE 10-1
Isaac Newton, 1642–1727. From a mezzotint executed in 1740 by James
Macardel from a portrait by Enoch Seeman. (Courtesy of the Ronan
Picture Library and the Royal Astronomical Society.)

times the gravitational force between them! Consequently, atomic physicists
justifiably ignore the gravitational force in their calculations.

The one exception is that of the Earth's gravitational pull on all objects
situated on it. In this case, we may set m_1 = mass of the Earth, and this mass
gives a measurably large force on m_2. This force is none other than the force of
gravity, which gives the feeling of "weight" to m_2. This exception brings out the
essential property of gravitation that makes it so important for astronomy.
Astronomy deals with massive objects for which either m_1 or m_2 or both m_1 and
m_2 are large. Even though r is also large in astronomy, the largeness of the
masses turns out to be dominant.

Compare now the other interactions vis-a-vis gravitation in the astronomical
situation. The strong and weak interactions are of short range. They will not
be important over interstellar or intergalactic distances; their main contribution
lies in the dense interiors of stars. The electrical interaction is not likely to be

important over a large scale because the heavenly bodies are electrically neutral; it is, however, important for producing radiation from these bodies under the various circumstances discussed in earlier chapters. When we are concerned with the large-scale motion of massive astronomical bodies, or when we are concerned with the equilibrium of such objects, we must turn to gravitation for a significant contribution. Motions of stars and planets, motions of galaxies and clusters of galaxies, and the large-scale behavior of the universe as a whole are examples of the importance of the role of the force of gravitation. The equilibrium configurations of various stars and the formation of black holes are cases where gravitation is pitted against other forces of nature; in many cases it asserts its dominance and supremacy.

Before considering gravitation in more detail, we shall look at the basic notions of dynamics—the science of motion. Again, it was Newton who laid the foundations of dynamics and used it mathematically to study the consequences of his law of gravitation. This Newtonian framework lasted for more than two centuries until it was totally revised by Albert Einstein.

§10-2. Motion

To anyone observing nature, motion is perhaps the most striking and all-pervading phenomenon. A changing rather than a stationary system attracts the observer's attention, and a close examination of every such situation reveals that something is moving. Even in so-called stationary systems, the component parts often move in such a way as to give the overall impression of no change in the system. On a windless day, for example, the air is not really at rest. Molecules comprising the air are in random motion, just as the molecules of water in a still lake are in unceasing motion.

Why do things move? How do they move? It is hardly surprising that people have asked these questions in several different contexts. The movement of an arrow, the flight of a bird, the propulsion of a cart, the flow of a river—from such terrestrial phenomena to the heavenly phenomena of motion of stars and planets, the Sun and the Moon—posed problems that demanded explanations. Documented history shows an interesting evolution of ideas with the interspersing of philosophical reasoning, actual observations, religious dogmas, and experiments, which began with the Greeks more than 2,000 years ago and which led to a first satisfactory solution of these problems in the form of Newton's celebrated laws of motion.

A FORCE IS ACTING ON A BODY ONLY WHEN ITS MOTION CHANGES

In the *Principia*, Newton gave a systematic discussion of the laws that govern the motions of material bodies. The groundwork for these laws had been laid earlier by Galileo (1564–1642), however, with what is now known as the *first law of motion: A body continues to move in a straight line with uniform speed if no force acts upon it.* This idea marked a revolution from age-old concepts dating back to the Greeks. For the first time, it was stated that a force was necessary, not for motion, but for a *change* of motion.

How much force is needed to produce a desired change of motion? This quantitative question is answered by Newton's *second law of motion,* which is often stated in the form,

$$force = mass \times acceleration.$$

Mass is a measure of the quantity of matter in the body. In the context of the second law, however, it also measures the *inertia* of the body. Inertia is the property that implies the resistance the body offers to any external agency (that is, a force) trying to change its state of motion. The larger the inertia (that is, the larger the value of m), the smaller will be the change of motion for a given force. This change is measured by *acceleration.*

What is acceleration? It is the rate of change of velocity. Now, velocity contains two bits of information. Velocity tells us the speed as well as the direction in which a body is moving. A change in either or both of these will result in an acceleration. For example, an automobile is moving at 50 miles per hour. The driver presses the gas pedal, and in 1 minute the speed changes to 60 miles per hour. This change in speed is made without changing the direction on the expressway. What is the acceleration?

The change in speed is $60 - 50 = 10$ (miles per hour). This change is produced in 1 minute $= \frac{1}{60}$ hour. Hence, the change produced per hour $= 10 \div \frac{1}{60} = 600$ miles per hour per hour, the acceleration in the direction of motion.

As another example, a stone attached to a piece of rope is whirled around in a circle of radius r with a constant speed v. Although the speed does not change, the direction of motion changes continually. Hence, the stone is accelerated. The magnitude of the acceleration is v^2/r and is directed towards the center of the circle. The force exerted by the rope on the stone is therefore also directed toward the center (see Figure 10-2).

The direction of motion of the stone turns through a small angle $v \, \delta t/r$ in a small time interval δt. This implies a change $v^2 \, \delta t/r$ in the component of velocity toward the center, and a change involving δt^2 in the velocity component along the tangent to the circle of Figure 10-2. For sufficiently small δt, the latter change can be neglected, whereas the change in the radial direction, when divided by δt, gives v^2/r. Acceleration being defined as the rate of change of velocity, this acceleration is toward the center.

Newton's *third law of motion* states: *Action and reaction are equal and opposite.* When an external agency exerts a force on a body, the body exerts an equal and opposite force back on the agency. A heavy meteorite falling on the Earth may disintegrate by the force of impact; at the same time, it produces a crater in the surface of the Earth. This is an example of action and reaction.

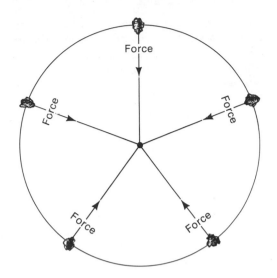

FIGURE 10-2
To keep a stone whirling around in a circle, a force must be applied to the stone, always in a direction *toward the center of the circle.*

§10-3. Dynamics

The subject of dynamics grew up with Newton's laws of motion. How will the body (or bodies) on which external forces act move under their influence? Dynamics has found applications not only in terrestrial phenomena but also in astronomy. Here we look at some of the important dynamical concepts that will be useful later.

In our everyday experience, we think of the surface of the Earth in our immediate locality as being flat; that is, we ignore the curvature of the Earth. We can also neglect changes of distance from the Earth's center, so that the gravitational acceleration on a small body of mass m produced by the Earth is simply a constant, which is usually written as g. This result follows because, treating the Earth as a sphere, the gravitational force exerted on the body by the Earth is

$$F = \frac{GmM}{R^2}$$

where M is the mass of the Earth and R is distance from the center. Now, because acceleration is force divided by the mass of the body on which the force acts, in this case m, the acceleration is $F/m = GM/R^2 = g$, say.

There is no gravitational force in the horizontal direction. A stone or ball thrown into the air moves with its horizontal component of velocity unimpeded by gravity. The vertical component of velocity is, however, subject to the downward acceleration g of gravity. By Newton's second law, an upward-moving ball will have its vertical velocity component checked at a time rate g, and a downward-moving ball will have its falling component of velocity augmented at a time rate g. If a ball is thrown at time $t = 0$ with an initial vertical component v_0, as time goes on the vertical component v behaves as in Figure 10-3. This figure is simply a graph of the line $v = v_0 - gt$. In any small time interval δt, the upward velocity is checked by an amount $g\,\delta t$. Adding the many small time intervals into which an interval t can be divided, we get $-gt$ for the reduction of the vertical component at time t. The ball ceases to move upward at $t = v_0/g$. Thereafter, it falls back to the ground, which it hits at $t = 2v_0/g$. If the ball is thrown with a horizontal velocity component u_0, and if we neglect air resistance, the ball will strike the ground at a distance $2u_0v_0/g$ from the point of projection (that is, $2v_0/g$ multiplied by u_0).

In the preceding example, we split the *net* velocity of the ball into two components: u_0 the *horizontal* component, and v_0, the *vertical* component. How are these components related to the *net* velocity?

Notice that the two components are in mutually perpendicular directions. Construct a *velocity diagram* along the following lines. In a horizontal direction, draw a line AB having a length that represents u_0 on some convenient scale. Draw a line AC perpendicular to AB (to correspond to the vertical direction), and make its length represent v_0 on the *same* scale. Now complete the rectangle $ABDC$ by drawing CD parallel to AB and BD parallel to AC.

The net velocity is represented, both in magnitude and direction, by the line AD, the diagonal of this velocity rectangle, from our starting point A.

Simple geometry, using the Pythagorean theorem, tells us that

$$AD^2 = AB^2 + BD^2 = AB^2 + AC^2$$

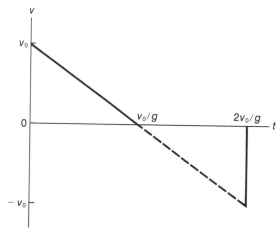

FIGURE 10-3
The graph of v against t shows how the vertical (upward) component of the velocity of the ball drops to zero at $t = v_0/g$ from a starting value v_0. The dotted continuation of the graph below the t axis indicates the downward motion of the ball. This downward motion continues until the ball hits the ground at $t = 2v_0/g$.

since $AC = BD$ for the rectangle. If the magnitude of the net velocity is w_0, the preceding relation tells us that

$$w_0{}^2 = u_0{}^2 + v_0{}^2.$$

The limitation in throwing a ball is determined by $w_0{}^2 = u_0{}^2 + v_0{}^2$, which, for a given person, cannot exceed an amount k, say, set by the muscles of the arm. For a specified value of k, how shall we best distribute u_0 and v_0 to throw the ball a maximum distance? We have to make the product $u_0 v_0$ as large as possible, subject to the restriction that $u_0{}^2 + v_0{}^2$ cannot exceed k. The answer is to throw in such a way that $u_0 = v_0 = \sqrt{k/2}$. From elementary algebra, we have

$$2u_0 v_0 = u_0{}^2 + v_0{}^2 - (u_0 - v_0)^2 = k - (u_0 - v_0)^2.$$

Thus, $u_0 v_0$ will be largest when $(u_0 - v_0)^2$ is least. This condition is met by $u_0 = v_0$, giving $u_0 = v_0 = \sqrt{k/2}$. The initial vertical and horizontal components of velocity thus being equal, the ball must be thrown at an angle of $45°$ to the ground. Remember that the horizontal range is always $2u_0 v_0/g$; thus, the maximum range given by $u_0 = v_0 = \sqrt{k/2}$ is k/g. Hence, a determination of the maximum range would decide the strength k of the arm.

The concern of a baseball player is not with maximum range but with making the time $2v_0/g$ of a throw as short as possible. If d is the horizontal distance required, the time of the throw is also d/u_0, which must be equal to $2v_0/g$ so that $2u_0 v_0 = g\,d$. Consequently, as the baseball player seeks to make $2v_0/g$ small by giving the ball only a small initial vertical component, he must make u_0 correspondingly large so that $2u_0 v_0$ will equal the required value of $g\,d$. Indeed, the greater the horizontal speed u_0 with which the player can fire the ball, the smaller v_0 can be and the shorter the time of the flight. This explanation shows why baseball pitchers have flat trajectories. But making u_0 large and v_0 small does not economize on k! We have $2u_0 v_0 = u_0{}^2 + v_0{}^2 - (u_0 - v_0)^2 = g\,d$. Writing $v_0 = \epsilon \sqrt{k}$ with ϵ small, $u_0{}^2 + v_0{}^2 = k$ gives $u_0 = (1 - \tfrac{1}{2}\epsilon^2)\sqrt{k}$ to sufficient accuracy, in which case $u_0{}^2 + v_0{}^2 - (u_0 - v_0)^2 = g\,d$ leads, after a little algebraic manipulation, to approximately $k = g\,d/2\epsilon$. Remember that d is a required distance, and, the quicker the throw (that is, the smaller ϵ), the greater the muscle power k that is needed. Flat throws over a considerable distance d require what is known in sports terminology as a great arm.

How high has the ball risen at time $t = v_0/g$? At time $t = 0$, the ball has the upward component of velocity v_0, and, at $t = v_0/g$, the upward component is zero (Figure 10-3). Because the upward component behaves linearly with time (that is, because the graph of Figure 10-3 is a straight line), the average upward component between $t = 0$ and $t = v_0/g$ is simply $\tfrac{1}{2}v_0$, and the height attained by the ball is $v_0{}^2/2g$ ($\tfrac{1}{2}v_0$ multiplied by the time v_0/g).

This problem can be generalized to obtain a result of considerable importance. At time t during the upward flight, the vertical component of velocity is

$v_0 - gt$, and the average upward velocity between time t and the initial moment is $\frac{1}{2}(v_0 + v_0 - gt)$. Hence, the height h attained by the ball at time t is given by $h = \frac{1}{2}t(2v_0 - gt) = v_0t - \frac{1}{2}gt^2$.

If we now denote the vertical component at time t by v so that $v = v_0 - gt$, we can express h in terms of v instead of in terms of t. Thus, $t = (v_0 - v)/g$, and $h = v_0t - \frac{1}{2}gt^2 = \frac{1}{2g}(v_0{}^2 - v^2)$. Multiplying this result by the mass m of the projectile, we can rewrite the last equation in the form

$$\tfrac{1}{2}mv^2 + mgh = \tfrac{1}{2}mv_0{}^2.$$

As the projectile falls back to the ground, the upward component of velocity reverses its sign. Instead of being v when the height is h, the upward component is $-v$ when the height is h. This change makes no difference to the preceding equation, however, since the velocity appears only in the form of a square, v^2. Thus, the same equation continues to hold good during the descent of the projectile. Moreover, we can add $\frac{1}{2}mu_0{}^2$ to both sides equally, giving

$$\tfrac{1}{2}m(u_0{}^2 + v^2) + mgh = \tfrac{1}{2}m(u_0{}^2 + v_0{}^2).$$

This result is the *equation of conservation of energy*. The right-hand side is the initial *kinetic energy;* $\frac{1}{2}m(u_0{}^2 + v^2)$ is the kinetic energy (the energy of motion) at height h, and the term mgh is the change in the gravitational *potential energy* occasioned by rising through the height h. This simple result is an example of a much more general result,

kinetic energy + potential energy = constant,

which holds good when the Earth's curvature is taken into account and when the elementary concept of height is replaced by that of distance from the center of the Earth.

Let us consider the curvature of the Earth in a case like that of Figure 10-2, but with gravity g holding a satellite in a circular orbit just above the Earth's surface instead of a stone being held on a rope. With V the orbital velocity of the satellite and R the radius of the Earth, the acceleration of the satellite toward the center is V^2/R. From Newton's second law, this acceleration must be equal to g so that $V^2 = Rg$. The time T required for the satellite to make a complete circuit of the Earth is given by $T = 2\pi R/V$, which is equal to $2\pi\sqrt{R/g}$. Putting $g = 32$ feet/second/second and R about 2×10^7 ft gives T about 5,000 seconds or about 80 minutes.

In the preceding discussion, forces of dissipation, such as friction and air resistance, were assumed to be small and ignorable. For dissipative forces, work has to be done against their resistance. This work usually appears as heat and is said to be lost because it cannot be recovered as kinetic energy, as in the case of the potential energy arising from the Earth's gravity. As an example, a spacecraft or a meteor falling through the Earth's atmosphere heats up due to the dissipative force of resistance by the air. But if we take care to include the heat in our energy balance, we regain the law of conservation of energy. This is the *first law of thermodynamics,* which we shall meet again in Chapter 11. Force fields like the Earth's gravity, or an electric or magnetic field, are *not* dissipative. Work done against them is converted entirely in the form of potential energy, and it is fully recoverable as kinetic energy if required. Such forces are said to be conservative.

§10-4. The Law of Gravitation

The law of gravitation was first published by Newton in its complete form in the *Principia,* although it seems likely that Robert Hooke (1635–1703) arrived at the correct form of the law,

$$F = \frac{Gm_1m_2}{r^2},$$

ahead of Newton.

Actually, the form of the inverse square law of gravitation was first deduced from the observed motions of planets. Suppose we have the Sun at S and a planet at P in Figure 10-4. Both are attracted toward each other along SP. By Newton's third law, the force of S on P is equal and opposite to the force of P on S. Consequently, neither the Sun nor the planet can remain at rest by the first law of motion. However, since the Sun is much more massive than the planet, the change in the Sun's motion will be much smaller than the corresponding change in the planet's motion. (For example, the Sun is about 300,000 times more massive than the Earth.) Thus, the second law of motion requires that the effect will be much greater on the motion of the planet than on the motion of the Sun. Indeed, the effect on the motion of the Sun is so small that it can for many purposes be ignored.

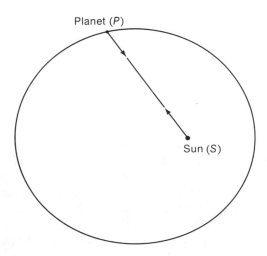

FIGURE 10-4
The gravitational force that the Sun exerts on a
planet is exactly balanced by the force that
the planet exerts on the Sun.

Johannes Kepler (1571–1630) charted the orbits of planets with painstaking accuracy, and he came up with three precise laws of planetary motion that described these orbits. The question was: What type of gravitational force could produce such orbits? It seems that Hooke arrived at the inverse square law from such a form of empirical reasoning. It was Newton, however, who formulated the laws of motion mathematically, deducing the shapes and other details of the planetary orbits, precisely as they had been obtained by Kepler from his analysis of the observed motions.

THE LAW OF GRAVITATION LEADS TO A DETERMINATION OF THE MASS OF THE EARTH

Once again, we write M for the mass and R for the radius of the Earth. The force on a body of mass m at the Earth's surface is GmM/R^2, and the acceleration g is GM/R^2, so that

$$M = \frac{gR^2}{G}.$$

All three quantities on the right-hand side of this equation can be determined by observation: g from the fall of a stone, R from a measurement of the curvature of the Earth's surface, and G from an actual laboratory measurement of the force between two masses m_1 and m_2 separated by a known distance. Such a laboratory measurement was first made by Henry Cavendish (1731–1810). Figure 10-5 shows the apparatus that Cavendish used. Today, g, R, and G are all known very accurately, and the mass of the Earth determined by the preceding equation turns out to be $(5.977 \pm 0.004) \times 10^{27}$ g.

FIGURE 10-5

In the time of Newton, the distance of the Earth from the Sun was not well known, so the constant G in Newton's gravitational formula was not known. One way to determine G was by an experiment in which the deflection of hanging pellets (x) toward the large known weights marked W was measured. Such an experiment was performed at the end of the eighteenth century, about 70 years after Newton's death, by Henry Cavendish.

THE LAW OF GRAVITATION DETERMINES THE LEAST SPEED THAT A ROCKET MUST HAVE IN ORDER TO LEAVE THE EARTH

If m is the mass of the rocket and M the mass of the Earth, then, at a separation distance r, the force acting on the rocket is GMm/r^2, which decreases as r increases, that is, as the rocket moves away from the Earth. The ability of gravity to restrain the rocket therefore weakens as the rocket moves outward, and if the rocket is fired with a sufficiently fast initial speed, it can move away so far that the Earth's gravitation no longer has a significant effect upon it. The rocket in such a case would have escaped from the Earth.

To estimate how fast the initial speed V must be for escape to occur, let us appeal to *the equation of conservation of energy,*

$$kinetic\ energy\ +\ potential\ energy\ =\ constant,$$

but using a more sophisticated formula for the potential energy than we gave before. Formerly, we had *mgh* for the change of potential energy in lifting a mass m through a height h. Remembering that $g = GM/R^2$ at the surface of the Earth, it is not hard to see, that for h small compared to R, *mgh* is effectively the same as

$$\frac{GmM}{R} - \frac{GmM}{R + h}.$$

(The slight difference between this expression and *mgh* comes from our former neglect of the change in g arising from variations in the distance from the center of the Earth.)

In the same way, if we lift a mass from height h to height $2h$ above the surface of the Earth, the change of potential energy is described by

$$\frac{GmM}{R + h} - \frac{GmM}{R + 2h}.$$

Suppose we go in such small steps of h out to a great distance r from the center of the Earth. Let us write

$$r = R + nh$$

where the number of steps n is large. The potential energy change in any intermediate step, say, the kth step, is given by an expression similar to that above:

$$\frac{GmM}{R + (k - 1)h} - \frac{GmM}{R + kh}.$$

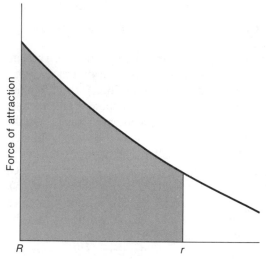

Force of attraction

R   r

Radial distance from the center of the Earth

FIGURE 10-6
The shaded area represents the change of potential energy in moving from a distance R to a distance r from the center of the Earth. The area is $GMm \left(\dfrac{1}{R} - \dfrac{1}{r} \right)$.

What is the net change in the potential energy from R to r? This net change is given by adding all the preceding step-by-step changes. It is easy to verify that the final answer is simply

$$\frac{GmM}{R} - \frac{GmM}{r}.$$

Our procedure here will be familiar to the student of calculus. In Figure 10-6, the force of attraction is plotted against r. The potential energy change between R and r is just the area under this curve between the ordinates at R and r.

Setting v for the speed of the rocket when it reaches distance r, the left-hand side of the energy equation is

$$\underset{\substack{(kinetic \\ energy)}}{\tfrac{1}{2}mv^2} + \underset{\substack{(change\ of \\ potential\ energy)}}{\frac{GmM}{R} - \frac{GmM}{r}},$$

whereas initially the left-hand side was simply $\tfrac{1}{2}mV^2$. Since the left-hand side is always the same, we can equate these two expressions to obtain the following result for v^2:

$$\tfrac{1}{2}mv^2 + GmM \left(\frac{1}{R} - \frac{1}{r} \right) = \tfrac{1}{2}mV^2. \tag{A}$$

What does the relation (A) imply? Notice first that, as we increase r, the corresponding v becomes smaller and smaller. The rocket loses speed because it is being continually opposed by the force of gravity. In relation (A), this opposition shows up through the potential-energy term, which, as we saw earlier, arises from the force of attraction.

Physicists use the word *potential barrier* to describe such situations. An athlete can easily jump a 2-ft fence; a professional high jumper can do much better, although everyone has his or her limit. In nature, the controlling forces often put limits on motion, and the potential barriers indicate how these limits operate.

In our example of the rocket, Earth's gravity erects a potential barrier. Unlike the high jumper's barrier, the gravity barrier extends all the way to infinity, although its "height" becomes progressively smaller until it drops to zero at infinity. How fast must the rocket be fired to surmount this barrier and reach out to infinity? It is easy to settle this question with the help of the relation (A). This relation tells us that, as r increases, v decreases. We do not want v to become zero at a finite value of r, for v becoming zero would imply that the rocket failed to surmount the gravity potential barrier. If the rocket did fail to surmount the barrier, it would again fall back toward the Earth. Can we arrange for the rocket to come to rest (if at all) only as r becomes infinite? To find the answer to this question, let r go to infinity in (A) and set $v = 0$. We then get the simple result,

$$\frac{GM}{R} = \tfrac{1}{2}V^2$$

or

$$V = \sqrt{\frac{2GM}{R}}.$$

This velocity is the one at which the rocket must be fired in order to just escape from the Earth. Putting in the known numbers for G, M, and R, we get the answer

$$V_{escape} \cong 7 \text{ miles/second.}$$

This notion of escape velocity will be encountered in Chapter 11 in connection with black holes. Basically, the larger the escape velocity, the taller the potential barrier to be penetrated for an escape from the gravitational pull. Thus, the strength of the gravitational control exerted by an object is indicated by the magnitude of the escape velocity from its surface.

The escape velocities of some astronomical objects are as follows:

The Moon	$\sim$ 1.5 miles/second
The Sun	$\sim$ 400 miles/second
Sirius B (a white dwarf)	$\sim$ 3,000 miles/second
Neutron stars	$\sim$ 100,000 miles/second.

The Newtonian laws of motion and gravitation served physicists well for two centuries. Numerous applications for them were found, not only in astronomy but also in the rest of science. Directly or indirectly, Newton's laws of motion inspired developments in other branches of physics, such as in the phenomena of electricity and magnetism. Nevertheless, these laws have undergone revolutionary changes in modern times. Why?

The success or failure of scientific laws, theories, and hypotheses is judged in the last analysis by how well they can account for natural phenomena. It is because of their failure to explain the observed results and because of their inconsistency with the developments in the rest of theoretical physics that Newtonian concepts had to be replaced by more sophisticated ones.

LIGHT DOES NOT OBEY NEWTON'S CONCEPT OF RELATIVE MOTION

In 1887, exactly two centuries after the publication of the *Principia,* a startling result emerged from an experiment performed by Albert A. Michelson and Edward W. Morley on Mount Wilson in southern California. The background to this experiment is briefly as follows.

Physicists of the last century believed that light, like sound, needed a medium to travel through. The supposed medium was called *aether*. Attempts to detect the existence of aether proved futile, however. The Michelson–Morley experiment was such an attempt; its aim was to detect the Earth's speed relative to aether.

It is common for a boat to travel faster downstream than it does upstream. If v is the speed of the stream and c the speed of the boat in stationary water, then the speed of the boat downstream will be

$$v + c,$$

while the speed upstream will be reduced to

$$c - v.$$

A calculation (see Figure 10-7) shows that a boat moving in a direction perpendicular to the stream has the speed

$$\sqrt{c^2 - v^2}.$$

For "boat" in this discussion, read "light"; for "stream," read "aether." With this substitution, the principle behind the Michelson–Morley experiment becomes clear. As the Earth rotates, say, with surface speed v, there should be an aether drift in the east-west direction (because this is opposite to the direction

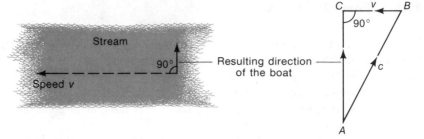

FIGURE 10-7

In order to go directly across the river, the boat must be turned at a slant along *AB*, and it must be given a velocity component *v* upstream to cancel the flow of the river downstream. Thus, in the triangle *ABC*, *BC* is equal to *v* downstream, *AB* equals *c*, and *AC* is perpendicular to *BC*. By the theorem of Pythagorus

$$AC^2 = AB^2 - BC^2 = c^2 - v^2,$$

so that the resulting required velocity of the boat is $\sqrt{c^2 - v^2}$ in the direction required. (Refer to the discussion of the velocity rectangle in Section 10-3.)

in which the Earth rotates). Thus, a return trip by light over a distance *l* in the east-west direction should take the time

$$\frac{l}{c + v} + \frac{l}{c - v} = \frac{2lc}{c^2 - v^2}.$$

A similar return trip in the north-south direction, perpendicular to the aether drift, should, however, take the time

$$\frac{2l}{\sqrt{c^2 - v^2}},$$

which is *less* than the east-west travel time by a factor

$$\sqrt{1 - \frac{v^2}{c^2}}. \tag{B}$$

Now, although their apparatus was sensitive enough to measure an effect 100 times smaller than (B), Michelson and Morley failed to observe *any* reduction whatsoever for the north-south direction.

This *null* result caused great consternation among scientists in the last decade of the nineteenth century; it cast doubt not only on the existence of aether but also on the basic concepts of motion that were regarded as well established since Newton. Interpretations of the Michelson–Morley result were attempted by H. Poincaré, G. F. Fitzgerald, and H. A. Lorentz, but they were of a somewhat makeshift character. A radically new interpretation was offered in 1905 by Albert Einstein (Figure 10-8).

FIGURE 10-8
Albert Einstein, 1879–1955, at his desk in the Patent Office, Bern, Switzerland, where he was working as a clerk during the period when he developed the special theory of relativity.

§ *10-6. The Special Theory of Relativity*

THE SPEED OF LIGHT IS THE SAME FOR ALL INERTIAL OBSERVERS

An inertial observer is one who is moving with uniform speed in a straight line, that is, one on whom no external force acts.

The face value of the null result of Michelson and Morley was that the speed of light was the same in the north-south and east-west directions. Einstein followed Poincaré in taking the constancy of the speed of light as a fundamental principle. He then faced up to the strange and seemingly paradoxical situations to which this principle appeared to lead.

Suppose two observers have a relative velocity v in a given direction. According to Newtonian ideas of motion, if one observer measures the speed of light traveling in the direction of the other observer to be c, then the other observer should measure the speed of light to be $c \pm v$. However, according to the postulate of the constancy of the speed of light, both observers should see light to be traveling with speed c. Clearly, the Newtonian law of addition of velocities had to be changed. Since velocity means the ratio of a space displacement to a time interval, the revolutionary implication of the fundamental principle was that our everyday ideas about the measurements of spatial displacements and of time intervals had to be wrong. Appendix A describes the changes in spacetime geometry necessitated by Einstein's fundamental principle of the constancy of the speed of light.

MEASUREMENTS OF THE SPACETIME POSITIONS AT WHICH PHYSICAL EVENTS OCCUR ARE NOT UNIQUE

The special theory of relativity, as Einstein's new ideas came to be called, abolishes entirely the apparently common-sense notion of a basic universal time, the so-called absolute time which, in Newtonian physics, is supposed to exist for *all* observers. Each inertial observer has his own time, called his *proper time*, which he can measure with his own clock. However, if he compares his clock with clocks of other inertial observers flashing past him, using an explicit method to be described in a moment, he will find that these other clocks run slower than he expects from the time readings of his own clock, slower for an observer moving past him with speed v by a factor

$$\sqrt{1 - \frac{v^2}{c^2}}.$$

This is the same factor noted earlier in connection with the Michelson–Morley experiment.

As the name *relativity* implies, these effects are not absolute but relative. Just the same effect will be seen from the point of view of the second observer in the preceding example! At first, this requirement appears paradoxical and impossible. However, with a little patience we can see that no contradiction really exists. To see how things work, consider two inertial observers A and B. In Figure 10-9, we have the spacetime diagram of observer A.

The time axis represents the so-called world line of A, that is, the curve that tells us where to find A in the spacetime diagram at any specified time. (Since we are measuring time t in the inertial frame of rest of A, A is obviously always in the same place at any given time.) Similarly, the straight line inclined to the time axis is the world line of B. A and B are taken to be at the same place when $t = 0$. Also, A and B set their watches to zero as they pass each other at this instant.

To find where B is at a later time, A sends out a light signal that leaves A at time t_1, is reflected by B, and returns to A at time t_2. Then, A argues that the

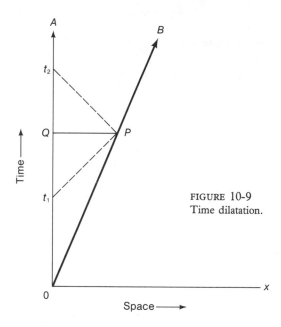

A

B

t_2

Q ———— P

Time →

t_1

0

Space →

x

FIGURE 10-9
Time dilatation.

light traveled a total distance $(t_2 - t_1)c$, half of it toward B and the other half back from B. Accordingly, he sets the distance of B at the instant when the signal is reflected as $(t_2 - t_1)c/2$. (The point of reflection is shown at P on B's world line in Figure 10-9. The broken lines are the light tracks.) By A's reckoning, the reflection happened at the half-way instant, $(t_1 + t_2)/2$. The observer A therefore concludes that B is moving away from him with a velocity

$$v = \frac{t_2 - t_1}{t_2 + t_1} c. \tag{C}$$

What should be the time recorded by B's watch at P? Should it be $(t_1 + t_2)/2$ as expected by A? To seek the answer to this question, we appeal to the basic symmetry between A and B. Notice that when light left A at t_1, it reached B at P. Suppose the time recorded by B is βt_1 where β is a constant factor. It is easy to see that, if A sent the light signal at $2t_1$, B would receive it at $2\beta t_1$, and so on. Now, the same factor β must clearly operate between B and A. That is, if B sends a light signal to A at a time τ as measured by B's watch, it must reach A at a time $\beta\tau$ as measured by A's watch. Remarkably, this symmetry between A and B enables us to fix β. We note that the time by B's watch at P is βt_1, so that the time by A's watch, when the return signal from B (sent at P) comes, must be $\beta \times \beta t_1 = \beta^2 t_1$. Thus, we have

$$\beta^2 t_1 = t_2.$$

Using formula (C), it is easy to see (by eliminating the ratio t_2/t_1) that β must be related to v by the equation

$$\frac{v}{c} = \frac{\beta^2 - 1}{\beta^2 + 1},$$

that is,

$$\beta = \sqrt{\frac{c + v}{c - v}}.$$

Thus, the time at P by B's watch must be

$$\tau = t_1 \sqrt{\frac{c + v}{c - v}}.$$

How does this value of τ for B compare with the time estimated by A? The latter was

$$t = \frac{t_1 + t_2}{2} = t_1 \frac{c}{c - v}.$$

Thus, we get

$$\tau = t \sqrt{1 - \frac{v^2}{c^2}}.$$

This means that, if B communicated her time τ at P along with the return signal to A, then A would discover that B's watch was slow compared to his own by the factor

$$\sqrt{1 - \frac{v^2}{c^2}}.$$

Notice that B can also perform the same type of reflection experiment as A did, and B would similarly conclude that A's watch was slow compared to her own by exactly the same factor! Remarkably enough, there is no paradox in this. A paradox would have arisen only if A and B were at the same place at all times and had been able to notice that each other's watches were slow.

It is worth emphasizing that these conclusions were arrived at on the assumptions of (i) the constancy of the speed of light ($= c$) and (ii) the symmetry between the inertial observers (A and B). Although what we have described is a thought experiment, it is possible to translate it into effects that are observable in practice. Take, for instance, the factor β that compares the time on A's watch at the time he sends a light signal to the time on B's watch at the time she receives it. If A keeps sending light waves of a certain fixed frequency ν to B, B will receive these waves, not with the same frequency ν, but with a *reduced* frequency ν/β because of the time-stretch factor β. This is the well-known Doppler effect that is discussed at length in Appendix C. The existence of the time dilatation factor

$$\sqrt{1 - \frac{v^2}{c^2}}.$$

can also be established experimentally. Such an experiment is described in Appendix A.

Nevertheless, effects like these were peculiar enough to make many leading physicists of the early twentieth century doubt the validity of the postulate of the constancy of the speed of light. The fact that spacetime measurements do not have an absolute status came as a shock to those who had been brought up in the Newtonian tradition; however, the mathematical elegance of the special theory of relativity became gradually apparent. One consequence of Einstein's theory was that physicists could no longer isolate purely spatial measurements from purely temporal ones; the two got mixed. It was Hermann Minkowski who first showed how to combine space and time in a new kind of geometry that we shall call the *geometry of special relativity*. Some details of this geometry are given in Appendix A.

THE MASS OF A BODY DEPENDS ON ITS MOTION

The new ideas of relativity also led to a revision of Newtonian dynamics. The mass of a body in the Newtonian system is always the same, whether or not the body is at rest. In special relativity, if we measure the mass of a body to be m_0 when it is at rest relative to the measuring apparatus, then its mass when it is moving with speed v relative to the apparatus will be found to be $m_0\gamma$, where γ is simply the *reciprocal* of the slowing-down factor just calculated, namely,

$$\gamma = \frac{1}{\sqrt{1 - (v^2/c^2)}}.$$

In actual experiments, we do not measure masses. We measure momenta. Thus, for a particle moving with velocity v and having rest mass m_0, the momentum is not $m_0 v$ as in Newtonian dynamics. It is

$$mv = \gamma m_0 v.$$

Measurements of momenta are possible through experiments involving collisions. In a typical collision, the total momentum of all participating particles is conserved; that is, the total momentum before collision equals the total momentum after collision. This result, which was a consequence of Newtonian dynamics, is carried over into special relativity with the preceding modification of the definition of momentum. Accelerator experiments involving collisions of fast moving ($v \simeq c$) particles have confirmed the existence of the factor γ.

Thus, as v increases towards c, γ rises rapidly and becomes infinitely large at $v = c$. Physically, γ becoming infinitely large at $v = c$ means that it is increasingly difficult to increase the speed of a body closer and closer toward c. By the second law of motion, the force required rises rapidly and reaches infinity as $v \longrightarrow c$. Thus, no technology, however powerful, can increase the speed of a body to that of light. The velocity of light provides the upper limit on any movement of matter, a limit that can never be attained.

What of light itself? Quantum theory has shown that light may be interpreted as a stream of quanta—packets of energy often called *photons*. How do photons thought of as particles manage to travel with the speed of light? The answer is that they have *zero* rest mass. Indeed, the general rule is that all particles of zero rest mass travel with the speed of light.

Scientists have conjectured whether there exists a third category of particles, called *tachyons*, that always travel faster than light (see Figure 10-10). If we

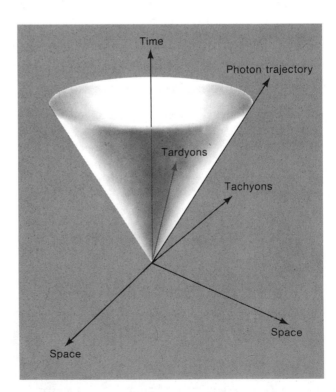

FIGURE 10-10
The light cone.

draw a spacetime diagram for light particles (photons) emitted from the origin, the light will reach a distance $r = ct$ from the origin at time t. The trajectories of the photons generate a hypercone in four-dimensional spacetime (three space plus one time). The trajectories of material particles (sometimes called tardyons) always lie inside this cone, which is called the light cone, whereas tachyon trajectories lie outside it. So far, experiments have failed to reveal the existence of tachyons.

MASS AND ENERGY HAVE AN EQUIVALENCE, $E = Mc^2$

In the prerelativity era, scientists had established two distinct conservation laws—one for mass and the other for energy. Special relativity suggested the possibility of converting mass into energy, the single equivalence popularly described by the relation $E = Mc^2$. For example, if we put $m = 1$ g and $c = 3 \times 10^{10}$ cm/second, the energy created by destroying 1 g of matter is 9×10^{20} ergs; it is sufficient to boil 30,000 tons of ice at normal atmospheric pressure.

This relation is no theoretician's pipe dream. It was translated into practice in the form of an atomic bomb in 1945. Today, nuclear reactors generate energy from the masses of the reacting particles. This principle is the same one that governs the generation of energy inside stars and permits them to shine for long periods of time.

If we consider the equivalent energy of a rest mass m_0 and a moving mass $m = m_0 \gamma$, the difference has the energy equivalent

$$(m_0 \gamma - m_0)c^2 = m_0 \left(\frac{1}{\sqrt{1 - (v^2/c^2)}} - 1 \right) c^2.$$

This energy equivalent is the kinetic energy, that is, the energy due to motion. For small v, it can be shown, by the binomial expansion of the square root expression (ignoring negligible contributions), that this formula for the kinetic energy simplifies to give $\frac{1}{2}mv^2$, the Newtonian expression we arrived at earlier. This derivation of the Newtonian expression is a verification of the equivalence of mass and energy.

For v small compared to c, the Newtonian laws are still approximately valid. Even for the speeds of spacecrafts leaving the Earth, the Newtonian laws still provide a good working theory.

Let us take one example. For the escape velocity from the Earth, the ratio $(v/c)^2$ is smaller than 14 parts in 10 billion, and the neglect of these 14 parts in 10 billion is a measure of the approximation involved in using the Newtonian laws. In astronomy, however, we do encounter very fast-moving particles. For example, the fastest moving cosmic-ray particles have speeds so close to the speed of light that they differ from it by less than 5 parts in 10^{25}. For them, γ is as high as 10^{12}. We cannot use Newtonian ideas for such particles, nor can we use them for the particles that gave rise to the emission of radio waves by the synchrotron process of Chapter 4, nor for the emission of x rays by the inverse Compton process of Chapter 7.

Einstein extended his ideas on special relativity to the following general principle: *all laws of physics look the same to all inertial observers*. This principle has not yet been thoroughly tested, but all experiments to date appear to be consistent with it. One important implication is that no physical signal (that is, a signal transmitted using material particles or zero rest mass particles such as photons) can travel faster than light. This fact caused Einstein to examine another well-established Newtonian concept—the law of gravitation. We will see how this led him to the general theory of relativity.

§10-7. The General Theory of Relativity

Although Newton's inverse square law of gravitation worked very well, there were certain conceptual problems with it when it was confronted by the special theory of relativity. Conversely, the phenomenon of gravitation posed conceptual problems for special relativity. Einstein sought to resolve both these difficulties by proposing, in 1915, the general theory of relativity. General relativity is a remarkable conceptual development—so remarkable that, when it was first advanced, very few people could grasp its full significance. We outline here the essential features of this theory without going into intricate mathematical details. We begin by discussing the difficulties already mentioned and the way they were dealt with by Einstein.

GRAVITATION IS A MANIFESTATION OF SPACETIME GEOMETRY

The force of gravitational attraction, according to the Newtonian inverse square law, acts instantaneously between two bodies, however far apart they may be. This idea is in conflict with special relativity, which says that no interaction of a material origin can travel faster than the speed of light. To see this conflict, consider the following thought experiment (see Figure 10-11). If the Sun were suddenly removed from the solar system, how soon would we on Earth discover the event? According to Newton's inverse square law of gravitation, the effect would be felt by Earth instantaneously. The Earth would take off from its elliptical orbit in a tangential direction and continue thereafter to move in a straight line. However, according to Einstein's theory of special relativity, this information should not reach us until at least $\sim$8 minutes after the Sun's removal, $\sim$8 minutes being the time it takes light to travel from the Sun to the Earth.

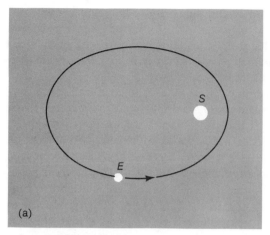 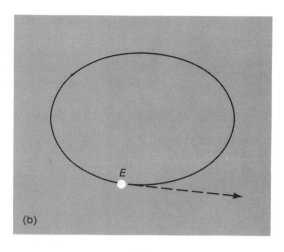

FIGURE 10-11

In (a), the Earth (*E*) is moving in an elliptical orbit around the Sun (*S*). If *S* were suddenly annihilated by some means, *E* would take off from the orbit in a tangential path as shown by the dotted line in (b). In Newtonian gravitation, this would happen instantaneously, but it takes ~8 minutes for light to travel from *S* to *E*. Thus, the disappearance of *S* will be "seen" by people on *E*, ~8 minutes after the event.

A further problem arises from the definition of an inertial observer. Such an observer is supposed to be free of all external forces, but is there a material object anywhere in the universe that is not acted upon by some external force? A little thought shows that all physical systems, animate or inanimate, are subject to the force of gravitation. This force is one that *cannot be switched off*. An electric force or a magnetic force can be switched off, or removed, by a suitable shielding mechanism, but gravity cannot be got rid of in any such way. Only by going far, far away from all matter can we hope to achieve something like a state of no gravitational force. But this prescription is of little use to scientists on Earth or to astronomers studying the universe. Thus, in the presence of gravity, even the special theory of relativity needed to be revised.

Einstein got around this difficulty of the inertial observer by taking gravity as a nonremovable feature of spacetime itself, something much more intrinsic than an entity occupying spacetime. His radical solution was to identify the presence of gravitation with the *geometric properties* of spacetime. The starting point of his theory of gravitation, the so-called *general theory of relativity,* was the following concept: *Due to the presence of matter, the geometry of spacetime is non-Euclidean. The non-Euclidean nature of spacetime manifests itself in the phenomenon of gravitation.*

Euclid (ca. 300 BC) appears to have been the first to lay down the systematic foundations of geometry that we learn at school. The geometry of Euclid is based on a set of axioms from which results are deduced about figures of various shapes and sizes. The theorems of Euclidean geometry have found numerous applications in everyday life, as, for instance, in surveying, engineering, and navigation. These applications led people, including scientists and mathematicians, to believe that Euclid's geometry was the only possible geometry, both as a mathematical system and as a fact of the real world.

These beliefs were shattered, at any rate as far as mathematicians were concerned, in the last century. Lobachevsky (1793-1856), Gauss (1777-1855), and Bolyai (1802-1860) demonstrated that, by changing the Euclidean axioms, it was possible to have other geometries that were also mathematically self-consistent. These other geometries came to be referred to as non-Euclidean geometries.

As an example of a non-Euclidean geometry, consider the spherical surface of the Earth (see Figure 10-12). Suppose that an albatross flying always at a constant height undertakes a journey starting from the North Pole, *N*. Moving in a straight southerly direction along the Greenwich meridian, it reaches the equator where it makes a left turn. It then proceeds straight along the equator, a quarter of the way around the Earth, where it makes another left turn. It is now on the 90° meridian along which it proceeds northward. When it reaches the North Pole *N*, it finds that it has arrived from a direction at right angles to the direction along which it had set out.

Now, take a look at the triangle *NAB* in Figure 10-12 that describes the track of the albatross. This triangle has *three* right angles. This situation is entirely different from Euclid's rule that the angles in every triangle must add up to two right angles. Does this situation mean that Euclid was wrong? He was right within the terms of reference of his own geometry. In Figure 10-12, *we are not using Euclidean geometry.*

In what way does the geometry on the surface of the sphere differ from Euclid's in its basic terms of reference? The difference lies in the so-called parallel postulate. Euclid assumed that, given a straight line *l* (see Figure 10-13)

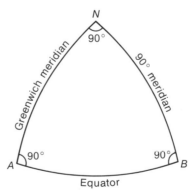

FIGURE 10-12
The triangle on the surface of the Earth, with the vertices *N*, *A*, *B* as shown, has three interior angles, each equal to a right angle.

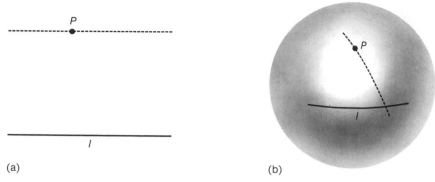

(a)   (b)

FIGURE 10-13

The parallel postulate of Euclid states that, through a point P not lying on a straight line l, one and only one straight line can be drawn parallel to l. This is true for a straight line l and the point P on a plane (a), but it is not true on the surface of a sphere (b). In the latter case, all straight lines through P meet l.

and a point P outside it, one and only one straight line could be drawn through P parallel to l. This is so self-evident to our everyday ideas that it is likely to be (and indeed, was) mistaken for the truth. On the surface of the sphere, however, Euclid's postulate is not true. All straight lines through P, which are arcs of great circles, meet l. Moreover, there are other non-Euclidean geometries, quite different from the geometry on a sphere, where *more than one* straight line can be drawn parallel to l through P.

It was Einstein who ingeniously exploited, for the first time, the potential of non-Euclidean geometries in describing gravitation.

§10-8. Gravitation According to Einstein

To see how the general theory of relativity goes about solving and interpreting gravitational problems, let us reconsider the simple example of a ball thrown vertically upward with a starting ground velocity v_0. We have seen that, according to Newton's second law of motion, the ball rises to a height of $v_0^2/2g$ and then falls (g is the acceleration due to Newtonian gravity). In Figure 10-14, we plot the height $h = v_0t - \frac{1}{2}gt^2$ attained by the ball as a function of the time t. The curve has the form of a parabola. If there were no gravity, the ball would have continued to move upward in the vertical direction with velocity v_0. The trajectory would then have followed the broken straight line in Figure 10-14. In the Newtonian framework, we argue that the dotted line is the case of no force, while the continuous line arises from the Earth's gravity, changing the state of motion according to the law that force equals mass times acceleration. We say that the trajectory is bent because of the force of gravity.

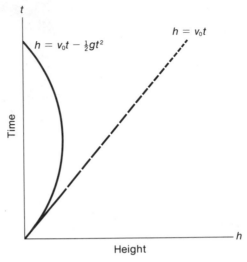

FIGURE 10-14
The continuous curve describes the world line of a particle thrown vertically with initial velocity v_0. In the absence of Earth's gravity, the particle would have moved along the broken, straight line (the first law of motion). According to the Newtonian system, the Earth's gravity supplies a force that bends the trajectory to the continuous form. According to Einstein, the geometry of spacetime is modified by Earth's gravity, so that the continuous line represents uniform motion in a straight line in non-Euclidean spacetime.

IN A PURELY GRAVITATIONAL SITUATION, PARTICLES FOLLOW THE STRAIGHT LINES OF A NON-EUCLIDEAN GEOMETRY

Einstein's point of view was quite different. He argued that it is meaningless to talk of the broken straight line of Figure 10-14 because a situation of no gravity cannot be attained in nature. The only real trajectory in nature is the continuous one of Figure 10-14. Thus, if the curved line is the only real one, why not regard it as describing uniform motion in a straight line?

This idea looks almost absurd at first sight, but let us examine it further. What do we mean by a straight line? The obvious definitions—"line of shortest distance" or "curve of unchanging direction"—depend on how we measure distances and directions. If we follow Euclid's rules, then the broken line is obviously straight and the continuous line is not. But what if we change the rules of geometry? In a non-Euclidean geometry, it might be possible to regard the continuous line of Figure 10-14 as straight and the broken line as curved, just the opposite from Euclidean geometry. This is the crux of Einstein's argument. The Earth's gravity makes the geometry in its vicinity non-Euclidean in just the way that the continuous trajectory of Figure 10-14 represents uniform motion in a straight line. Notice that we are now back to the first law of motion because gravity has been eliminated as a force. It has been completely reinterpreted as a property of the geometry of spacetime, which is non-Euclidean.

The same reinterpretation is to be given to all phenomena operating under the force of gravitation. Thus, the geometry of spacetime in the neighborhood of the Sun is to be non-Euclidean in just the way that the planetary orbits around the Sun can be looked upon as straight trajectories of uniform motion.

THE PRECESSION OF THE PERIHELION OF MERCURY VERIFIES THE GENERAL THEORY OF RELATIVITY

To find out how planets move around the Sun, Newton wrote certain *equations of motion* that gave the acceleration in terms of the inverse square law of force. By solving these equations, he determined the orbits of the planets. In Einstein's theory, the procedure is different. First, we write Einstein's mathematical equations, which describe how the non-Euclidean geometry is to be determined by the presence of matter—in this case, the Sun. This latter problem was first solved in an approximate way by Einstein in 1915, and then in a complete way by Karl Schwarzschild in 1916. The next step is to calculate *geodesics,* the technical name given to the straight lines of a non-Euclidean geometry. A selection from these geodesics is then used to describe the motions of planets in Einstein's general theory of relativity.

To all intents and purposes, the practical answers for the orbits of the planets came out to be the same as in Newton's theory. The very slight differences were significant only for the planet Mercury, and then only in one small respect. Figure 10-15 shows the orbit of Mercury schematically. The point P is the closest point on the orbit to the Sun S, and it is called the perihelion. In the Newtonian theory, if we ignore the gravitational effects of other planets, the planet Mercury should move in the same elliptical orbit over and over again. By contrast, the orbit in the Einstein theory rotates slowly. That is, after completing one orbit, the perihelion point shifts from P to P' as shown in Figure 10-15.

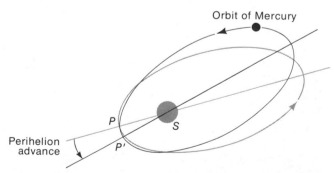

FIGURE 10-15

The orbit of Mercury, with the eccentricity much exaggerated. In the nineteenth century, the French astronomer U. J. J. Le Verrier found the orbit to be turning around by an amount that could not be wholly accounted for in terms of the gravitational forces exerted on Mercury by the other planets. (From J. C. Brandt and S. P. Maran, *New Horizons in Astronomy,* First Edition. W. H. Freeman and Company, Copyright © 1972.)

This shift is known as the precession of the perihelion of Mercury. It has been grossly exaggerated for illustrative purposes in Figure 10-15. The actual amount predicted by the general theory of relativity is an angular shift of the direction *SP* of 43ˢ (that is, about 1.2 per cent of a degree) per century.

It is worth remarking here that observations over many years had in fact shown the perihelion of Mercury to be precessing by about 575ˢ per century. Of this precession rate, all but 43ˢ per century could be accounted for by taking into account the effects of the other planets on Mercury. A net discrepancy of about 43ˢ per century therefore still remained to be accounted for. The fact that the general theory of relativity predicts the exact shift anomaly unexplained by Newtonian theory is taken as a great triumph for the seemingly strange ideas from which Einstein developed his theory.

LIGHT ALSO PROPAGATES ALONG THE GEODESICS OF A NON-EUCLIDEAN GEOMETRY

Another way of verifying Einstein's idea of a non-Euclidean geometry is the following. Suppose we draw a triangle around the Sun, with its three sides grazing the Sun's surface. As shown in Figure 10-16, the three sides of the triangle are now drawn in a non-Euclidean geometry. Will the three angles add up to 180°? Einstein's equations predict that the sum of the three angles should be slightly in excess of 180°.

In practice, it is not possible to perform an actual experiment for adding the angles of Figure 10-16, but a variant of such an experiment has been performed several times. How do we draw straight lines in space? We use an important result of Einstein's theory, namely, that light travels in a straight line. In Figure 10-16, the lines appear to be bent if we interpret them in the Newtonian way.

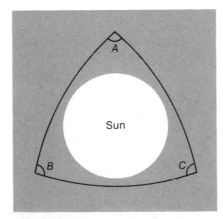

$$A + B + C > 180°$$

FIGURE 10-16
The triangle described by the tracks of light rays around a massive object like the Sun should have its three interior angles add up to more than 180°.

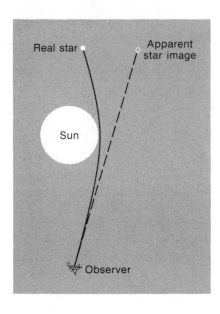

FIGURE 10-17
When a star is observed as it is close to occultation by the Sun during a solar eclipse,
its direction appears to change. This is the
effect of gravitation on the track of light.

For light following these lines, the apparent bending is due to the Sun's
gravitational force, which attracts photons as well as it attracts other particles. A
practical scenario is described in Figure 10-17, where light from a distant star
grazes the solar limb. As it is bent by the Sun, the line of sight from the star
appears to change its direction (shown by a broken line in Figure 10-17). The
predicted value of this angular bending according to the theory of relativity is
1.75ˢ (that is, about 5/10,000 part of a degree), and half this value according to
Newton's theory.*

Because of many experimental difficulties, measuring the deflection of Figure
10-17 for visible light has been inconclusive. Similar recent measurements using
radio-wave and microwave techniques have confirmed the Einstein prediction of
1.75ˢ, however, to within a small margin of error (currently less than 1 percent).
Here, then, is another triumph for general relativity and one for the new techniques of measurement.

§ 10-9. Gravitation in Relation to Astronomy

We have now seen how the two ways of describing gravitation work. Newton's
way is easier to understand and apply. It has, however, certain conceptual
difficulties that render it inconsistent with special relativity. General relativity is

---

*This statement refers to a *modified* Newtonian theory in which it is assumed that photons, the
carrier particles of light, are subjected to the Newtonian law of gravitation. The *original* Newtonian
theory did not make any such claim, and according to it there should be *no* bending of light.

free from these difficulties, but it deals with gravitation in a more subtle and indirect way. Which theory should astronomers use?

A rule of the thumb is the following. Suppose we are discussing the gravitational effects of a mass M at a distance r from it. Construct the dimensionless parameter

$$\alpha = \frac{2GM}{c^2 r},$$

where c is the speed of light and G the constant of gravitation. If α is very small ($< \sim 10^{-3}$), we can safely use Newtonian theory. If, on the other hand, α is comparable to 1, Einstein's theory must be used. In the intermediate range of α, a compromise—called the post-Newtonian approximation—is used.

The post-Newtonian approximation draws on the essential features of relativity but simplifies them to look like the Newtonian theory. The following table lists a few values of the parameter α for some astronomical systems.

TABLE 10-1
*Theories of gravity appropriate for describing
astronomical systems*

System	α	Type of theory used
Sun's gravity at the Earth	10^{-8}	Newton
On and inside a typical star	10^{-6}	Newton
On and inside white-dwarf stars	10^{-3}	post-Newtonian approximation
On and inside a neutron star	10^{-1}	post-Newtonian approximation
Black hole	1	Einstein
The universe	$10^{-2} - 1$	Einstein

In the next chapter we discuss black holes and white holes, an important field of study in which the general theory of relativity has recently made dramatic contributions that are on the speculative frontiers of physics.

General Problems and Questions

1. A train is moving at 80 mph. A car moving on a parallel expressway is moving at 55 mph. What is the speed of the train as seen by the driver of the car when (i) the train is moving in the same direction as the car and when (ii) the train is moving in the opposite direction relative to the car?
2. Design an experiment to measure the acceleration due to the Earth's gravity.

3. Read Chapter 9 of *Astronomy and Cosmology* by Fred Hoyle (W. H. Freeman and Company, 1975) for a review of the historical development of the ideas on gravitation.

4. If Newton's first law of motion is valid, why does a car running on a horizontal track eventually come to rest if no power is supplied?

5. If a cup is dropped from rest from a height of 10 ft and falls on a tiled floor, it breaks. Will the impact be lessened if the cup is thrown upward to start?

6. How will you move on a perfectly smooth, horizontal ice rink?

7. What are Kepler's laws? How will you deduce Newton's law of gravitation from them? (See *Astronomy and Cosmology* for a discussion).

8. How will you calculate the height of a synchronous satellite?

9. Use Newton's laws to calculate the period of a satellite moving above the Earth's surface at a height of 400 miles.

10. Write an essay on Newton's laws of motion.

11. In the albatross's trajectory described in Figure 10-12, imagine that it is carrying a rod that is initially at right angles to its direction of motion. It carries this rod throughout its travel without changing the rod's direction, although it itself turns. When it arrives back at N, by what angle has the rod turned?

12. The mass of a cosmic ray electron has increased one thousandfold over its rest mass. What is its speed?

13. Derive the formula (in the text) where a boat with intrinsic speed c moving perpendicular to a stream of speed v has an effective speed $\sqrt{c^2 - v^2}$.

14. What did the Michelson–Morley experiment try to measure? Why was its null result so disconcerting?

15. Calculate the escape velocity for the Earth and the Moon.

16. Show that, when the escape velocity is equal to the velocity of light, the parameter mentioned at the end of the chapter is unity.

17. In what way was Einstein's handling of the phenomenon of gravitation different from Newton's?

18. Verify, by constructing a saddle-shaped surface, that if we draw a triangle, its three angles add up to less than 180°. (To draw a triangle between any three points, stretch strings between them while constraining them to lie on the surface.)

19. By stretching strings on a globe between London and New York, convince yourself that the shortest distance between these cities is not along the line you would get by joining these two points by a straight line on a flat map. (This illustrates the difference between Euclidean and non-Euclidean geometries. The airlines follow the non-Euclidean route.)

20. Repeat Problem 19 for London and Los Angeles.

21. Discuss the phenomenon of the motion of planets within the frameworks of Newtonian gravitation and Einstein's general theory of relativity. When does the difference between the two become apparent?

22. Discuss the phenomenon of the bending of light. How do you reconcile the bending with the notion that light travels in a straight line?

23. If, by a thought experiment, we reduce the Earth's radius but keep its mass constant, how small must it become in order for the gravitational field on the surface of the Earth to be considered strong?

Chapter 11
Black Holes

§11-1. Introduction

As we saw in Chapter 10, terrestrial and solar system phenomena are adequately described from a practical point of view by Newtonian ideas of motion, taken together with the spacetime geometry of special relativity. Indeed, the reasons for abandoning Newtonian gravitation in the solar system are conceptual rather than practical. The theory of general relativity becomes substantially different from the Newtonian theory only when the gravitational distortion of spacetime is large. In this chapter, we consider two objects where this happens—black holes and white holes. Both types of objects are concepts within the brains of theoreticians at present. At the time of writing, there is no direct and conclusive evidence for their existence in the universe. These objects are the outcome of pushing the implications of Einstein's theory to its very limit, and as such they are speculative. They (especially the black holes) have fired the imagination of theoretical astronomers in recent years. Indeed, it is difficult to find areas in modern astronomy where a black hole has not been mentioned for some reason or other.

The considerable literature on black holes is of two types. One type represents the many advances made on the theoretical front and is highly technical. The other is written at a nontechnical level and leads to a superficial description that raises as many questions as it attempts to answer. This book is not written at the level of the first type, but we attempt to go beyond the level of the second type by introducing elementary mathematics. (The concepts developed in the last chapter will be useful.)

§*11-2. Escape Velocity*

BLACK HOLES ARE BLACK BECAUSE LIGHT CANNOT ESCAPE FROM THEM

The concept of a black hole dates back to 1799 when Laplace* (see Figure 11-1) proved the following theorem: *the attractive force of a heavenly body could be so large that light could not flow out of it.*

In Section 10-4, we discussed the concept of an escape velocity. In the case of the Earth, the minimum velocity required for a body projected from its surface to leave the confines of Earth's gravity is about 7 miles/second. In general, for a mass M and a radius R, the escape velocity is given by the formula

$$v = \sqrt{\frac{2GM}{R}}.$$

What happens if $v = c$? In such a case, even a particle projected at the speed of light c could not escape from the gravitational field of mass M. Since light (photons) as well as material particles experiences the effects of gravitation, the mass M would be invisible to a distant observer. Neither its own light (if it had any) nor light from another luminous source scattered or reflected by its surface could leave its neighborhood. *Such an object is called a black hole.* Being black does not imply the absence of light for an observer within the object. It does imply that the object cannot be seen by a distant observer.

From the preceding formula, which is essentially the same as that used by Laplace, the radius of a black hole of mass M cannot exceed

$$R_s = \frac{2GM}{c^2}$$

---

*In the year 1799, Laplace published a proof of the result that a sufficiently massive and condensed body would appear invisible. A translation of this proof, which appeared in a German astronomical journal, can be found in the appendix to *The Large Scale Structure of Space-Time* by S. W. Hawking and G. F. R. Ellis (Cambridge University Press, 1973). It appears, however, that even Laplace was anticipated in this concept of black holes by an English physicist, John Mitchell. On November 27, 1783, Mitchell read, before the Royal Society, a paper entitled, "On the Means of Discovering the Distance, Magnitude, etc. of the Fixed Stars, in Consequence of the Diminution of the Velocity of the Light, in Case of such a Diminution should be found to take place in any one of them, and such other Data should be procured from Observations, as would be further Necessary for that Purpose." In this paper (published in *Philosophical Transactions of the Royal Society,* 74, 35, 1784), Mitchell gives, among other stellar calculations, the basic principles of black holes within the Newtonian framework. (For information about Mitchell and his work, see a short article by G. Gibbons in *The New Scientist,* June 28, 1979, p. 1101.)

FIGURE 11-1
Pierre Simon Laplace. (Reprinted from *Celestial Mechanics*, by the
Marquis de Laplace, translated by Nathaniel Bowditch, New York,
Chelsea Publishing Company, 1966, with the permission of the
publishers.)

If M is put equal to the mass of the Earth, 5.977×10^{27} g, R_S is about 1 cm,
whereas if M is put equal to the mass of the Sun, R_S is about 3 km. The actual
radius of the Earth is $\sim 6,400$ km, and the radius of the Sun is about
700,000 km. These numbers give some idea of the tremendous shrinkage
involved before a familiar object can become a black hole. Can such shrinkage
take place? It is interesting to view this question in both Newton's framework
and Einstein's.

§11-3. *Gravitational Collapse in Newtonian Gravitation*

GRAVITATION BECOMES STRONGER IF A SYSTEM GIVES IN TO IT

The force of gravitation is peculiar in many ways, and we have already discussed
certain aspects of its peculiarity in Chapter 10. Here we describe and elaborate
on the phenomenon of gravitational collapse. To contrast gravitation with other
forces found in nature, consider the following example.

Imagine two lumps of matter connected by an elastic spring. In Figure 11-2, the two lumps are shown with the spring stretched beyond its natural length. The elastic force in the spring tends to contract it, with the result that the two lumps are attracted toward each other. If they are released, they will move toward each other in submission to the elastic force of the spring. The situation a little later is shown in Figure 11-3. The lumps are still moving toward each other, but the spring has become shorter and acquired its natural length. The tendency to contract is no longer there, and the attractive force has disappeared. The motion of the two lumps will continue, however, in the same way until the spring shortens and develops an elastic *repulsive* force. This repulsive force will eventually make the lumps come momentarily to rest, after which they will move apart again.

Figure 11-4 shows what happens when we have a gravitational force acting between two lumps. If their masses are M_1 and M_2 and their initial separation is R, the force of attraction is

$$\frac{GM_1M_2}{R^2}.$$

FIGURE 11-2
The two lumps of matter are attracted toward each other as the spring, stretched beyond its natural length, tends to contract in the direction of the arrows.

FIGURE 11-3
The spring has momentarily acquired its natural length, and there is now *no* force between the two lumps. However, they continue to move in the direction of the arrows because of the velocities they have acquired.

M_1 GM_1M_2/R^2 M_2

FIGURE 11-4
The inverse square force of gravity between M_1 and M_2 *grows* as they give into it and approach each other. For example, if the distance between M_1 and M_2 is halved, this force will be quadrupled.

As they give in to this force and move toward each other, the force does not decrease. If, at a later stage, their separation has decreased to r, the force has increased to

$$\frac{GM_1M_2}{r^2}.$$

If r is 10 times smaller, the force is 100 times larger!

Most forces in nature decrease when the system on which they act yields to them. This behavior is of the type shown by the elastic spring. Gravitation, on the other hand, increases in strength when the system yields to it, with the consequence that the system is obliged to yield yet more and more. This finally leads to the catastrophic situation known as *gravitational* collapse.

We now discuss the spherical mass shown in Figure 11-5. Imagine this mass to be a ball of dust with no pressure inside it. All particles of the dust ball will be attracted toward each other, with the result that the ball, as a whole, will begin to shrink. This shrinkage will bring the dust particles closer to each other and *increase* their mutual forces of attraction. The ball will shrink further and faster. How long will this continue? Until all the dust particles come so close that the ball has shrunk to a point! This process—a phenomenon known as *implosion* (the reverse of explosion)—will take place with increasing inward speed.

WITHOUT INTERNAL PRESSURE, THE SUN WOULD COLLAPSE INTO A BLACK HOLE IN 29 MINUTES

Imagine what would happen if the pressure inside the Sun were suddenly removed. The Sun would become like the dust ball and begin to shrink, at first slowly, and then with increasing rapidity. Finally, it would shrink to a point, with the entire process taking less than half an hour. In Figure 11-6, the Sun's collapse in these hypothetical circumstances is shown graphically. The radius of

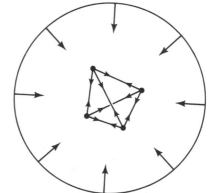

FIGURE 11-5
All constituents of the dust ball attract each other as shown for a few typical representatives. The outcome is an overall shrinkage of the dust ball.

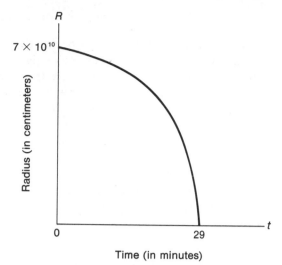

FIGURE 11-6
Under a pressure-free implosion induced by
gravity, the Sun would collapse from its
present state to a point in less than half an
hour.

the Sun is plotted against time, with $t = 0$ being the moment when contraction begins. Notice how rapidly R decreases to zero as the end is approached. This is the phenomenon of gravitational collapse.

In reality, the Sun is a remarkably stable object. Although it has a tremendous tendency to contract under its own gravitation, strong internal pressures prevent this contraction. An equilibrium exists between these internal pressure forces and the forces of gravitation. The pressures are maintained by the heat generated by thermonuclear reactions inside the Sun (see Chapter 8). As long as a star has nuclear fuel to burn, it can usually generate sufficiently strong pressures to withstand gravity. But what happens when the nuclear fuel is exhausted? We return to this critical question after discussing gravitational collapse as seen by Einstein's theory of gravitation.

§11-4. Gravitational Collapse in General Relativity

We saw in Chapter 10 that gravitational effects manifest themselves in general relativity through a form of non-Euclidean geometry determined by the presence of matter. (The examples discussed in Section 10-8 used non-Euclidean geometry.) We now consider how non-Euclidean geometry affects time-keeping devices, that is, clocks, and we do this first for the simple case of a noncollapsing body.

Imagine two observers, A and B. Observer B is located on the surface of a noncollapsing spherical body of mass M and radius R (see Figure 11-7), whereas observer A is located far away. We expect intuitively that the gravitational effects of the body will be strong near B and very weak near A, and we take the effects near A to be negligible. Suppose clocks of identical structure have been provided to A and B. Would the clocks run at the same rate in both locations?

To test this question, A and B make an arrangement. Observer A is to send signals of light to B at specific intervals, say every hour on the hour, and B is to send signals to A in exactly the same way. Will the signals reach their destinations at hourly intervals? The answer to this question is illustrated in Figure 11-8. A does not receive B's signals at hourly intervals. Instead, they come at slightly longer intervals, of an amount

$$\frac{1}{\sqrt{1 - (2GM/c^2R)}} \text{ hour.}$$

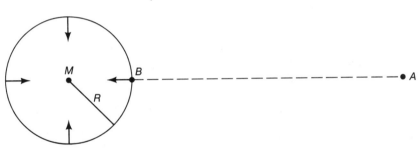

FIGURE 11-7
Observer B is located on the surface of an initially noncollapsing object, whereas observer A is located far away along the radial direction from the center of the object through B.

FIGURE 11-8
If observer B sends out a wave frequency ν_B, observer A receives it with frequency ν_A, and vice versa. In the initial stages, the frequencies are related by

$$\nu_A = \sqrt{1 - \frac{2GM}{c^2R}}\, \nu_B.$$

Similarly, B receives A's signals at shorter intervals of

$$\sqrt{1 - (2GM/c^2R)} \text{ hour.}$$

In practice, this prediction of Einstein's theory has been verified in the following way. Instead of terrestrial, man-made clocks, astronomers use the frequencies of spectral lines (see Chapter 3), the frequencies being determined by atomic and molecular processes. If we follow Einstein's ideas, such processes should run the same way on the surface of a massive star as they do on the Earth, the only difference being that the departure from Euclidean geometry is more marked in the case of the star than it is on the Earth. As we have already seen, this departure shows in the different rates of running of the clocks. In terms of the frequencies of spectral lines, the preceding result can be described in the following way.

Suppose B sends a signal in the form of a spectral line of frequency ν_B and wavelength λ_B, between which there is the usual relation

$$\nu_B \lambda_B = c.$$

Thinking of B's clock as determined by the oscillation of this spectral line, and taking into account that B's clock appears to run slow to A, we see that A has to wait for a longer period to receive one complete wave emitted by B, than B does to receive one from A. The result is that A's measured frequency ν_A is smaller by the factor already noted,

$$\nu_A = \nu_B \sqrt{1 - \frac{2GM}{c^2R}}.$$

Since the wavelength λ_A measured by A is related to ν_A by the usual formula,

$$\nu_A \lambda_A = c,$$

observer A will find a longer wavelength λ_A than that sent out by B, with λ_A given by

$$\lambda_A = \frac{\lambda_B}{\sqrt{1 - (2GM/c^2R)}}.$$

If A is looking at the entire visible frequency spectrum of a source of light at B, he will find all the wavelengths systematically increased by the same factor. The result is a shift in the spectrum toward the red end, that is, the long wavelength end. The factor $1/\sqrt{1 - (2GM/c^2R)}$, by which the wavelength appears to have increased, is usually written as $1 + z$, and z is said to be the *redshift* of the light from B.

In the same way, B will find the wavelength of any light received from A to be reduced. This effect is known as a *blueshift* since the spectrum in this case would be shifted toward the blue.

GRAVITATIONAL REDSHIFTS HAVE BEEN OBSERVATIONALLY VERIFIED

Clearly, for the redshift to be large, the parameter $\alpha = 2GM/c^2R$ must not be small. Yet, for the Sun, $2GM/c^2R$ is only about four parts in a million, which is difficult to measure. (The spectrum of the Sun is very complex, and it is awkward to disentangle the many solar spectrum lines to an accuracy of a few parts in a million.) For white-dwarf stars, however, the parameter α is on the order of 10^{-3} (see Table 10-1), so that the redshift effect is much larger for this special class of star than it is for the Sun. The first claimed astronomical observation of the gravitational redshift was for the white dwarf Sirius B, the binary companion to the bright star Sirius A.

A terrestrial verification of the gravitational redshift has been made. To understand the method used, we recall an important conclusion of quantum physics, that light is made up of photons with a relation between the energy E and frequency ν of a particular photon expressed by

$$E = h\nu,$$

where h is Planck's constant, which has a value depending on the units of time and energy (see Chapter 2). Suppose a photon is dropped from a height H. By the time it hits the ground, it has more energy than it started out with, the gain being an amount equal to its change of gravitational potential energy (Section 10-3). How do we calculate this quantity? We do so in an intuitive but correct way by combining ideas from Newton's theory of gravitation and Einstein's theory of relativity. Newton's theory of gravitation tells us that a mass m has a potential energy mgH that is released when the mass falls to the ground, whereas the theory of special relativity tells us that an energy E is equivalent to a mass:

$$m = \frac{E}{c^2}.$$

Setting $E = h\nu$, we see that the gravitational potential energy of a photon of frequency ν, released in falling a height H, is

$$\frac{h\nu}{c^2} \times gH.$$

When such a photon hits the ground, it has a new energy given by

$$E' = E + \frac{h\nu}{c^2}gH = h\nu\left(1 + \frac{gH}{c^2}\right).$$

Hence, its new frequency is given by

$$\nu' = \nu \left(1 + \frac{gH}{c^2} \right).$$

Thus, light in falling in a terrestrial environment through a height H is blue-shifted by the factor $(1 + gH/c^2)$. In the same way, light climbing upward against the Earth's gravity would be redshifted. The blueshift of freely falling protons was first measured by R. V. Pound and G. A. Rebka in 1960 by an experimental method illustrated in Figure 11-9. The result of the experiment was in agreement with the preceding formula.

Thus, we can argue in general that a gravitational redshift arises when light travels from a strong to a weak gravitational field, and a blueshift arises when light propagates in the opposite direction. The change in frequency arises because of the work done in climbing or descending a gravitational potential. In one case (redshift), the work is done against gravity, whereas in the other case (blueshift), work is done by gravity.

FOR A COLLAPSING OBJECT, AN ADDITIONAL REDSHIFT ARISES FROM THE DOPPLER SHIFT

Let us return to the case of the gravitational collapse of a ball of dust, viewed in the framework of general relativity. Einstein's equations are more complicated than Newton's, but surprisingly, in this particular case they admit a solution that

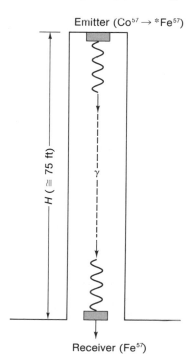

Emitter (Co^{57} → *Fe^{57})

$H (\cong 75 \text{ ft})$

γ

Receiver (Fe^{57})

FIGURE 11-9
In the experiment by R. V. Pound and G. A. Rebka, the emitter of photons was a nucleus of cobalt, Co^{57}. Co^{57} decays into an excited state, *Fe^{57}, of iron. The excited nucleus emits a γ-ray photon that travels downward. If the photon does not gain energy in free fall, it is absorbed by the receiver (Fe^{57}) at the bottom of the tower. However, the absorption depends sensitively on frequency of the photon. Since the frequency has in fact increased in free fall, absorption cannot take place unless the receiver is moved downward with appropriate velocity. A downward motion of the receiver generates a Doppler redshift that cancels the gravitational blueshift. By noting the requisite downward velocity of the receiver. Pound and Rebka were able to calculate the gravitational blueshift.

at first looks very similar to Newton's theory of gravitation. There are, however, significant differences that we now consider in a qualitative way.

We take observer B to be stationed on the surface of the collapsing dust ball and observer A to be far from the center of the ball and at rest relative to it. We also consider a third observer C who is momentarily coincident with B, but like A, C is at rest relative to the center of the ball. We consider light emitted by B to be passed first from B to C, and then from C to A. Since B and C are positionally coincident, light passes between them without any gravitational change. Nevertheless, observer C finds that a spectrum line of frequency ν_B emitted by B is changed to some frequency ν_C by the Doppler effect of the motion of B with respect to C—that is, by the collapse of the object. The corresponding wavelength λ_C can be written in terms of λ_B by introducing a Doppler factor,

$$\lambda_C = \lambda_B(1 + z_{\text{Dopp}}),$$

where z_{Dopp} depends on the speed of collapse of B with respect to C. The light now passes from C to A in exactly the same way as before, with the wavelength λ_A measured by observer A given in terms of λ_C by the same expression as before,

$$\lambda_A = \frac{\lambda_C}{\sqrt{1 - (2GM/c^2R)}},$$

in which R is the radius of the dust ball at the moment of emission of the spectrum line from B. Combining these two equations gives the following relation between λ_A and λ_B.

$$\lambda_A = \frac{1}{\sqrt{1 - (2GM/c^2R)}} \cdot (1 + z_{\text{Dopp}}) \cdot \lambda_B.$$

Since z_{Dopp} is a positive number depending on the speed of the collapse, the redshift effect is larger for the emission of light from the surface of the object than the purely gravitational factor depending on the parameter $2GM/c^2R$ that we studied before.

AN EVENT HORIZON DEVELOPS AT $2GM/c^2R = 1$ FOR GRAVITATIONAL COLLAPSE ACCORDING TO GENERAL RELATIVITY

Four stages of collapse of a spherical dust ball according to Einstein's theory of general relativity are shown in Figure 11-10. In stage I, the ball is distended with $\alpha = 2GM/c^2R$ small, when the main contribution to the redshift comes from the Doppler factor $1 + z_{\text{Dopp}}$. Even so, the redshift is small, and hourly

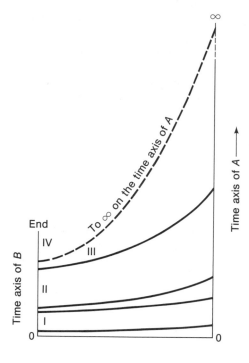

FIGURE 11-10

In stage I, the signals from observer B are redshifted by small amounts. In stage II, the redshift is noticeable. In stage III, the redshift rises rapidly so that there is a terminal signal (at the boundary of III and IV) that never reaches observer A even if he waits forever. Beyond this (into stage IV), no signals can come out of what has become a black hole of the dust ball. Stage IV ends in a catastrophic singularity for B. The lines connecting the two time axes are not light tracks; they simply establish the correspondence between emission and reception times on the clocks of B and A.

flashes of light from B would be received at intervals only slightly longer than an hour on A's clock. In stage II, the ball has contracted so much that the redshift factor,

$$f = \frac{1}{\sqrt{1 - (2GM/c^2R)}} \cdot (1 + z_{\text{Dopp}}),$$

is significantly greater than 1. If f were 24, A would receive B's hourly flashes a whole day apart. Indeed, as the ball shrinks, R decreases while M remains unchanged, and the ratio of corresponding time intervals at A and B steadily increases. This is shown in Figure 11-10, where we show a few typical transmission intervals on B's time axis against the growing intervals of reception on A's time axis. At the end of stage III, the radius has shrunk to the critical value,

$$R = R_s = \frac{2GM}{c^2},$$

and the ratio f is infinite. This means that a signal sent by B at the precise moment when $R = R_s$ would never reach A, even if A were immortal. Also, all light rays sent by B at this instant become infinitely redshifted, that is, the photons have zero frequency and zero energy. What this implies is that no information about B, at or beyond stage III, is ever accessible to A.

This information barrier at $R = R_S$ is called the *Schwarzschild barrier,* named after K. Schwarzschild, who first obtained it in 1916 as a consequence of the mathematical solution of Einstein's equations for the non-Euclidean geometry produced by a spherical mass M.

If the collapsing body were emitting light, it would appear to A to be getting fainter and fainter. As the outer surface of the body approached the Schwarzschild barrier, the faintness would increase remarkably rapidly. The characteristic time scale for this phenomenon is on the order of

$$\tau_S = \frac{R_S}{c}.$$

For a body of mass equal to that of the Sun, τ_S is on the order of a few microseconds. Thus, as far as the external appearance is concerned, the collapsing body will practically disappear from A's view as its external surface approaches R_S. For all practical purposes, it will have become a *black hole.*

In an exact mathematical sense, however, the situation is quite the contrary! In the mathematical language of general relativity, the collapsing object becomes a black hole when its external surface actually *reaches* the Schwarzschild barrier. As we just saw, even if A were to wait forever, he would not learn that this stage had been reached.

Popular literature abounds in statements like: "such and such star has become a black hole," or "when the supermassive object becomes a black hole. . . ," and so forth. These statements have to be treated with caution. We must appreciate the fact that in our role as outside observers like A, we can never claim that a collapsing object *has become* a black hole because information transmitted from the object at the stage of its crossing the Schwarzschild barrier can never reach us. Statements like these *imply* that these objects have become so faint that they are no longer detectable with our telescopes. In this practical sense, they are *black.*

The Schwarzschild barrier is also called an *event horizon.* Just as we cannot see events that occur beyond the horizon of the ocean, so we cannot see events that occur at or beyond the event horizon. In the collapse of a spherical body, the part of the collapse in the range $R < R_S$ is invisible to A. We can say that the view of the future of the collapsing body at or beyond $R = R_S$ is cut off from us by the event horizon.

§11-5. How Are Black Holes Formed?

SOME HIGHLY EVOLVED STARS ARE CANDIDATES FOR BLACK HOLES

We saw in Chapter 8 that stars evolve by burning a sequence of nuclear fuels: first, hydrogen to helium; then, helium to carbon, oxygen, and neon; then, neon to magnesium, silicon, and sulfur; and finally, magnesium, silicon, and sulfur into the iron-group elements. Massive stars may lose excess material, either smoothly and continuously, to leave a white-dwarf residue, or violently and explosively, to leave in some cases a neutron-star residue. In other cases, however, the residue may exceed the largest mass that a neutron star can have. In the latter case, the residue then collapses toward a black hole. The unseen companion of the binary Cygnus X-1, discussed in Chapter 7, is believed to be an example of evolution to a black hole. Many other similar cases probably exist.

THE PROGNOSTICATION FOR THE FUTURE OF AN OBSERVER SITTING ON A COLLAPSING OBJECT IS POOR

What is the situation for observer B of the preceding discussion? As far as B is concerned, there is a future beyond stage III of Figure 11-10. Observer B will continue to fall inward with the object until the object becomes a point mass at stage IV. In Newtonian physics, observer B would simply become infinitely compressed at this stage, but general relativity implies something much more sinister! In general relativity, the observer reaches a *singularity,* a singularity being a general breakdown in Einstein's mathematical rules governing the geometry of spacetime. In Einstein's theory, the future of B beyond stage IV does not exist. The life of B is finished, not just for a practical reason like becoming too compressed, but finished in principle. Even the matter of which B has hitherto been composed ceases to exist at the singularity.

Is this a defect of the theory? We could argue that, as outside observers, we ourselves never see this singularity and that the fate of B is not our concern. This point of view is not satisfactory, however, because one of the claims of physics is to be able to discuss *all* observers.

A second reaction to the situation might be to argue that, since we ignored pressures throughout the preceding discussion, perhaps we can prevent the singular future in store for B by applying suitable pressures determined in a normal physical way. Important results based on Einstein's theory have been derived by R. Penrose, S. W. Hawking, and R. P. Geroch, and these results show that no amount of pressure of the types known to physicists will prevent the incidence of stage IV. Thus, a spacetime singularity seems inevitable unless some very weird and unknown physics operates after stage III.

SOME BLACK HOLES MAY BE PRIMORDIAL

A possibility to be considered is that objects surrounded by event horizons existed when the universe was formed (if the universe had an origin—see Chapter 13). Such objects may have any mass, unlike the black holes formed by stellar evolution. (Stellar black holes must have masses above the limit for neutron stars—see Chapter 8.)

Although black holes cannot be detected through the emission of radiation from within the event horizon, black holes retain a gravitational influence on their surroundings from which their presence can be inferred. Figure 11-11(a)

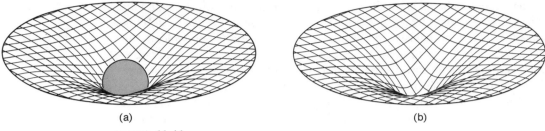

FIGURE 11-11

(a) The warping of spacetime produced by gravitating matter shown schematically. (b) The warping produced by a black hole. Even if the black hole is invisible, the warping of spacetime produced by it is measurable in principle. Through such measurements, black holes could be detected.

shows how spacetime geometry becomes warped during stage II of the gravitational collapse of an object, whereas Figure 11-11(b) shows stage III when spacetime becomes so warped that the object itself becomes invisible to an external observer. This warping is in principle detectable, for example, through the passage of external light through the distorted region.

SUPERMASSIVE BLACK HOLES MAY BE PRESENT IN MANY ASTRONOMICAL OBJECTS, INCLUDING QUASARS AND THE CENTERS OF GALAXIES

Many astronomers believe that very massive black holes exist in quasars and in the centers of many galaxies. Whereas evolutionary astrophysical processes are often considered to have led to the formation of these black holes, we ourselves prefer the concept that they are residues from a primordial state of the whole universe, a problem that we considered in detail about 15 years ago.

Even though the black holes in quasars and in the centers of galaxies would have masses upward of 1 million times the Sun, gravitational effects along the lines of Figure 11-11 do not provide the best evidence for their existence. The best evidence has already been given in earlier chapters, evidence based on the explosive emission of the fast-moving particles that give rise to the outbursts of radio sources (Chapter 4) and to intense x-ray emission (Chapter 7). Evidence from x-ray emission has also suggested that there may be black holes at the centers of the dense star clusters known as globular clusters. Halos of some hundreds of globular clusters surround many of the galaxies, including our own.

The reader may wonder how it happens that fast-moving particles and x rays can provide evidence for the presence of black holes if visible light cannot emerge from the interior of the event horizon of a black hole. The answer is that the fast-moving particles and x rays come from the environs of black holes, *outside their event horizons*. They come by processes associated with the presence of black holes, processes that we study in the remainder of this chapter. Although the basic concept of a black hole was known to physicists and astronomers as long ago as 1916, the concepts relating to the environs of a black hole are post-1963. In 1963, R. Kerr discovered the solution of Einstein's equation for a *spinning* black hole, and it was this discovery that led to modern developments of the subject.

§11-6. A Black Hole Has No Hair

"A black hole has no hair!" This celebrated statement by J. A. Wheeler implies that when a body undergoes gravitational collapse to form a black hole, very few items of information survive to tell outside observers what the physical characteristics of the black hole are. The basis for this remark is a theorem by R. H. Price, the gist of which is explained next.

A BLACK HOLE CAN BE DETECTED IN THE EXTERNAL WORLD BY ITS MASS, ITS SPIN, AND ITS ELECTRIC CHARGE

In Section 11-4, we looked in a qualitative way at the gravitational contraction of a spherical ball of dust. Our discussion was based, however, on an exact solution (in the general theory of relativity) of the problem of what happens to a homogeneous spherical ball of dust without internal pressures. What would happen if the body were neither spherical nor homogeneous? What would happen if the body were spinning as it contracted? What would happen if it had magnetic fields and electric currents circulating within it? Figures 11-12, 11-13,

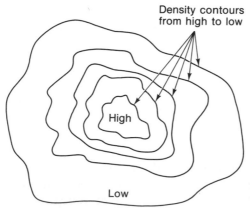

FIGURE 11-12
A nonhomogeneous, nonspherical body.

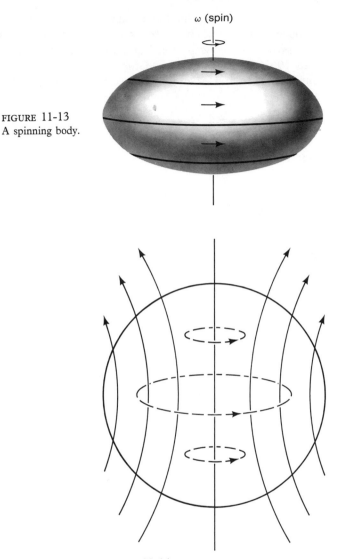

FIGURE 11-13
A spinning body.

FIGURE 11-14
A body containing electric currents (dashed lines)
and magnetic fields (continuous lines).

and 11-14 show such objects. We mentioned black holes being formed from stars. Collapsing stars may have all these various characteristics to begin with. How many of them do they retain during collapse? Unfortunately, general relativity does not (or has not so far been able to) answer these questions. Price's theorem, however, looks at a somewhat simpler state of affairs.

Imagine a body that has all the characteristics of Figures 11-12 to 11-14, but in small amounts. If it is flattened at the poles because of spin, the flattening is small. The body may have an electromagnetic field emerging from it because of its internal electric charges and currents, but these effects are too small to produce any appreciable effect on the non-Euclidean nature of the geometry *outside* the object. Price's theorem describes what happens to such a slightly but generally disturbed body as it collapses. First, its various built-in irregularities are noticeable to the outside observer if the outside observer performs suitable experiments. However, when the stage comes for a black hole to be formed, that is, when an event horizon is about to form, most of the information concerning these irregularities is lost for the external observer. The initially observable irregularities are radiated away. The only information that survives is the mass of the body, its electric charge, and its spin (the so-called angular momentum of the body).

In technical terms, Price's theorem states that if a physical property of the slightly disturbed body is describable by a field of spin s, then only moments up to one order lower, that is, up to order $s - 1$, survive.

For example, electromagnetic information is carried by a photon which has spin 1. Price's theorem therefore allows only electrical information of zero spin to survive. This information is represented by the electric charge of the body. The gravitational information is carried by a so-called graviton that has spin 2 (see Appendix B). Hence, the only surviving gravitational information is of order zero and one. These bits of information are conveyed by the mass ($s = 0$) and by the angular momentum ($s = 1$), respectively.

Because classical physics depends only on gravity and electromagnetism, we can argue that, so far as the known interactions of classical physics are concerned, the most general black hole is characterized by its mass, charge, and angular momentum. This is a conjecture prompted by Price's theorem. So far, a general proof that any body, however large its initial irregularities, will reach this state has not been given.

§11-7. The Kerr–Newman Black Hole

THE STRUCTURE OF A BLACK HOLE IS MODIFIED BY SPIN AND BY ELECTRIC CHARGE

The simple black hole formed by the collapse of a perfectly spherical body is completely characterized by its mass M. As we found in Section 11-4, it has a horizon at a radius

$$R_S = \frac{2GM}{c^2}.$$

The only information contained in this Schwarzschild black hole for the outside observer is M. We also saw in the previous section that Price's theorem suggests the most general black hole would be characterized by just three quantities: mass M, charge Q, and angular momentum H. An exact mathematical description of this more general black hole is available in general relativity, and it is known as the Kerr-Newman black hole. Although the details of collapse to this final state are not known, the Kerr-Newman solution has played an important role in our understanding of the way black holes behave. We now describe briefly some properties of this more general black hole.

First, it is a combination of two earlier solutions. In 1916, H. Reissner and G. Nordström independently obtained a solution of Einstein's newly formulated mathematical equations of general relativity that described the gravitational field of a static electric charge. The *Reissner-Nordström black hole* therefore has $M \neq 0$, $Q \neq 0$, and $H = 0$. As was already mentioned, in 1963 R. P. Kerr obtained a solution for the spinning, uncharged black hole: $M \neq 0$, $Q = 0$, and $H \neq 0$. The Kerr-Newman black hole combines this Kerr solution with the Reissner-Nordström solution; it has M, Q, and H all nonzero.

The horizon of the Kerr-Newman black hole occurs at a radial coordinate given by

$$R_+ = \frac{GM}{c^2} + \frac{1}{c^2} \sqrt{G^2M^2 - GQ^2 - h^2},$$

where $h = cH/M$. Notice that, for R_+ to be a real number, the quantity under the square root must be positive; that is,

$$G^2M^2 - GQ^2 - h^2 \geqslant 0.$$

If this quantity *is* positive, there appears from the mathematics to be another horizon at

$$R_- = \frac{GM}{c^2} - \frac{1}{c^2} \sqrt{G^2M^2 - GQ^2 - h^2}.$$

However, since $R_- < R_+$, the external observer is concerned only with R_+. In the particular case when the quantity under the square root is zero, we have $R_+ = R_-$.

There is *no* horizon if the quantity under the square root is negative. Unlike the situation already discussed for the Schwarzschild black hole, there is in this case the intriguing possibility of the external observer being in a position to witness the final state of the gravitational collapse—the singularity. The singularity is then called *naked*. Do black holes of this type exist? Or is there a *cosmic censorship* that permits only those black holes to exist that have horizons concealing the singular fate of the collapsing body from the external observer? This is as yet an open question.

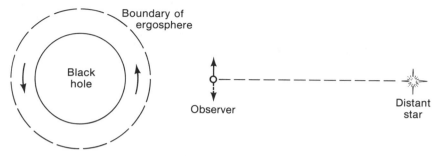

FIGURE 11-15

A rotating black hole viewed along the axis of rotation. An external observer at a typical latitude will feel as if he is being carried away (continuous arrow) along with the rotating black hole. As long as he is outside the broken circle, he can apply external forces to maintain himself at rest relative to the distant star background (against which the black hole's rotation is measured). The broken circle is the latitude section of the boundary of the black hole's *ergosphere*. Once the observer is on or inside the ergosphere, he cannot prevent himself from being dragged along by the black hole. The continuous inner circle represents the event horizon, $R = R_+$.

AN EXTERNAL BODY CAN GAIN ENERGY FROM A ROTATING BLACK HOLE

Is there any way in which an external observer can feel the presence of a rotating black hole? Suppose the observer comes closer and closer to the black hole (see Figure 11-15) while keeping an eye on the distant stars in the universe. The distant stars provide a background against which the rotation of the black hole can, in principle, be measured. Will the observer be able to arrange it so that the distant stars do not appear to rotate? As the black hole is approached, the observer will find an increasing tendency to get carried away in the same sense as that in which the black hole is rotating. To keep stationary, he will need to apply a force against this tendency, a force that increases as the black hole is approached. A stage known as the *static limit* will come when he will be swept away by the black hole no matter how hard he tries to counteract this rotational sweeping force. When this happens, he has entered a zone called the *ergosphere*.

A section showing how the extent of the ergosphere changes with a change in the latitude is given in Figure 11-16. The poles are (as usual) along the axis of rotation of the black hole. Here, at the poles, the ergosphere touches the horizon. Then, as we move toward the equator, the ergosphere extends beyond the horizon. At a latitude l, the radial coordinate for the ergosphere is given by

$$R_l = \frac{GM}{c^2} + \frac{1}{c^2} \sqrt{G^2M^2 - GQ^2 - h^2 \sin^2 l}.$$

In Figure 11-15, we already saw a section taken at a latitude $l < 90°$. The boundary of such a section of the ergosphere is circular and concentric with the boundary of the horizon.

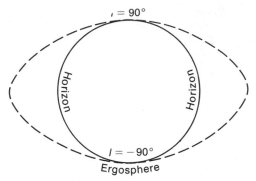

FIGURE 11-16
A meridional section of the rotating black hole. The poles
($l = \pm 90°$) are on the axis of rotation. The event horizon is
the continuous circle, whereas the ergosphere has the outer
boundary shown by the broken curve.

Why the name *ergosphere?* The nomenclature depends on the possibility that
a black hole may afford energy extraction in this region (the Greek word *ergo*
means "work"). R. Penrose suggested that a projectile fired from outside into the
ergosphere begins to rotate with the black hole so that the projectile acquires
more rotational energy than it possessed originally. The projectile can now
break up into two pieces. Of these, one piece may fall into the black hole
singularity, whereas the other may come out of the ergosphere. The piece
coming out may then have *more* energy than the energy of the original projectile.
This is illustrated in Figure 11-17.

FIGURE 11-17
In the Penrose process, a particle falling into the
ergosphere is dragged along by the black hole's rotation. At point A, it breaks into two components,
one of which falls into the black hole. The other
escapes with greater energy.

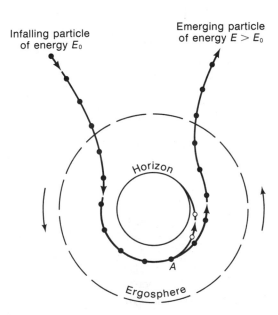

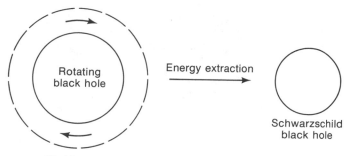

FIGURE 11-18
Scenario describing the conversion of a rotating black hole into a
nonrotating one after maximum available energy has been extracted.

In the Penrose mechanism, the black hole contributes part of its energy of
rotation to the projectile so that the black hole itself is slowed down, a process
that can continue until the black hole has given away all its energy of rotation.
The ergosphere then no longer exists. Its outer boundary is now coincident with
the horizon. Starting from a Kerr black hole, we arrive back again at the
Schwarzschild black hole. When this state is reached, no more energy can be
extracted from the black hole. The Schwarzschild black hole represents the final,
irreducible state in which external processes can only *increase* the energy of the
black hole instead of decreasing it (see Figure 11-18).

This example is a particular case of a general set of rules that govern the
behavior of black holes. These rules have emerged after many theoretical
investigations into general relativity, and they are known as the laws of black-
hole physics.

§11-8. The Laws of Black-Hole Physics

The example of energy extraction just described suggests a strong analogy with
thermodynamics, a science developed in the nineteenth century, notably by
Rudolf Clausius, Lord Kelvin, and Ludwig Boltzmann. The *first law of thermodynamics* recognizes that heat is a form of energy and simply restates the law
of conservation of energy that we encountered in Section 10-3.

A heat engine is a physical system that goes through a number of cycles, the
state of the system at the end of each cycle being the same as it was at the
beginning of the cycle. During some parts of the cycle, heat is absorbed from its
surroundings by the engine, and during other parts, heat is given up to the
surroundings. Writing Q_1 for the total heat absorbed and Q_2 for the total heat
given up in each cycle, the difference $Q_1 - Q_2$ is the mechanical work done per
cycle by the engine. The name "engine" implies that Q_1 is larger than Q_2 so that
the mechanical work derived from the engine is a positive quantity. A physical

system for which Q_2 is larger than Q_1 is known as a refrigerator, and to operate a refrigerator, mechanical work must be done on the system. (The work is usually done by an electric motor.) These are examples of the first law of thermodynamics.

Suppose a heat engine is designed to absorb a certain fixed quantity of heat Q_1 in each cycle. The smaller Q_2 can be made, the greater the efficiency of the engine. But how small can Q_2 be? In the eighteenth century, engineers had already discovered that the naive idea of trying to make Q_2 zero could by no means be achieved. Indeed, in the eighteenth century, Q_2 was not much smaller than Q_1, so the mechanical work derived from early engines had $Q_1 - Q_2$ considerably smaller than Q_1. A limitation is imposed on Q_2 by the *second law of thermodynamics*. It can be simply stated for an engine that absorbs Q_1 at a constant temperature T_1 and gives up Q_2 at a constant lower temperature T_2. For such an engine, the smallest possible value of Q_2 is $T_2/T_1 \times Q_1$, which implies that the largest fraction of the heat absorbed at T_1 that can go into mechanical work is

$$1 - \frac{T_2}{T_1}.$$

The trouble with early engines was that T_2 was not much smaller than T_1. The aim of the modern engine designer is to make T_2 considerably smaller than T_1. This aim is more readily achieved for a gasoline or diesel engine than it is for a steam engine.

The result for the smallest value of Q_2 was derived in the nineteenth century by introducing a concept known as *entropy*. The entropy of a system is a measure of its disorderliness. In a state of low entropy, the constituents of the system (the constituents being atoms, molecules, or larger subunits) are well ordered, whereas in a state of high entropy, the constituents are considerably disordered. When a coffee cup falls to the ground and breaks, the cup starts with a state of low entropy and ends with a state of much higher entropy. The *second law of thermodynamics* states: *in any physical process, the entropy of all participating systems taken together can never decrease.*

BLACK-HOLE PHYSICS HAS ANALOGIES IN THERMODYNAMICS

The *first law of black-hole physics* states: *energy and momentum are conserved in every physical process.* This first law is hardly surprising since black holes are creatures of the general theory of relativity, and this theory rigidly adheres to the laws of conservation of energy and momentum. What is surprising and interesting is that black holes appear to behave analogously to the second law of thermodynamics.

Before turning to the second law of black-hole physics, let us go back to the special case of the Schwarzschild black hole. We have already noted that the Schwarzschild black hole absorbs ambient matter and radiation without giving out any matter or radiation. Thus, it will always increase its energy and hence its mass M. The event horizon acts as a one-way membrane, and this suggests at first that M for a black hole must always increase. Is this analogous to the increase of entropy in the second law of thermodynamics?

A more careful analysis shows that the answer to this question is no. Indeed, we have already seen the example of the Penrose process that *extracts* energy from a rotating black hole and hence decreases M. Thus, the mass (or energy) of a black hole does not necessarily have the property of being able to increase.

Nevertheless, there is another physical parameter associated with a black hole that does have a nondecreasing property. It is the area of the black hole—more precisely, the surface area of the event horizon. For a Schwarzschild black hole, this is simply

$$A = 4\pi R_s{}^2 = \frac{16\pi G^2 M^2}{c^4}.$$

(Thus, both A and M increase for a Schwarzschild black hole.) For the Kerr-Newman black hole, the area is given by

$$A = 4\pi\left(R_+{}^2 + \frac{h^2}{c^4}\right) = 4\pi\left[\left(\frac{GM}{c^2} + \frac{1}{c^2}\sqrt{G^2M^2 - GQ^2 - h^2}\right)^2 + \frac{h^2}{c^4}\right].$$

Note that this more involved expression for area is an indication of the fact that the non-Euclidean geometry of the Kerr-Newman black hole is much more complex than that of the Schwarzschild black hole. If we now consider what happens to the area of the rotating black hole in the Penrose process, we will find that although M decreases in the process, the area never decreases. Qualitatively, this may be seen in the following way. In the Penrose process, the incoming particle acquires energy by taking part of the angular momentum of the black hole. Thus, for the black hole, both M and h decrease. Quantitative calculations show that, whereas the decrease of M leads to a *decrease* of area, this decrease is more than compensated for by the *increase* of area caused by the decrease of h. At best, if the process is performed with perfect efficiency, the area of the black hole stays constant while it gradually sheds its angular momentum. As shown in Figure 11-19, this process must end when all angular momentum has been lost. Thereafter, any external interaction with the nonrotating black hole (Schwarzschild black hole) leads to an increase of area. We have assumed throughout this particular example that $Q = 0$. If $Q \neq 0$, we can extract further energy by extracting electric charge until finally $Q = 0$.

This example brings out the central role played by area in determining the behavior of black holes. In 1971, Stephen Hawking gave a formal reasoning for the nondecreasing nature of the area of a stationary black hole. His result is expressed as *the second law of black-hole physics: in all physical processes involving black holes, the total surface area of the participating black holes can never decrease.*

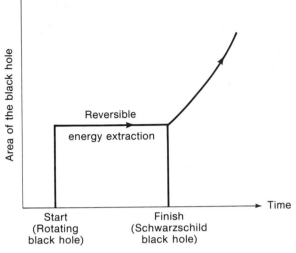

FIGURE 11-19
If we work with perfect efficiency, we can extract energy from a rotation black hole by keeping the area constant at all times. The process ends inevitably at the Schwarzschild black hole. Thereafter, any interaction with the black hole will only increase its area.

THE SURFACE GRAVITY OF A BLACK HOLE IS ANALOGOUS TO TEMPERATURE

With the analogy between the surface area of a black hole and entropy established, the next question depends on another thermodynamic quantity, the temperature. In thermodynamic equilibrium, we can ascribe a certain temperature to the system. Can we ascribe an analogous parameter to the black hole in a stationary state, that is, in a state not changing with time?

The answer is yes; the requisite quantity is the surface gravity κ. In a crude analogy with the acceleration due to gravity g on the surface of the Earth, we can look upon κ as measuring the strength of the black hole's attraction. *The zeroth law of black-hole physics* states: κ *is a constant over the event horizon of a stationary, axially symmetric black hole.* Thus, stationaryness and axial symmetry play a role analogous to equilibrium in the thermodynamic definition of temperature.

Finally, in analogy to the third law of thermodynamics, which states that the absolute zero temperature is unattainable by any finite series of operations, we have *the third law of black-hole physics,* which states: *by no finite series of operations can we make the surface gravity κ of a black hole zero.* For example, for a Schwarzschild black hole of mass M, the surface gravity is given by $\kappa = c^4/4GM$, which would require M to tend to infinity for κ to tend to zero. The mass cannot be made infinite in a finite series of physical operations.

If we push the analogy with thermodynamics further, we encounter a curious situation. When a body with a given temperature is placed in an environment of lower temperature, the body radiates energy in the form of heat. If we place a black hole with a given surface gravity in a vacuum, should it also radiate? We have just seen that the black hole does not allow any matter to escape outward from its horizon. How, therefore, can a black hole radiate energy? The very notion of a radiating black hole appears paradoxical. Yet, in 1974, Stephen Hawking suggested a novel way of resolving this apparent paradox. He argued that, whereas a black hole cannot radiate in classical physics, it can in quantum physics. Many examples exist in classical physics where a particle cannot surmount a potential barrier, whereas in quantum mechanics the barrier can occasionally be surmounted (see Figure 11-20). Hawking's idea, which has been quantitatively studied by him and others, is qualitatively described next.

In quantum theory, a vacuum is not just a region devoid of anything; it is full of what are called *virtual pairs* of particles that are spontaneously created and destroyed. A pair consists of a particle together with its antiparticle, and pairs in the vacuum are said to be *virtual* because they do not last long or produce any directly observable result (see Figure 11-21). They do play an important role, however, as unobserved intermediaries in many quantum phenomena.

Suppose we now have a black hole in a vacuum. It may capture both members of a pair of virtual particles, or it may capture only one of them (see Figure 11-22). In the former case, the pair is gobbled up; in the latter case, however, when one particle is eaten up by the black hole, its mate is let loose. The impression created outside the black hole would be that the black hole had

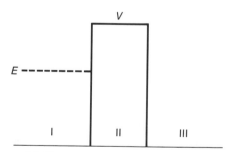

FIGURE 11-20
Classically, a particle of total energy E moving in region I will not be able to surmount the high potential barrier V of region II. The particle will be reflected at the boundary of I and II. Quantum mechanically, however, there is a finite probability of the particle penetrating into region II and coming out into region III. Many experiments with atoms have conclusively established this quantum mechanical prediction.

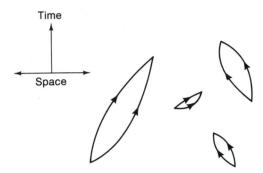

FIGURE 11-21
The creation and annihilation of virtual pairs
of particles and antiparticles in a vacuum.

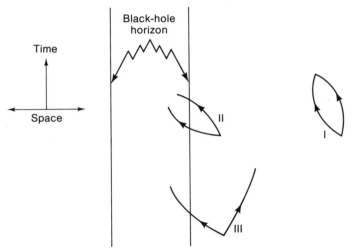

FIGURE 11-22
Three types of effects a black hole can have on pairs created in vacuum. In
I, there is no noticeable effect; the pair is simply annihilated. In II, both
members of the pair are swallowed by the black hole. In III, one member is
eaten up and the other let loose. This loose member appears to be emitted
by the black hole.

somehow created the emerging particle. Whenever the particle pair is spontaneously created in a vacuum, one of these two possibilities might occur, and the
probabilities of their occurrence can be computed. Such a calculation leads to
the result that a black hole of surface gravity κ radiates like a *black body* (see
Chapter 3) with temperature

$$T = \frac{h\kappa}{4\pi^2 ck},$$

where k is the Boltzmann constant and h is Planck's constant.

For a Schwarzschild black hole of mass M, the temperature comes out to be

$$T = \frac{hc^3}{16\pi^2 GkM} \cong 6 \times 10^{-8} \frac{M_\odot}{M} \, \text{K}.$$

This is very small for astrophysical black holes that have, say, $M > 4M_\odot$. But the temperature can be high if the black hole was created at the time the universe was created, if it is a primordial black hole of small mass. At high temperature, the black hole may radiate and lose its mass and hence radiate faster and faster until it is evaporated away entirely! The time scale of evaporation turns out to be about

$$10^{76} \left(\frac{M}{M_\odot} \right)^3 \text{ seconds}$$

for a starting black hole mass M. Thus, for the lifetime of a black hole to be greater than the age $\sim 10^{17}$ seconds of the universe (see Chapter 12), the black hole mass must be greater than $\sim 10^{15}$ g. Primordial black holes of mass less than $\sim 10^{15}$ g will not have survived to the present epoch. It has been suggested that recently observed bursts of γ rays may be primordial black holes in their final stages of evaporation.

The Hawking process prompted several interesting investigations. Many of the questions it raised are conceptually difficult to understand and tackle, and some are still highly speculative.

§11-9. The Detection of Black Holes

With so many advances on the theoretical front in black-hole physics, the question naturally arises: has a black hole been seen? Let us reconsider this question in more detail. If by seeing we mean the classical process of detection of an object by using light (or electromagnetic radiation in general), then the answer must necessarily be no. For as we have already pointed out, according to classical physics, a black hole cannot radiate or scatter radiation. If, on the other hand, we take recourse in the Hawking process involving quantum physics, then a black hole radiating significant energy must be much *smaller* in mass than the Sun. However, in those cases in which we have the strongest reasons for thinking that black holes might exist, as in stellar evolution, the mass must be *greater* than the mass of the Sun. How then should more massive black holes be detected?

One answer is by gravitation. Imagine what would happen if the Sun were to become a black hole. (This is just a thought experiment! Current ideas do not indicate that the Sun is massive enough to become a black hole.) The Earth would still continue to go around the solar black hole in its usual elliptical orbit.

People on Earth would be able to deduce, from Kepler's and Newton's laws, that an attracting body was present at one of the foci of the orbit. By the same token, we could look for black holes through their gravitational influence on their neighbors.

X RAYS AND HIGH-SPEED PARTICLES CAN BE EMITTED FROM THE ERGOSPHERES OF BLACK HOLES

Because planets are not self-luminous, the type of Earth-Sun arrangement just described is not of much practical use for the detection of black holes, but interesting developments can take place in binary star systems, such as the one shown in Figure 11-23. If in such a system one member is a black hole, and if it is situated close enough to its companion to exert appreciable tidal forces on it, such forces can pull matter from the surface of the companion toward the black hole, forming a thin disc as in Figure 11-24. We can suppose that new matter is continually being pulled in, while previously accreted matter circulates in the disc for a while and then eventually falls into the black hole. Such *accretion discs* can form whenever one member of a close binary system is a compact object like a white dwarf, a neutron star, or a black hole. Due to friction within the material of the disc, the circulating particles heat up and begin to radiate. The result is mostly radiation in the x-ray region, together with the emission of high-speed particles. The process is very similar to the one we

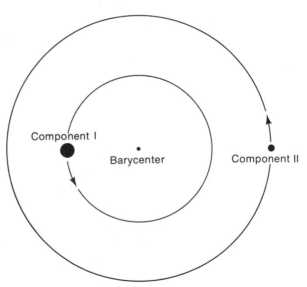

FIGURE 11-23
In a binary system two stars pursue orbits about a common barycenter.

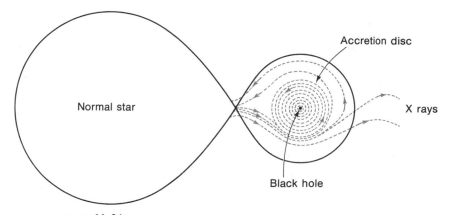

FIGURE 11-24
The double star configuration, with one member a black hole, can give rise to x rays from the accretion disc formed around the black hole by infalling matter from the companion star.

studied at the end of Chapter 7, and in the case of a spinning black hole, a gain of energy in the ergosphere can be particularly effective in producing an intense outburst.

OUTBURSTS FROM THE ERGOSPHERES OF SPINNING BLACK HOLES OF VERY LARGE MASS CAN EXPLAIN RADIO SOURCES AND QUASARS

In Chapter 10, we saw that the energy associated with a mass M is given by Einstein's famous relation, $E = Mc^2$. A spinning black hole can store rotational energy up to an amount of this general order, Mc^2, where M is the mass of the black hole. When M is on the order of 1 million times the mass of the Sun, the energy thus available in the ergosphere of the black hole is enormous, $\sim 10^{60}$ ergs. This is adequate to explain the energy emission from radio galaxies and quasars. No other process that has yet been studied in mathematical detail rivals a spinning black hole in this respect. No other process appears capable of meeting the energy requirements of radio galaxies and quasars. *It is for this reason that these objects are believed to contain black holes of very large mass.*

§*11-10. White Holes*

In Section 11-4, we found that the general theory of relativity implies an end to the life of observer B, who is falling freely with a collapsing object. The end comes when observer B hits a spacetime singularity. General relativity is, however, a *time symmetric* theory. This means that if the theory predicts a chain of events happening in succession, it also predicts another chain of events happening in succession. This second chain of events is formed from the first chain in *reverse chronological order*. To put it another way, if we take a movie of the first happening and run the film backwards, what we see in reverse is also a possible happening according to the theory of relativity. Hence, in the case of a collapse to singularity, we can envisage a new possibility by time reversal—that of emergence of the object from a singularity (see Figure 11-25). This phenomenon is known as a *white hole*.

It has been suggested that when observer B arrives at a singularity at the end of the collapse, he then emerges, along with the object, into another universe as a white hole. Thus, the *implosion* in a black hole is followed by an *explosion* of a white hole into another universe connected to the first one.

It must be emphasized here that the general theory of relativity does not provide any basis for such a linkage. According to this theory, the two pictures are separate; a white hole and a black hole could exist in isolation. In our example, the future of observer B ends at the singularity. Similarly, in a white hole, the life of B begins in a singularity. In a later chapter, we show how a natural connection between black holes and white holes can be established in a theoretical framework other than general relativity.

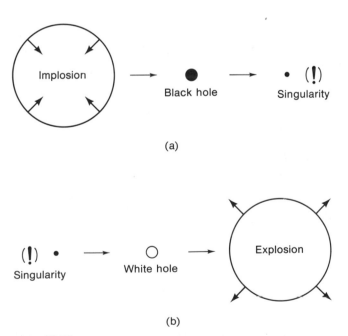

FIGURE 11-25
According to general relativity, (b) is the time-reversed version of (a), and vice versa.

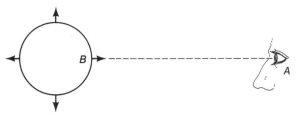

FIGURE 11-26
To remote observer A, the white-hole surface, on the near side, is approaching. Hence, there is a Doppler blueshift. In the early stages, the blueshift is much stronger than the gravitational redshift.

RADIATION FROM A WHITE HOLE IS BLUESHIFTED

Why the name white hole? Unlike a black hole, a white hole is easy to see. This is shown in Figure 11-26, where we have a white hole in its initial stages of explosion. To a remote observer A, the expanding surface of the white hole (on its front side) appears to be approaching. Whenever a source of light approaches an observer, the frequency of the light appears to the observer to have increased. This phenomenon is the Doppler *blueshift** that takes place for a white hole. The blueshift is to some extent countered by the gravitational redshift discussed earlier because the gravitational field near a white hole is very strong. In general, for a white hole, the Doppler blueshift wins over the gravitational redshift. In the early stages of expansion, the blueshift is so large that a photon in the visual wavelength range comes to observer A as an x ray or a γ-ray photon. For this reason, white holes have been suggested as likely causes of γ-ray bursts and x-ray sources of short-term duration. Over the last few years, many such sources have been detected in our galaxy by satellite-based detectors. A typical γ-ray burst lasts for a second or so.

In optical astronomy, observations of radial velocities have indicated several examples of explosions in the nuclei of galaxies (see Figures 4-6 and 4-24). It is perhaps possible to think of white holes as being involved here too, although a theoretical framework wider than general relativity would then be needed, as we shall discuss in Chapter 14.

It is worth emphasizing that, unlike the black-hole event horizon, the Schwarzschild barrier (in the case of a white hole) does *not* prevent light from the white hole crossing outward to an external observer. The strong blueshift is able to carry light across the potential barrier. Thus, in principle, it is possible to see a white hole growing from an initial point source, the singular beginning. Hence, the white hole, if it exists, should turn out to be a remarkable object, at least in its early stages.

---

*For a discussion of the Doppler effect, see Appendix C.

In spite of its spectacular properties, the white hole has remained a poor relation to the black hole in the field of relativity. Many theoreticians find it unattractive because of its singular beginning (although they do not mind a singular end to the black hole), whereas others feel that it may not survive long enough to be observationally relevant. Until we understand better the physics close to a spacetime singularity, it seems to us that these objections to the white-hole concept are somewhat premature. Moreover, astronomers do not seem to mind the biggest conceivable model of the white hole—the model of the big-bang universe that we consider in the next chapter.

General Problems and Questions

1. Verify, by direct substitution, that the value of R_S for the Earth is less than 1 cm, and for the Sun, less than 3 km.

2. What is the law of attraction for two masses connected by an elastic spring stretched beyond its normal length? Check to see that the attraction vanishes at some stage of the oscillation described in the text.

3. Do the same as in Problem 2 for gravitating masses. Where exactly do you notice the difference?

4. Show that a dust ball of mass M, collapsing under its own Newtonian gravity, has a radius R that satisfies the following equation of motion:

$$\frac{d^2R}{dt^2} = -\frac{GM}{R^2}.$$

5. Solve the equation of Problem 4 and show that, if R starts with a value R_0 from rest, it reaches the value zero in a time

$$t_0 = \frac{\pi}{2}\sqrt{\frac{R_0{}^3}{2GM}}.$$

6. In Problem 5, substitute for the mass M and the radius R_0 the values from the Sun. See if you get the value quoted in the text.

7. Write an essay on the way the Sun maintains a nearly stationary shape in spite of its own strong gravitational force.

8. Estimate the gravitational redshift from:
   a. The Sun's surface.
   b. The surface of a white dwarf of mass equal to the mass of the Sun but with an average density of 10^6 g/cm^3.
   c. From the surface of a neutron star of mass $M = 2M_\odot$ and radius $R = 20$ km.

9. Observer B is on a *slowly* contracting object of mass $M_\odot$ and sends a signal to A every second. Suppose the radius of the object was $R_\odot$ initially. On graph

paper, try to plot the interval between successive signals received by *A* against *B*'s clock time until *B* almost reaches the horizon.*

10. Explain why *no* external observer can truly claim that he has observed or detected a black hole.

11. Explain the words "event horizon" by an analogy with the horizon on the Earth.

12. What do you understand by *singularity* of spacetime? Try to visualize what would happen to an extended object falling into a singularity. Its front end will be moving faster than its rear end, and this will lead to growing tidal forces.

13. Write an account of the way black holes may form by stellar evolution.

14. Comment on the statement: "A black hole has no hair."

15. Show that the area of the extreme Kerr black hole is half the area of the Schwartzschild black hole of the same mass.

16. Explain cosmic censorship. Which type of singularity will violate it?

17. What is the physical property of the ergosphere?

18. Develop the analogy between black-hole physics and thermodynamics.

19. A Kerr black hole of mass *M* and angular momentum *H* is reversibly transformed into a Schwarzschild black hole. What will be the mass of the new black hole? If the transformation is *not* reversible, will the black hole have a greater or lesser mass? Or will it be the same?

20. In the Hawking process, estimate the time taken for the evaporation of a black hole whose mass is equal to 1 million tons.

21. Write an essay on Cygnus X-1 and its role in the detection of black holes.

22. A photon from a white-hole surface was emitted at a wavelength of 5,000 Å. It was blueshifted by 1,000. What will be its wavelength at reception? What type of radiation does it represent?

23. How could a white hole arise in the universe? What are likely places suggested as white holes?

24. Explain how a white hole can give rise to very high energy radiation.

---

*There is an amusing analogy with everyday life. Imagine *A* to be an applicant, *B* an inefficient bureaucrat. *A* will always feel that *B* delays in answering correspondence where *B* does not think so since the time scales are different!

Chapter 12
Introduction
to Cosmology

§12-1. What Is Cosmology?

A one-sentence answer to the question "What is cosmology?" might go as follows: cosmology is the study of the structure of the universe taken as a whole. But this definition is too wide-ranging for our present discussion since it would involve both animate and inanimate. If pursued consistently, it would involve both large and small—past, present, and future. Consequently, we must restrict ourselves to more modest aims. What do astronomical observations tell us about the large-scale structure of the universe? Are the laws of physics, as we know them today, able to explain the patterns that emerge from such a study? The investigation of these questions is what we mean by cosmology.

To put the adjective "large-scale" into perspective, we consider the content of spacetime within the range of the most powerful present-day telescopes, and we do so with respect to various well-known units of length, mass, and time.

THE LIGHT-YEAR OR THE PARSEC IS THE LENGTH UNIT MOST APPROPRIATE TO COSMOLOGY

Judged subjectively, the Earth seems to be a big object, especially if we have to work over any appreciable bit of it. Our planet is approximately a sphere with radius $6{,}400 \text{ km} = 6.4 \times 10^8$ cm. The distance of the Earth from the Sun is

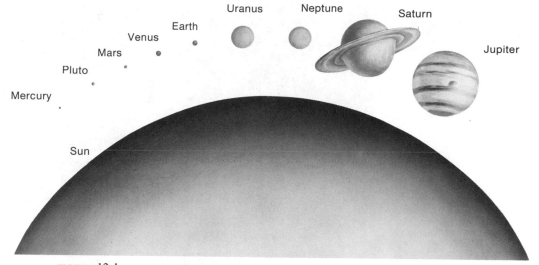

FIGURE 12-1

The nine planets of the solar system in scale relative to the Sun. (From J. Gilluly, A. Waters, and A. Woodford, *Principles of Geology*, 3d ed. W. H. Freeman and Company, Copyright © 1968.)

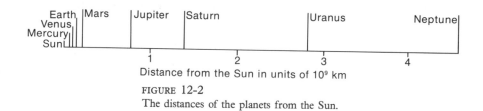

Distance from the Sun in units of 10^9 km

FIGURE 12-2

The distances of the planets from the Sun.

about 1.5×10^{13} cm, nearly 23,500 times greater than its radius. Pluto, the outermost planet of our solar system, is about 40 times farther away from the Sun than this, roughly 1 million times the radius of the Earth. Figure 12-1 illustrates the relative sizes of the Sun and the planets, whereas Figure 12-2 shows the relative distances of the planets from the Sun.

Our galaxy contains approximately 10^{11} stars distributed in a lenticular volume, as shown in Figure 12-3. As the figure indicates, the solar system is located some two-thirds of the distance from the galactic center. It is no longer convenient to describe this much bigger distance in terms of the length units used here on Earth. Two much larger units are now used. The more physically meaningful of these units is the *light-year*, the distance traveled by light in a year. In terms of the centimeter, we have

$$1 \text{ light-year} = 9.46 \times 10^{17} \text{ cm.}$$

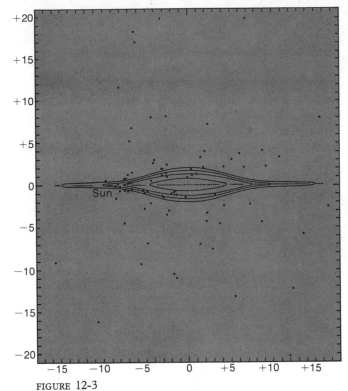

FIGURE 12-3
A schematic representation of our galaxy, showing the location of
the Sun. The unit of distance here is the kiloparsec, about 3,260
light-years.

The distance of the solar system from the center of our galaxy is about 30,000
light-years.

The other suitable unit in which to express such large distances is the *parsec*
(pc), with the approximate equivalence

$$1 \text{ pc} = 3.26 \text{ light-years.}$$

The parsec is the distance at which the radius of the Earth's orbit around the
Sun subtends an angle of 1s of arc. Historically, the parsec arose as a unit of
length in the nineteenth century in connection with a trigonometric method then
employed for determining the distances of the stars nearest to the solar system.

Until the present century, most astronomers believed that our galaxy was all
there was to the whole universe. The data obtained with powerful telescopes
using primary mirrors of large aperture gradually changed this view, however,
by revealing that our galaxy was only one of many. Other galaxies of various
shapes and forms are found all through the universe, some resembling ours and

FIGURE 12-4

The galaxy M31, with a diameter of ~100,000 light-years, situated in the constellation of Andromeda. This galaxy, sometimes referred to as the Andromeda nebula, is the nearest of the large galaxies. The central regions are visible to the naked eye. (Courtesy of Hale Observatories.)

others very different. There are isolated field galaxies as well as galaxies that occur in groups or clusters, numbering from about 10 galaxies for a small cluster to about 1,000 for a large cluster. Figures 12-4 to 12-10 illustrate the various galactic forms and clusters. The separation between neighbors is usually expressed in a unit of 1 million parsecs (1 mpc). The galaxies in large clusters lie in regions with spatial diameters up to about 5 mpc. Several large clusters are often seemingly associated together in so-called *superclusters* with diameters up to about 50 mpc.

FIGURE 12-5
The galaxy IC 1613, about 10,000 light-years in diameter, is a diffuse dwarf member of the local group. (Courtesy of Hale Observatories.)

FIGURE 12-6
The symmetric spiral galaxy M81 is seen near the center of a small group of galaxies in the constellation of Ursa Major. The group is about 500,000 light-years in size. (Courtesy of Hale Observatories.)

FIGURE 12-7
A cluster of galaxies about 500,000 light-years in diameter, in the southern sky, RA 10h30m, Dec −27°. (Courtesy of Dr. V. C. Reddish, Science Research Council.)

FIGURE 12-8
A rich cluster of galaxies in Corona Borealis, located about 1,200 million light-years away from us. (Courtesy of Hale Observatories.)

344

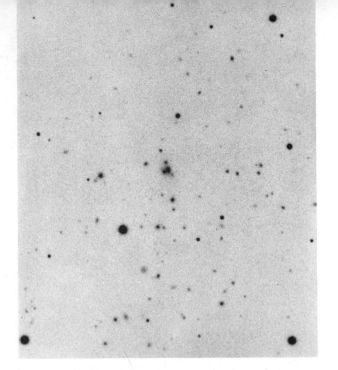

FIGURE 12-9
In the depths of space at a distance ∼3,600 million light-years, many galaxies are faintly seen. This cluster is in the constellation of Hydra. (Courtesy of Hale Observatories.)

FIGURE 12-10
Another deep photograph. The marked galaxies are at distances of about 6,000 million light-years away from us. (Courtesy of Hale Observatories.)

The region accessible with a powerful telescope extends to about 3,000 mpc, that is, to a distance of some 10 billion light-years. This distance of 10 billion light-years, as we see later in this chapter, represents the characteristic scale of the universe.

MASSES ON A LARGE SCALE ARE CONVENIENTLY SPECIFIED IN TERMS OF THE MASS OF THE SUN AS UNIT

Let us again ascend the cosmic ladder with respect to mass, from the lowly rung of the Earth up to the mass of the whole observable universe.

The Earth has a mass of about 6×10^{27} g, and the Sun has a mass of about 2×10^{33} g. As in the case of the distance scale, the local unit of mass, the gram, is too small to be convenient in cosmology. Here we take the mass of the Sun, denoted by $M_\odot$, as our unit. In terms of this unit, the mass of the Earth is about $3.10^{-6} M_\odot$. The mass of our galaxy, shown schematically in Figure 12-3, is estimated to be about $2.10^{11} M_\odot$. Galaxies more massive than ours exist. Thus, the giant elliptical galaxies like M87 that we found to be an intense emitter of x rays (Chapter 7) have masses that are several times $10^{12} M_\odot$, and a cluster of galaxies may have a total mass as large as $10^{14} M_\odot$.

What is the estimated mass of the whole of the observable universe? The number of observable galaxies lies in the range from 10^9 to 10^{10}, so that a rough estimate for the observable universe would be some $10^{21} M_\odot$ (on the order of 10^{54} g). This estimate does not include unseen matter that may exist in non-visible forms. Astronomers differ greatly in their opinions of how much unseen matter there might be. Some think the amount comparatively small; others think the amount might be as much as 100 times greater than the visible galaxies, totaling about $10^{23} M_\odot$ for the whole of the matter within a distance of 10 billion light-years.

THE SECOND AND THE YEAR ARE THE TIME UNITS MOST OFTEN USED IN COSMOLOGY

The time scales covering the life histories of astronomical objects vary considerably, and no single unit is convenient for all of them. The periods of pulsars and the time durations of x-ray and γ-ray bursts involve intervals on the order of a second or less. The explosions of supernovae also occur in time scales of a few tens of seconds. On the other hand, the evolutionary history of a solar-type star, beginning at the main sequence, covers a time span of billions of years.

Turning to galaxies, many have rotation periods of tens or even hundreds of millions of years, as is illustrated for our own galaxy in Figure 12-11. Even longer time scales prevail in the evolution of galaxies. Although the detailed processes of the formation and evolution of galaxies are still not very well understood, a galaxy made up largely from stars of small mass—for example, an elliptical galaxy like M87—is believed to be more than 10 billion years old.

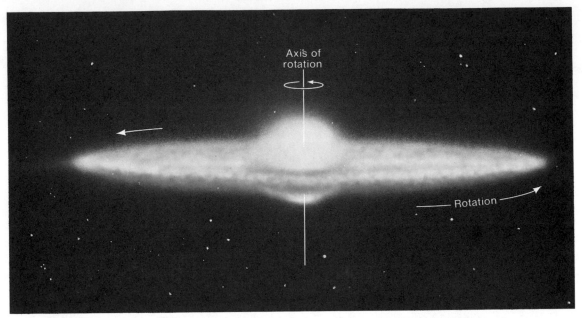

FIGURE 12-11
The galaxy rotates about the axis shown. The solar system completes one revolution in about 200 million years. (Adapted from Blaauw and Schmidt, *Galactic Structures,* University of Chicago Press © 1965.)

Is there a characteristic time scale that can be specified for the whole universe? The answer to this question is yes, and the characteristic time probably lies in the range from 10 to 15 billion years. To understand the basis of this affirmative answer, we consider next a critically important discovery due to Edwin Hubble that provided the foundation for the development of modern cosmology.

§12-2. Hubble's Law

With the 60-inch and 100-inch telescopes at Mount Wilson, California, Edwin Hubble made a systematic investigation of galaxies out to distances of a few hundred million parsecs. The aims of the investigation were:

1. To study the structural forms of the galaxies.

2. To estimate the apparent luminosities of the galaxies.

3. To measure the redshifts of the spectra of the galaxies.

The first determinations of the redshifts of galaxies had already been made by V. M. Slipher, but these determinations were made for a more limited sample of comparatively nearby, bright galaxies. Hubble's survey extended Slipher's work to more distant, fainter galaxies. The 100-inch telescope on which most of Hubble's work was done is shown in Figure 12-12.

Slipher had found the lines in the spectrum of a galaxy to be systematically at longer wavelengths than those observed in the laboratory. Thus, if the laboratory wavelength for a particular line was λ_0, the measured wavelength λ for the same line in the spectrum of a galaxy could be written in the form

$$\lambda = (1 + z)\lambda_0,$$

where the number z (positive in almost all cases) was the same for all the lines of the spectrum belonging to a particular galaxy. The number z was called the redshift of the galaxy in question. Different galaxies in Slipher's sample had different values of z, and so had the galaxies in Hubble's wider sample.

In relation to this work, we might wonder how the nature of the atoms responsible for a particular line in the spectrum of a galaxy is determined. As long as the spectrum has more than one line, ratios of the measured wavelength

FIGURE 12-12
The 100-inch telescope at Mount Wilson. This is the instrument on which Hubble did most of his important work on nebular redshifts. (Courtesy of Hale Observatories.)

of a line to the measured wavelengths of other lines can be obtained. Such ratios are unaffected by the redshift. Thus, if $(\lambda_A)_0$, $(\lambda_B)_0$ are the laboratory wavelengths of two lines, one produced by atoms of kind A and the other by atoms of kind B, we have

$$\lambda_A = (1 + z)(\lambda_A)_0 \quad \text{and} \quad \lambda_B = (1 + z)(\lambda_B)_0$$

for the measured wavelengths λ_A, λ_B in the spectrum of a galaxy of redshift z. Taking the ratios of these equations, the factor $(1 + z)$ disappears, and we have

$$\frac{\lambda_A}{\lambda_B} = \frac{(\lambda_A)_0}{(\lambda_B)_0}.$$

Now the ratio $(\lambda_A)_0/(\lambda_B)_0$, known from laboratory work, is highly specific to the lines in question. Only very rarely does it happen that the same precise ratio turns up in connection with a line involving atoms of a third kind, say, C. The essential uniqueness of $(\lambda_A)_0/(\lambda_B)_0$ means that, should measurements of two lines in the spectrum of a galaxy lead to just the same unique ratio, we can safely infer that the lines in question arise from atoms of kinds A and B. Association of λ_A with $(\lambda_A)_0$ and of λ_B with $(\lambda_B)_0$ then gives z, from $1 + z = \lambda_A/(\lambda_A)_0$, $1 + z = \lambda_B/(\lambda_B)_0$.

The crucial discovery made by Hubble was that a strong statistical connection exists between the luminosities of galaxies and their redshifts, in the sense that the smaller the luminosity, the greater the redshift. Since smaller luminosities meant greater distances, this discovery implied that the redshifts of the galaxies increase with their distances. Would the redshifts increase still more for galaxies more distant than those in Hubble's first sample? The first sample had consisted of individual field galaxies. To extend the distance range, Hubble turned to galaxies in clusters because the brighter galaxies in clusters are more uniform than field galaxies, so that statistical scatter was thereby much reduced. This uniformity permitted the sample of galaxies in distant clusters to be smaller in number than the sample of field galaxies had been, a condition essential to the work since the more distant galaxies now under investigation were fainter than before. Therefore, they were much more difficult to observe, each one requiring an increased number of hours at the telescope. The work indeed became so demanding that Hubble proceeded in collaboration with Milton Humason, Hubble being responsible for the luminosity determinations and Humason for the redshifts.

The results obtained by Hubble and Humason are shown in Figures 12-13 and 12-14. Nowadays it would be usual to present such results more directly, as a logarithmic plot of luminosities L against redshifts z. In the 1930s, however, luminosities were converted into magnitudes in accordance with the procedure of Figure 12-15, whereas redshifts were interpreted as Doppler velocities. Magnitudes and velocities appear in Figures 12-13 and 12-14.

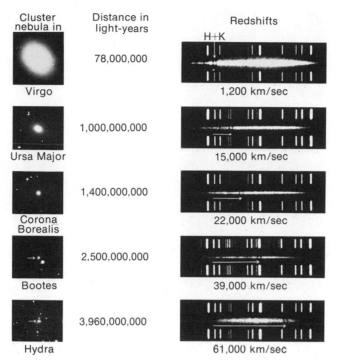

Cluster nebula in	Distance in light-years	Redshifts

FIGURE 12-13

The relationship between redshift and distance for extragalactic nebulae. The redshift of the spectrum lines from specific atoms, for example, the H and K lines of calcium (indicated by the arrows), when interpreted in terms of the Doppler effect, gave the speeds shown in this figure. Redshifts are expressed as velocities, $c \Delta \lambda / \lambda$. A light-year equals about 9.5 trillion (9.5×10^{12}) km. Distances are based on an expansion rate of 50 km/second/mpc, 1 pc being 3.26 light-years. (Courtesy of Hale Observatories.)

The Doppler velocity v is related to the redshift z by the formula

$$1 + z = \sqrt{\frac{c + v}{c - v}},$$

where c is the speed of light. This formula, which is discussed and proved in Appendix C, can readily be inverted to the form

$$v = c \cdot \frac{z^2 + 2z}{z^2 + 2z + 2}.$$

The speed of light is close to 300,000 km/second. Using this value for c and putting z equal to the measured redshift of a galaxy gives v expressed in kilometers per second. The ordinate of Figure 12-14 is the logarithm (to base

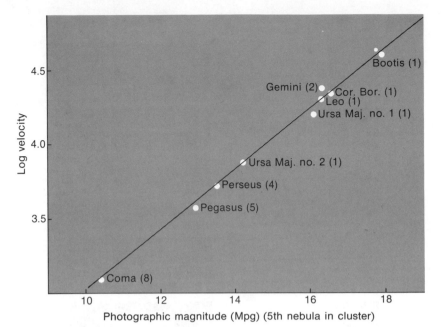

FIGURE 12-14

The observed results of Hubble and Humason support the linear relationship between m and log v.

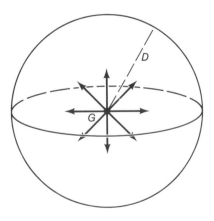

FIGURE 12-15

If galaxy G emits L units of energy per unit time, these units will cross a sphere of radius D centered at G in an *isotropic* way so that a unit area on the spherical surface receives $f = L/4\pi D^2$ units per unit time. The magnitude in this figure is defined by a logarithmic relation:

$$m = -2.5 \log f + \text{constant.}$$

The constant (not important here) is chosen in relation to the brightness of first magnitude stars.

10) of the value of v obtained in this way. For small redshifts, the preceding

351

§ *12-2. Hubble's Law*

relation simplifies to the approximate form

$$v = cz.$$

**THE OBSERVATIONAL RESULTS OF HUBBLE AND HUMASON
SUGGEST THAT THE REDSHIFTS OF THE GALAXIES
INCREASE PROPORTIONATELY TO THEIR DISTANCES**

Figure 12-14 contains two items of information. First, there is a linear relationship between log v and the magnitude m. Second, the *slope* of the line in the figure requires that the relation be of the form

$$\log v = 0.2m + \text{constant},$$

that is, the coefficient 0.2 is determined by the slope of the line as may readily be verified from an actual measurement of the figure. For the galaxies in question, it is sufficiently accurate to use $v = cz$, so that

$$\log (cz) = 0.2m + \text{constant}.$$

Now $m = -2.5 \log f +$ a constant, where f is the observed energy flux from the galaxy. Hence,

$$\log (cz) = -0.5 \log f + \text{some constant},$$

and writing

$$f = \frac{L}{4\pi d^2}$$

where L is the intrinsic luminosity of the galaxy, we have

$$\log (cz) = \log d - 0.5 \log (L/4\pi) + \text{some constant}.$$

If the observed galaxies all have the same luminosity, the $-0.5 \log (L/4\pi)$ term in this last equation is the same for all of them, in which case z for each galaxy is proportional to its distance d. Indeed, taking the antilog of the last equation then leads to an equation

$$cz = Hd,$$

where H is a constant depending on L and the other constants appearing in the preceding equations. This last result is Hubble's law, and H is known as the *Hubble constant*.

From the observations, Hubble arrived at $H = 530$ km/second/mpc, which means that a galaxy has a value of v that increases by 530 km/second for each megaparsec in its distance. This implies an increase of about 0.00177 in z for each million parsecs in distance, so that a galaxy at a distance of d megaparsecs has $z = 0.00177d$, provided d is not too large.

More recent observations, involving redeterminations of the luminosity values of galaxies, have revised H substantially downward, to about 75 km/second/mpc. The revisions came about for reasons discussed in Chapter 9, and even today there are disagreements about what the "true" value of H should be. Some astronomers set the value as low as 50 km/second/mpc, while others set H as high as 100 km/second/mpc.

The linear character of Hubble's law is considered well established for galaxies having redshifts less than about 0.3. Such galaxies have distances up to about 1,500 mpc. There could be a departure from $cz = Hd$ for still more remote galaxies. As we shall see later, the nature of such a departure, if it could be determined from observation, would give important information concerning the large-scale structure of the universe.

§12-3. The Expanding Universe

Hubble's law requires our galaxy to be located so that, with the exception of a few immediate neighbors (with very small negative z), all other galaxies are receding from us. At first, this seems to require a special position in the cosmos, as if we were positioned at a privileged center. A little thought shows, however, that such a deduction is in error. If we imagine ourselves located on any other galaxy, we would see just the same large-scale picture; the other galaxies would obey $cz = Hd$ with z and d *measured from our new vantage point*. This aspect is an important general property of the universe, known as *homogeneity*, that we consider in more detail in Section 12-4.

HUBBLE'S LAW DOES NOT DEMAND A PRIVILEGED POSITION FOR THE OBSERVER

The following two examples make clear the fact that Hubble's law does not demand a privileged position for the observer.

The first example is a rubber balloon in the process of inflation, as in Figure 12-16. Suppose there are dots on the balloon to serve as markers. No dot is in a privileged position. Yet, as the balloon expands, *all* the dots move away from one another.

As a second example, imagine a cubic, metallic grid of wires being heated in an oven, as in Figure 12-17. The wires expand in length with the heat, causing all the lattice points to move away from one another. Here, again, there is no particular privileged lattice point, although we might think of lattice points in

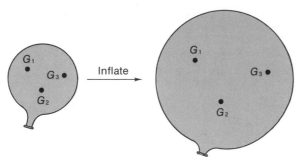

FIGURE 12-16
As we inflate the balloon, the dots G_1, G_2, and G_3 on its surface move away from each other. Yet, no dot can claim a privileged site on the spherical surface.

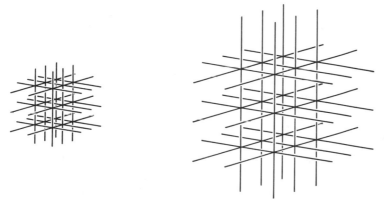

FIGURE 12-17
A cubic grid of metal wires expands on being heated. The lattice points all move away from one another. Here, again, no one point can claim a privileged position.

the interior of the grid as being different from those on the boundary of the grid. But by making the grid bigger and bigger, the proportion of such boundary points is reduced. As the grid becomes infinitely large, even this distinction between interior and boundary points disappears.

The first example is analogous to the closed, finite models of the universe that we consider in the next chapter, while the second example is analogous to open, infinite models. We can imagine, as with the heated wires, that the *space* in which the galaxies are embedded *expands,* causing intergalactic distances to increase in a uniform way. Figure 12-18 shows what happens in the course of time to three galaxies G, G_1, and G_2. At the earlier time, the galaxies form a smaller triangle than at the later time, but the *shape* of the triangle remains the same because all three sides change in the same proportion.

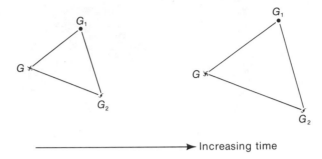

FIGURE 12-18

The triangles formed by our galaxy G and two other galaxies, G_1 and G_2, drawn for two different moments of time, have different scales.

This rule, applied generally, that *all distances between galaxies change with time in the same proportion* is often referred to as the *expansion of the universe*. Figure 12-19 shows three cases (I, II, III) of the way such distances can change with the time, the scale Q being applicable to all the interconnections between galaxies. Suppose the time associated with point P corresponds to the present epoch. If from the present we go back into the past along the straight line I of Figure 12-19, we reach point O at which $Q = 0$, meaning that *all* the observed galaxies were then at *one* point. At this time, the whole observable universe was shrunk to a single point.

The straight line I corresponds to a linear expansion of the universe with respect to time. What would happen if, instead of such a linear expansion, the expansion were faster in the past than it is now? Then we would have a curve like II in Figure 12-19, lying below the straight line I. The situation with $Q = 0$ is then reached from P over a shorter, backward time interval. This time interval from the present epoch back to $Q = 0$ is called the *age of the universe*. The age

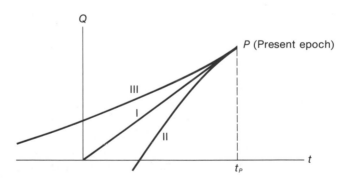

FIGURE 12-19

The scale factor Q plotted as a function of time t for three different cases of the expanding universe. P denotes the present epoch and the curves can be continued to the future of P to predict how the universe will behave in the future. For the straight line I, distance is proportional to t and velocity is constant. The Hubble constant is therefore simply $1/t$ at time t.

is evidently shorter in case II than it is for the linear expansion of case I. If, on the other hand, the expansion rate were smaller in the past, as for case III in Figure 12-19, the age would be longer. Indeed, for an expansion rate that slows more and more back into the past, it is possible that there is no finite age; as we go back into the past, Q becomes smaller, but the condition $Q = 0$ is never reached. This situation is the so-called steady-state model of the universe, whereas case II, with a finite age, occurs for the so-called big-bang models discussed in Chapter 13.

THE STAGE $Q = 0$ REPRESENTS THE ORIGIN OF THE UNIVERSE

Astronomers interpret $Q = 0$ as the condition occurring at the origin of the universe. The universe "began" at the epoch at which Q was zero. The speeds of recession of the galaxies we observe now are relics of this earlier condition, and they are considerably slower than they were in the past. Astronomical opinion favors a curve like II of Figure 12-19. In the next chapter, we consider whether this opinion can be proved by astronomical observations.

§*12-4. The Symmetries of the Universe*

In Section 12-2, we emphasized that Hubble's law holds good for an observer situated in *any* galaxy. This property of homogeneity of the universe is believed to apply in the following more general sense. It is possible to identify one's position in space *only from local details, not from the large-scale structure of the universe.* This belief cannot, of course, be proved by experimental test because, in practice, we cannot change our position in space to a significant degree. We can say, however, that no aspect of our observations is inconsistent with it. How might an observation be inconsistent with homogeneity?

Figures 12-20 and 12-21 show, in a schematic way, a homogeneous and an inhomogeneous distribution of galaxies. Figure 12-20 has local variations, but on a larger scale, the distribution is uniform. Figure 12-21, on the other hand, is markedly nonuniform on the scale of successive shells concentric with respect to our own galaxy. Both distributions refer to a particular epoch. If we ignore light travel times for a moment, we can see that observations that were consistent with Figure 12-20 would not be consistent with Figure 12-21, and vice versa. Observations of the densities of galaxies at increasing distances can therefore rule out at least one of these two situations. In practice, it turns out that Figure 12-21 is contradicted by observation, while Figure 12-20—the homogeneous situation—is consistent with such density determinations.

Light travel times are longer from distant galaxies than they are from nearby ones. We therefore observe the densities of galaxies at earlier and earlier epochs as the shells of Figure 12-21 increase in their radii. For this effect to be

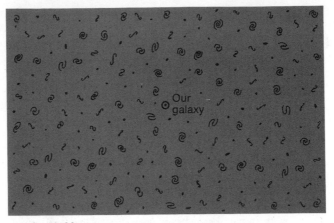

FIGURE 12-20
A uniform distribution of points (galaxies in space). There is
no region that is especially dense or rarified provided we do
not look at the distribution on a local scale.

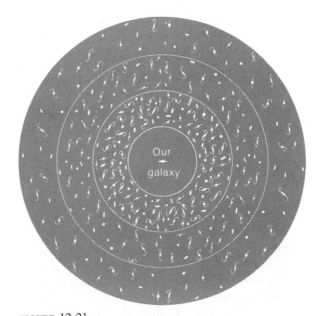

FIGURE 12-21
Although different concentric shells contain different densities
of galaxies, the situation still appears isotropic to an observer
at the center, but it would not appear isotropic to an observer
not at the center. (The density of galaxies is greatly exaggerated in this drawing.)

ignorable, it is necessary that the densities do not change much during the light travel times, a condition that is met if the expansion factor Q does not change appreciably. Reference to Figure 12-19 shows that, for this to be the case, light travel times must be small compared to the age of the universe—that is, small compared to 10 billion years. We are therefore limited to distances of a few hundred million light-years in making observations of the kind described here. Only if we already had knowledge of the precise dependence of Q on the time, which we do not have, could the observations be extended to greater distances. Because of this uncertainty, we can say that the observations are consistent with a homogeneous structure for the universe *as far as they go,* but the observations do not *prove* that this must be so. Homogeneity on a very large scale remains an article of faith among astronomers.

THE UNIVERSE IS ISOTROPIC ON THE LARGE SCALE WITH RESPECT TO OUR OWN GALAXY

When the positions of the nearest and brightest galaxies are plotted on the celestial sphere (for example, the galaxies of the New General Catalogue), the distribution is found to be nonuniform, over both adjacent areas of the sky and widely separated areas of the sky. Similar plots for galaxies down to fainter and fainter magnitudes are still nonuniform over adjacent small areas because of clustering, but when such local irregularities of detail are smoothed, the distribution becomes increasingly uniform over the whole sky. At the range of the most powerful telescopes, the distribution is very uniform. As seen from our galaxy, the universe on a large scale lacks any preferred direction. The situation appears to us to be *isotropic.*

Would an observer in any other galaxy find a similar result? There is no way in which this question can be answered directly from observation. Both situations of Figures 12-20 and 12-21 appear to us to be isotropic in the large, but, whereas for Figure 12-20 an observer in any galaxy would also find isotropy, observers in other galaxies for the situation of Figure 12-21 would *not* find isotropy.

Unlike the state of affairs in pre-Copernican days, when people were happy to think of the Earth occupying a central position with respect to the universe, scientists today find this point of view repugnant. That the large-scale features of the universe are arranged isotropically only with respect to us seems ridiculous to the modern scientific mind. The situation of Figure 12-21 is therefore rejected on intellectual grounds, *in principle,* as cosmologists say. In principle, we decide that observers in other galaxies would also find isotropy. In this case, it can be proved mathematically that the universe must be homogeneous on the large scale, a scale larger than that which is guaranteed by the observations just discussed.

The large-scale properties of homogeneity and isotropy greatly simplify the building of mathematical models of the universe as we find in the Chapter 13. Such models are said to obey the cosmological principle.

COSMOLOGISTS ALSO ASSUME THE EXISTENCE OF A
SYNCHRONOUS SYSTEM OF UNIVERSAL TIME

Throughout the preceding discussion, we used the concept of time in an essentially Newtonian sense. Yet, in Chapter 10, we saw that there is no unique system of time recognized by all observers. Each observer carries a clock that measures proper time only for that individual observer. How are we to reconcile this apparent discrepancy between Chapter 10 and the present discussion?

The observers of Chapter 10 were chosen in an abstract way, whereas we are concerned now with observers attached to galaxies. This restriction in the choice of observers is not in itself sufficient to resolve the problem, but it provides the basis for a resolution. We must add the further point that the galaxies are related to each other in accordance with Hubble's law. The effect of Hubble's law, combined with homogeneity and isotropy, is to make the proper times measured by all galactic observers synchronous with each other in a system of universal time analogous to Newtonian time. This important simplification of cosmological models arises in the following way.

Figure 12-22 explicitly illustrates the situation for case I of Figure 12-19. By drawing the world lines of the galaxies in suitably organized curves, we can discuss cases II and III of Figure 12-19 in the following way.

Figure 12-22 is a spacetime diagram in which one particular galaxy—our own—is at rest. A similar diagram could be drawn in which any other galaxy was considered to be at rest. The diagram is schematic in the sense that the abscissa x stands for all three dimensions of space. The time t is proper time for

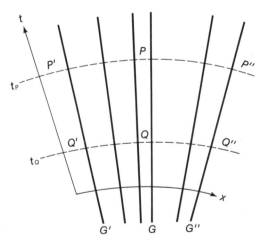

FIGURE 12-22
The world lines of all the galaxies represented in a single diagram. Subject to the assumption that the universe on a large scale is homogeneous and isotropic, the world lines of the galaxies have a highly systematic distribution like the uniformly divergent fan in this figure.

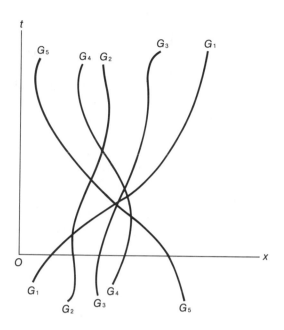

FIGURE 12-23
A turbulent universe in which the world lines of galaxies G_1, G_2, and so forth are all tangled up. Fortunately, our universe does not happen to be like this.

an observer in our galaxy. This means that the time difference $t_P - t_Q$ between the time sections through any two points P, Q on the world line of our galaxy is determined physically as the time interval measured by a clock carried from Q to P by the observer in our galaxy. What we would not generally expect, and what would certainly not be found if the other galaxies had the chaotic motions of Figure 12-23, is that $t_P - t_Q$ is also the proper time interval measured by every observer in every galaxy. Yet, in view of the restrictions just discussed, this situation is so. An observer in galaxy G', for example, carrying a similar clock from Q' to P', would also measure $t_P - t_Q$; so would an observer in G'' carrying a similar clock from Q'' to P''. It is this remarkable property of the world lines of the galaxies that permits us to treat time in a universal sense.

A point of some subtlety and interest remains. Of course, the statements of the last paragraph cannot be subjected to an actual practical test since there is no operational way in which such clock measurements can be carried out. We are not in a position to collate the measurements of observers in all the galaxies. It is just here that the distinction between the actual universe and a mathematical model of the universe is important. The mathematical model permits us to calculate what the time measurements must turn out to be *if the model correctly represents the universe*. Astronomers believe that the model correctly represents the universe provided it obeys Hubble's law and the cosmological principle, and provided that the calculations are made according to Einstein's general theory of relativity. These criteria are all be satisfied for the big-bang model considered in Chapter 13. Models for which a synchronous time system exists are said to obey the *Weyl postulate*, named after Hermann Weyl (1885-1955), who first discussed the general mathematical properties of such a time system.

§*12-5. Olbers' Paradox*

As long ago as 1826, the Viennese physician Heinrich Olbers asked the seemingly innocent question, "Why is the sky dark at night?" A satisfactory answer was hard to find, however. Indeed, until the present century, the question defied a proper solution. The argument to which it led, and which will now be discussed, became known as Olbers' paradox.

The spin of the Earth turns us away from the Sun each day, and a simple attempt at an answer would be that the cutting off of the sunlight makes the sky dark at night. This answer would be fine if there were nothing to the universe except the Sun and the Earth. But there are stars and galaxies, so the real question is why the stars and galaxies produce only a faint glow in the night sky, not a blaze of light as the Sun does in the daytime. To see that the question in this form is not a simple matter, let us consider the argument as it was given by Olbers (and even before Olbers by Swiss astronomer J. P. L. Cheseaux).

Olbers knew nothing about the galaxies being separated one from another, so he assumed the solar system to be embedded in a uniform distribution of stars. Since we can readily imagine the stellar contents of the galaxies to be smoothed throughout space, our modern picture can be made to conform to this situation postulated by Olbers. We can also average the luminosities of the actual stars to a standard value, say, L, to be used for all the hypothetical stars of our smoothed distribution, which we take to have n stars per unit volume.

The flux f from a star at a distance D is given by $f = L/4\pi D^2$, the flux being the energy received per unit time across a unit area taken perpendicular to the direction of the star. Evidently, the flux becomes very small as the distance becomes large, so that at first we might think that the contribution of faraway stars to the light of the night sky must be small. But consider the situation of Figure 12-24, in which there are many stars in the thin spherical shell of thickness a. The volume of this shell is given by multiplying the thickness by the surface area, $a \times 4\pi D^2$. Thus, the shell contains $4\pi a D^2 n$ stars, each contributing the flux $L/4\pi D^2$. Hence, the total flux from all the stars in the shell is

$$\frac{L}{4\pi D^2} \cdot 4\pi a D^2 n = anL,$$

which, surprisingly, does not depend on D. The faintness of an individual star caused by great distance is exactly compensated by the increasing number of stars of uniform distribution. What is the total flux from all the stars in the whole universe? The answer is given by considering a series of shells of increasing D, but all with the same thickness a. Each shell gives a contribution anL to the total flux.

Olbers assumed the universe to be spatially infinite, as indeed the universe was always taken to be in Newtonian physics. This assumption meant that the number of shells like those of Figure 12-24 was infinite, in which case the flux should be infinite and the sky should be infinitely bright, irrespective of whether

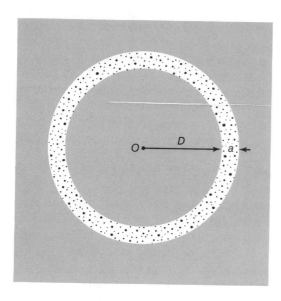

FIGURE 12-24
According to the argument given by Olbers, the total contribution to the brightness of the sky at point O from all stars in the shell of thickness a is the same, however large D may be.

or not we are turned toward the Sun! Clearly, there had to be something very seriously wrong in this argument, and the problem for nineteenth century astronomers was to find where the error lay.

There is no possibility of evading this conclusion by arguing that stellar radiation is absorbed en route to the Earth by material in space. This argument simply transfers the difficulty from one place to another. The absorbing material would heat up and itself begin to radiate. After a while, the material would radiate as much energy as it absorbed, and exactly the same argument would then apply.

The Olbers' argument, as it is presented here, tacitly assumes the stars to be point sources of radiation, whereas stars have a finite size. Nearby stars must therefore tend to block radiation from distant stars, and this effect was omitted in our calculation. A little thought shows that the blocking of distant stars by nearer ones avoids the infinite result just obtained. But the finite flux that remains is still enormously large, so large that the whole sky would have a surface brightness comparable to that of a typical star. Indeed, the Earth would be embedded in a radiation bath just as intense as the radiation at the photosphere of a star. Therefore, this modification of the argument does not lead to a successful resolution of the paradox.

OLBERS' PARADOX CAN BE RESOLVED IN SEVERAL WAYS

The paradox is properly resolved if the universe has a sufficiently short finite age. Light emitted by a star at distance D has a travel time D/c, and if the universe has age T—that is, if stars did not exist at a time greater than T before the present—then light from stars at a distance greater than cT cannot yet have reached the Earth. The sequence of the shells we just considered must therefore be cutoff at a distance cT. The number of shells of the type shown in Figure 12-24 contributing to the flux at the Earth is then finite; for T not too large, the flux is small, and the night sky is effectively dark.

This resolution of the paradox was accessible to astronomers of the nineteenth century, but it appears to have been less favored than the concept of an *island universe*. If the material content of the universe consisted of a finite cloud of stars embedded in infinite space, then much the same situation would arise. There would be no contribution to the flux at the Earth from beyond the boundary of the cloud, and provided the cloud were not too large, the flux would again be small; the night sky would be effectively dark.

Not until the early 1920s did astronomers in general accept that galaxies were systems quite independent of our own Milky Way. Many had clung tenaciously to the idea that galaxies were small, nebulous objects lying inside our own system (see Chapter 9). The need to resolve the Olbers' paradox doubtless had an influence in promoting this erroneous point of view.

In modern times, Olbers' paradox has presented no difficulty. It is "over resolved" today in the sense that it can be dealt with in more than one way. A finite age for the universe resolves the problem. So does the redshift phenomenon. The flux formula $f = L/4\pi D^2$ applies to the Newtonian universe of nineteenth century astronomers. It is also approximately valid in modern cosmological models provided D is not so large that the redshift z becomes appreciable. For increasing D, however, the redshift itself introduces a cut off in the flux according to the modified formula $f = L/4\pi D^2(1 + z)^2$. Working with the extra $(1 + z)^2$ factor, calculations using the known density of galaxies lead to a total flux at the Earth that is adequately small. The redshift effect on this flux is itself sufficient to resolve the Olbers' paradox.

General Problems and Questions

1. Calculate the length of a light-year in kilometers.

2. Suppose, for this problem, that the Earth moves in a circular orbit of radius $R \cong 1.5 \times 10^{13}$ cm around the Sun. A star is said to be located at 1 pc when the angle subtended by the radius of the Earth's orbit at the star is 1 arc second (1/3,600 of a degree). Estimate the distance of the star (roughly) in centimetres from these data.

3. Make the same assumption about the Earth's orbit as in Problem 2. Taking the gravitational constant to be about $\frac{2}{3} \times 10^{-7}$ cm^3/second^2/g, estimate the mass of the Sun. (*Hint:* Use Newton's law of gravitation and the fact that the Earth takes one year to complete the orbit.)

4. The Lyman α line has a wavelength 1216 Å. At what redshift will it begin to be seen in the visual part of the spectrum that is supposed to begin at $\sim$3600 Å?

5. An object has a redshift $z = 2$. If it is moving away, what is its speed of recession?

6. Verify that, if z is very small compared to 1, the formula for redshift simplifies to $v = cz$. Why is it incorrect to argue that an object of redshift 2 is moving with twice the speed of light?

7. By imagining yourself on another galaxy receding from ours according to Hubble's law, show that you would see the same Hubble law operating from your new location.

8. You have two electric bulbs, one of power 10, the other of power 1,000. The first one is kept at a distance of 10 ft from a light detector. Where would you keep the second to have it produce the same flux at the detector? What does your answer imply for the interpretation of distance of a galaxy in terms of its faintness?

9. Write an essay on Hubble's law.

10. Construct an example to illustrate the expansion of the universe.

11. If the scale factor Q obeys a linear law in Figure 12-22, estimate the time interval that has elapsed since the epoch of $Q = 0$ to the present day in terms of the present value of Hubble's constant.

12. Can you construct an example of a universe that is homogeneous but *not* isotropic?

13. Discuss the cosmological principle and the Weyl postulate.

14. Explain Weyl's postulate, and construct examples of motions of terrestrial systems that do and do not appear to show the regularities of Weyl's postulate.

15. In trying to get around Olbers' paradox, if we assume radiating objects to have radius a, how will we reduce the final answer in Olbers' calculation? (The answer is now reduced to the surface brightness of the typical star, so that if each star were like the Sun, the night sky would be so bright as to have a temperature of $\sim 5500°$C. Verify this.)

16. Convince yourself that, if the universe were not infinitely old, the Olbers' paradox would be resolved.

Chapter 13
The Big-Bang Universe

§13-1. Cosmological Models

The study of cosmology consists of finding mathematical models that simulate the observed behavior of the actual universe and of making predictions to be tested by future observations. In addition to these clear-cut aims, we can add the further and more subtle aim of extrapolating the known laws of physics as far as possible in both space and time. The first of these objectives is essential for any theory; the second stimulates new observations; the third may contribute to our basic understanding of science.

GRAVITATION IS THE CONTROLLING INTERACTION FOR COSMOLOGY

At first, all theories must have their roots in what we know already about physics. In the course of this book, we have seen that there are four fundamental physical interactions, and the one most likely to determine the large-scale structure of the universe is gravitation. In this chapter, we find that the dynamic behavior of the universe is indeed determined by the gravitational interaction. However, the other interactions are not excluded altogether. The electrical interaction provides us with information about the remote parts of the universe. Without the existence of light and other forms of radiation, all cosmological models would be only speculative exercises. We also see how our knowledge of the strong and weak interactions can tell us about the primordial condition of matter and how relics of primordial matter may still be found today.

EINSTEIN'S GENERAL THEORY OF RELATIVITY IS THE PHYSICAL BASIS OF COSMOLOGICAL THEORY

Although Newton's theory of gravitation is easier to grasp than Einstein's general theory of relativity, Newton's theory has certain conceptual inconsistencies with the rest of modern physics. We have seen that the solar system tests of the perihelion precession of the planet Mercury and the bending of light past the Sun are explained by Einstein's theory but not by Newton's. For these reasons, it seems desirable, in cosmology, to put our faith in general relativity rather than in the older theory of Newton.

There are other reasons particular to cosmology that support this point of view. In cosmology, we inevitably deal with very large distances—distances over which light takes billions of years to travel. One would need to be extremely cautious in attempting to apply the Newtonian concept of instantaneous action-at-a-distance over such vast distances. Einstein's theory does not suffer from this difficulty. Another problem arises when we attempt to calculate the Newtonian gravitational force exerted by one part of the universe on another part. Because of the infinity of Newtonian space, the calculation of such forces can be very ambiguous due to what mathematicians refer to as the form of the *boundary conditions at infinity*.

SENSIBLE COSMOLOGICAL MODELS CAN BE CONSTRUCTED IN NEWTONIAN THEORY

In 1934, E. A. Milne and W. H. McCrea showed how to circumvent the aforementioned difficulties in the special case of the homogeneous and isotropic universe discussed in Chapter 12. With a suitable interpretation of the behavior of light with respect to Newtonian theory, the models of Milne and McCrea turned out to be surprisingly similar to those of general relativity.

In our discussion of black holes (Chapter 11), we encountered a similar situation. We saw that the gravitational collapse of a spherically symmetric dust cloud, when calculated according to Newton's theory, has features in common with the calculation according to general relativity. Newton's theory fails, however, to reproduce the essential features of a spinning black hole; likewise, it would fail for more complex cosmological models that are without homogeneity and isotropy.

From what has been said, we clearly prefer to use general relativity in the present chapter. This does not establish with certainty, however, that general relativity is the perfect theory for describing cosmology. Preferable though it is to Newton's theory of gravitation, it too has certain drawbacks. We shall return to a discussion of these drawbacks in the next and final chapter.

We saw in Chapter 12 that models that satisfy the cosmological principle and also satisfy Weyl's postulate (see Section 12-4) possess the simplification that a synchronous time t can be defined for all galaxies. The spacetime sections determined by $t =$ constant are homogeneous and isotropic, as is illustrated schematically in Figure 13-1.

THE SPACES OF CONSTANT UNIVERSAL TIME HAVE CONSTANT CURVATURE

Mathematicians have another way of describing the homogeneous isotropic sections of spacetime. They are three-dimensional spaces of constant curvature. The curvature can be of three types: zero (A), positive (B), or negative (C).

For instance, we can understand the concept of positive and negative curvature with the aid of simple examples of two-dimensional spaces. These examples are none other than familar surfaces of special types. A flat surface—that is, the wall of a room or the surface of water at rest in a jar—has *zero* curvature. The surface of a sphere has *positive* curvature that is constant all over the surface. (The surface of an egg also has positive curvature, but it is not constant everywhere). The surface of a saddle, on the other hand, has *negative* curvature that is not constant all over the surface. A surface of constant negative curvature can be obtained by rotating a rectangular hyperbola about one of its axes to produce the hornlike shape illustrated in Figure 13-2.

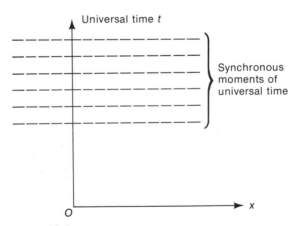

FIGURE 13-1
A system of universal time can be determined provided each observer sets the zero point of his clock in an agreed upon way. The dotted lines are spatial sections of spacetime and are homogeneous and isotropic.

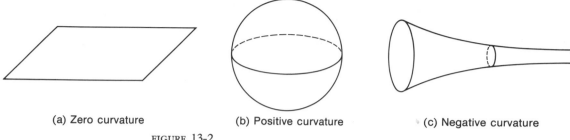

(a) Zero curvature   (b) Positive curvature   (c) Negative curvature

FIGURE 13-2
Examples of two-dimensional spaces of constant curvature.

There is a simple experiment for testing the kind of curvature a surface has at any given point on it. Take a small piece of paper, place it on the point in question, and try to cover the neighboring area of the surface entirely by the paper. If the paper neatly fits the area, we have case A. If the paper is wrinkled, we have case B. If the paper is torn, we have case C.

It is difficult to visualize the curvature of higher-dimensional spaces, but it can be readily described mathematically. We can appreciate, with the help of the preceding examples, that the three possibilities A, B, and C exist. Of these, the first possibility gives simple Euclidean geometry on the spatial sections of Figure 13-1 determined by $t =$ constant, whereas possibilities B and C require non-Euclidean geometries (see Chapter 10) on these spatial sections. Returning to Figure 13-1 and recalling the observational fact that the universe is expanding, we now have the following problem: we need to find the scale factor $Q(t)$ that tells us how the geometry within the spatial sections changes with time. Note that the phenomenon of expansion does not alter the type of space. A space belonging to a particular type, A, B, or C, will continue to belong to the same type. In B and C, the curvature decreases in magnitude as the space expands. In A, the curvature always remains zero, but the distance between any two galaxies increases as the space expands.

THE GENERAL THEORY OF RELATIVITY GIVES INFORMATION ABOUT THE EXPANSION FACTOR Q

In 1924, the Russian astronomer A. Friedmann considered the problem of determining the function $Q(t)$ with the help of Einstein's equations of gravitation. He assumed that the matter in the universe had negligible pressure. This assumption would be justified if we were in the well-ordered universe of Figure 12-22 rather than in the turbulent universe of Figure 12-23. Pressure-free matter is often called dust, and for this reason the cosmological models of Friedmann are often called *dust models*. The aim of the models was to obtain complete information on the large-scale structure of the universe. We discuss how far this aim was successful.

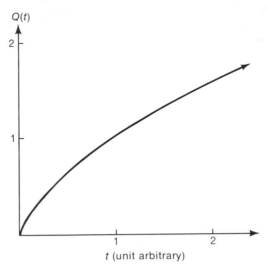

FIGURE 13-3
The behavior of $Q(t)$ according to Einstein's gravitational theory for a large-scale geometry of type A. The unit for $Q(t)$ is chosen so that $Q = 1$ at $t = 1$.

Einstein's theory places restrictions on the large-scale structure of the universe, but it still does not suffice to lead to a unique choice for the geometry of the universe. Particularly, it does not settle the form of geometry to be applied to the spaces $t = $ constant, the forms we called A, B, and C ($A \equiv$ Euclid). It does show, however, that if A is the appropriate form, the galaxies continue to expand away from each other forever. Indeed, in this case Einstein's theory determines an explicit result for $Q(t)$, shown in Figure 13-3. *By using the moment when $Q = 0$ as the zero point for fixing the clocks of all observers in all galaxies,* which is equivalent to the procedure suggested in Section 12-4, it can be shown that Q is proportional to $t^{2/3}$ [that is, to the square of the cube root of t: $(t^{1/3})^2$].

THE INFORMATION ABOUT Q IS BY NO MEANS COMPLETE

Einstein's theory is less explicit for the more difficult geometries denoted by B and C, but it does show that in B, the curve of $Q(t)$ turns over in the manner of Figure 13-4. The curve has a symmetrical form. Following the maximum of the scale factor $Q(t)$, the intergalactic distances shrink, the shrinkage being exactly the reverse of their earlier expansion. For the remaining geometrical possibility C, there is no reversal of $Q(t)$. The three cases are shown together in Figure 13-4, it being supposed that the convention of setting the zero points of all clocks so that $t = 0$ corresponds to the moment when $Q = 0$ is adopted, and that this is done irrespective of whether the geometry is A, B, or C. Notice that,

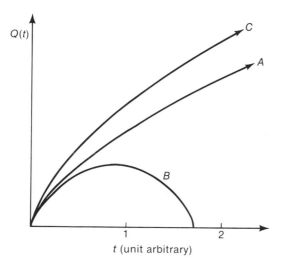

FIGURE 13-4

The behavior of $Q(t)$ according to Einstein's gravitational theory for the three forms of large-scale geometry: A, B, and C. The unit for $Q(t)$ is chosen so that $Q = 1$ at $t = 1$, but this situation will require a different scale on the vertical axis for each of the three cases.

whereas the behavior of Q for geometry A is definitely known, many possibilities remain for geometries B and C. Although we can say that B and C have the general properties shown in Figure 13-4, Einstein's theory does not permit us to distinguish between the possibilities shown, for example, in Figure 13-5.

We shall restrict much of our discussion to case A, partly because this case has a unique behavior of the scale factor Q, partly because its Euclidean treatment is simple, and partly for certain very general reasons that emerge in Chapter 14. This model for the universe is often referred to as the *Einstein–de Sitter model*, and we usually refer to it by this name.

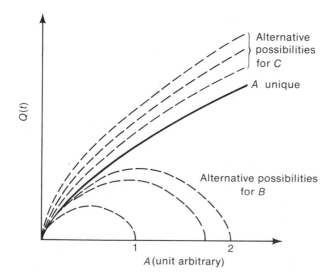

FIGURE 13-5

The behavior of $Q(t)$ is not uniquely determined when the large-scale geometry is either of type B or of type C. The unit for $Q(t)$ is chosen so that $Q = 1$ at $t = 1$.

The Big-Bang Universe

We now consider the possibility that new observations, going outside our local region of the universe, might give further information on the large-scale structure of the universe. Since in principle there is a way to distinguish between the geometrical forms *A*, *B*, and *C* using the observations of very distant galaxies, this topic is important for both theoretical and practical reasons. We continue this chapter by considering the situation in some detail.

THE GREATER OBSERVED DISTANCES BECOME, THE MORE IMPORTANT IT IS TO KNOW THE CORRECT GEOMETRY OF THE UNIVERSE

In Section 12-2, we saw how to represent galaxies in a diagram with observed apparent magnitudes plotted horizontally and with the Doppler velocity v plotted logarithmically as ordinate. It is undesirable now to plot Doppler velocities, however, because the Doppler interpretation of the observed redshift of the galaxies belongs to the geometry of special relativity. We wish now to extend our discussion to much greater distances over which the simple geometry of special relativity may not hold. All we need do to avoid using $v = cz$ is to plot the observed redshift z itself.

The nature of the redshift has been discussed previously, but it will be useful to discuss it again. Characteristic line radiations from atoms—the H and K lines of singly ionized calcium atoms, for example—are observed for a particular galaxy. Denote the measured frequencies by v'_H and v'_K, respectively. These frequencies are different from those observed for ionized calcium atoms in the terrestrial laboratories, say, v_H, v_K. However, the ratio v'_H/v'_K is the same as v_H/v_K. Define $1 + z = v_K/v'_K = v_H/v'_H$. Since v_K is larger than v'_K, and v_H is larger than v'_H, the redshift z is a positive number.

On the assumption that all the galaxies under investigation are intrinsically similar to one another, the Einstein–de Sitter model leads to the relation between z and apparent magnitude M shown in Figure 13-6.* In the next chapter, we see how to obtain this particular relation. At small values of z, the 45° line of Figure 13-6 is essentially the straight line of Figure 12-14, but, at large z, the curve leans to the right of this line.

There are corresponding curves for geometries of classes *B* and *C*. Those of class *C* lean still farther to the right of the 45° line, between the curve of Figure 13-6, drawn again in Figure 13-7, and a second curve also drawn in Figure 13-7. Curves for geometries of class *B* take the form shown in Figure 13-8. At large z, such curves lie markedly above that of the Einstein–de Sitter model. Thus, the curve of the Einstein–de Sitter model separates those of class *B* from those of class *C* as shown in Figure 13-9.

---

*In earlier chapters, *m* was used to denote an apparent magnitude. Here we switch to *M* so that *m* can later be used to denote the mass of a particle.

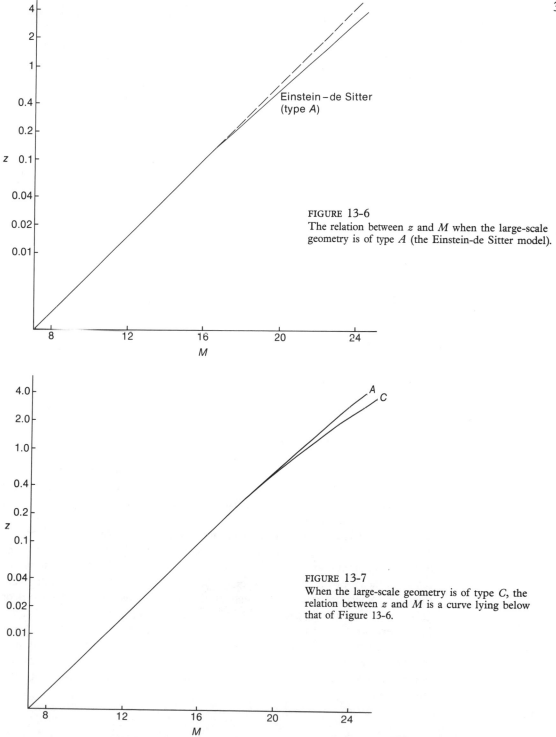

FIGURE 13-6
The relation between z and M when the large-scale geometry is of type A (the Einstein-de Sitter model).

FIGURE 13-7
When the large-scale geometry is of type C, the relation between z and M is a curve lying below that of Figure 13-6.

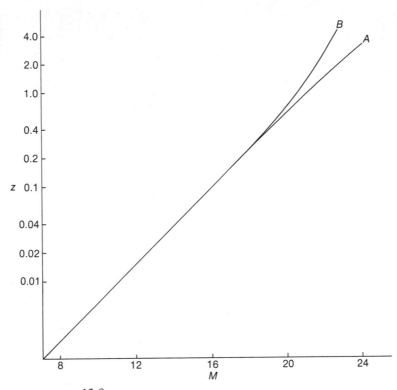

FIGURE 13-8
When the large-scale geometry is of type B, the relation between z and M is a curve lying above that of Figure 13-6. Otherwise, the relation is not determined in detail.

The 45° part of Figure 13-9 applies within the limited range of local geometry. At larger distances, there is in general a departure from this straight line, depending on the nature of the large-scale geometry. If indeed the galaxies observed at great distances were found to lie on the 45° line as appears to be indicated by Figures 13-10 and 13-11, and if the galaxies were indeed intrinsically similar to each other, then the world geometry would have to be of class B. Some astronomers do indeed believe Figures 13-10 and 13-11 show the large-scale geometry of the universe to be of class B. This is the basis of statements that are sometimes made to the effect that the expansion of the system of galaxies will eventually cease and be replaced by a contraction of the universe. However, many other astronomers feel that the difference between the 45° line and the curve of Figure 13-6 is so small, even for z as large as one-third, that this conclusion cannot be considered reliable. Small intrinsic differences between the observed galaxies in Figures 13-10 and 13-11 would vitiate the conclusion.

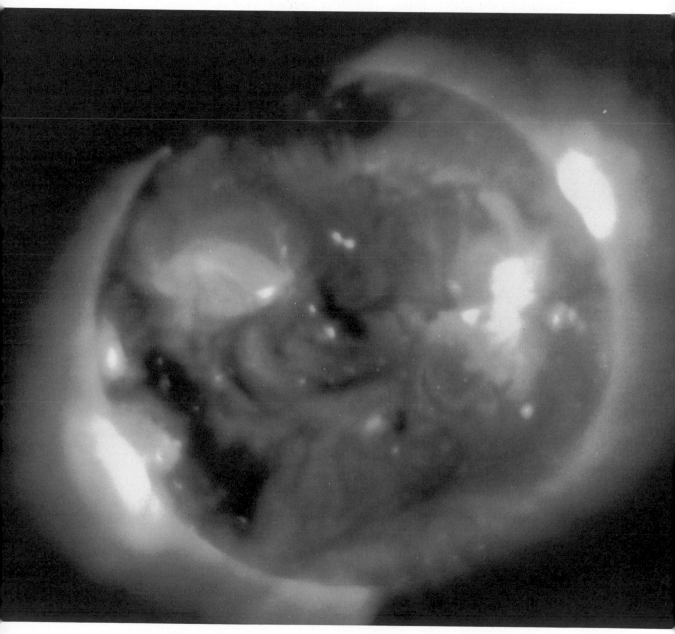

PLATE VI.
X rays emitted by the hot corona of the Sun.
(Courtesy of American Science and Engineering, Inc.)

PLATE VII.
The Fermilab particle accelerator at Batavia, Illinois.
(Courtesy of the Fermilab.)

A

B

PLATE VIII.
The Fermilab particle accelerator (A)
and the central building (B).
(Courtesy of the Fermilab.)

PLATE IX.
Energy from the Sun maintains the biosphere of the Earth.
(Apollo 16, courtesy of NASA.)

PLATE X.
The life-covered Earth. (Courtesy of NASA.)

PLATE I.
The Crab nebula. (Courtesy of the Lick Observatory.)

PLATE II.
The Orion nebula, a cloud of gas in which stars are now forming.
(Courtesy of the Lick Observatory.)

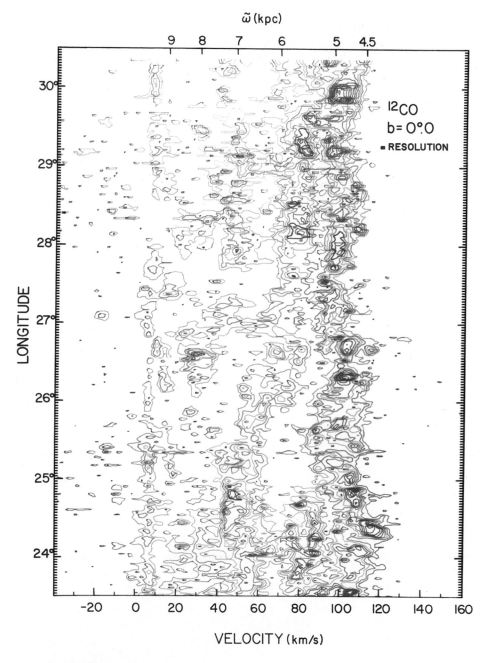

PLATE III.
Radiation from carbon monoxide at a wavelength of 2.6 mm reveals molecular clouds with density levels indicated by contours. The top scale is distance from the galactic center in kiloparsecs, and the bottom scale is the Doppler velocity of the clouds. The scale at left is longitude in galactic coordinates (galactic latitude is zero). (Courtesy of P. M. Solomon, D. Sanders, and N. Z. Scoville.)

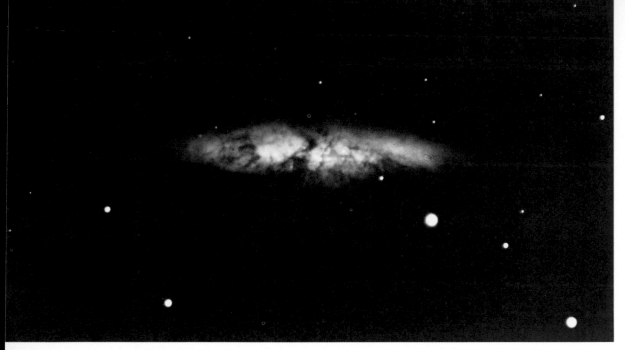

PLATE IV.
The galaxy M82. (Courtesy of the Hale Observatories.)

PLATE V.
The Ring nebula (NGC 6720), a planetary nebula.
(Courtesy of the Hale Observatories.)

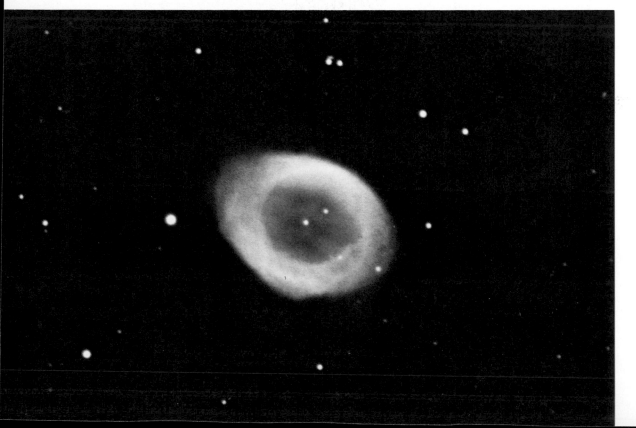

If observations could be carried to values of z significantly larger than those of the galaxies in Figures 13-10 and 13-11, it would be less difficult to decide this matter because the curve of Figure 13-6 for the Einstein–de Sitter model leans more and more away from the 45° line as z increases. If the 45°-line dependence of Figures 13-10 and 13-11 were found to be maintained up to $z = 1$, for example, the argument for regarding the universal geometry as class B would be strong. Such an extension to large z is technically difficult to achieve for galaxies, simply because galaxies at large z are very distant and therefore exceedingly faint.

When the quasi-stellar objects (QSOs) were first discovered, it was thought that they would soon lead to a resolution of this issue, because many QSOs were found to have the required large z values, z greater than unity in many cases. It will be clear from Figure 13-12, which gives the z-M relationship for some 250 QSOs, why this hope has not been realized. This figure shows that the QSOs have a very large variation among themselves and thus do not satisfy the essential condition of intrinsic similarity. In spite of this disappointment, it is likely that attempts will be made to push the observations of Figures 13-10 and

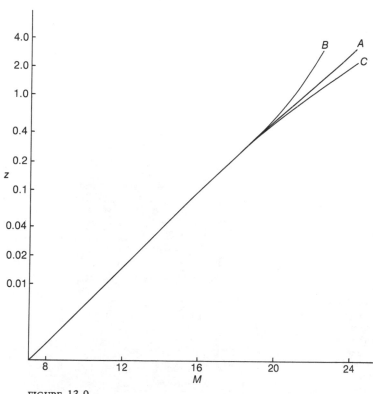

FIGURE 13-9
The form of the z-M relation for the three types of world geometry.

374

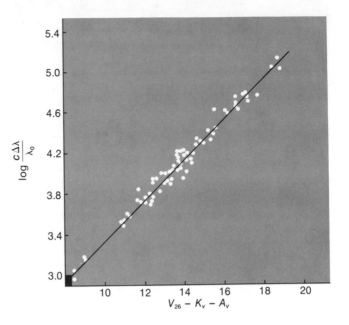

FIGURE 13-10

Since $z = \Delta\lambda/\lambda_0$, the left-hand scale is log (cz). The bottom scale gives the apparent visual magnitude of the galaxies (the brightest galaxy in each of 84 clusters) after certain corrections represented by K_v, A_v have been applied. (Reprinted courtesy of A.R. Sandage and *The Astrophysical Journal*, published by the University of Chicago Press; © 1972 The American Astronomical Society.)

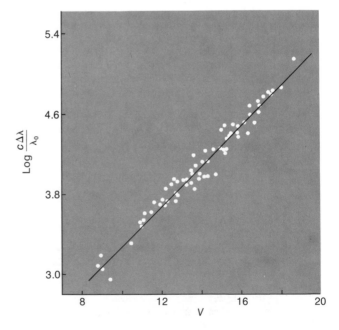

FIGURE 13-11

A diagram similar to Figure 13-10 for radio galaxies. For certain technical reasons, the corrections K_v, A_v have not been applied in this case. (Reprinted courtesy of A.R. Sandage and *The Astrophysical Journal*, published by the University of Chicago Press; © 1972 The American Astronomical Society.)

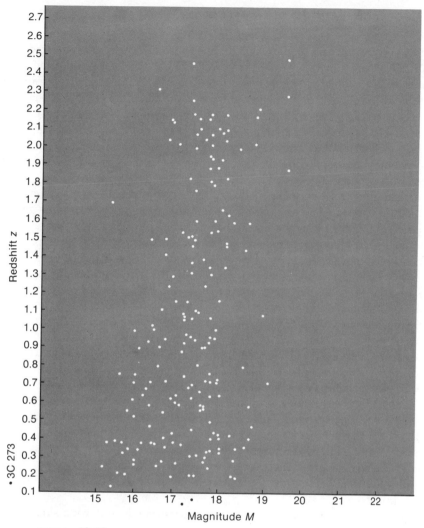

FIGURE 13-12

The early redshift-magnitude relation for some 250 quasi-stellar objects. A later plot for over 700 QSOs exhibits a similarly large scatter. Notice that z is plotted directly, not logarithmically.

13-11 to larger z values. Many new, large telescopes are at present under construction throughout the world, and more observing time will become available. New methods will become more sensitive than older methods. The prize of determining the nature of the universal geometry is a rich one, and astronomers are not likely to be deterred for very long by technical difficulties from attempting to seize it.

Just as optical astronomers can measure the apparent magnitude of an object, so can radio astronomers measure the power from a radio source by means of a radio receiver tuned at some chosen frequency. We are all familiar, from our everyday experience, with the general procedure to be used. We find that any practical receiver does not tune at just one exact frequency; it is always possible to hear a radio program even though the receiver is mistuned by a small amount. This small amount is known as the *bandwidth* of the receiver. What radio astronomers do is measure the power from a source using a receiver with a certain standard bandwidth.

In practice, radio astronomers do not have a free choice of the frequency at which they work, simply because so much man-made radio emission is generated all the time all over the Earth. There would be little chance of measuring the radio waves from a cosmic object if there were even very slight competition from a man-made source, since by everyday standards the power received from a cosmic source is very small indeed just because astronomical distances are so very large. It has been estimated that all the radio power received by all the world's radiotelescopes operating for a decade would not raise the temperature of a spoonful of water by as much as a millionth of a degree. So what radio astronomers must do is work at some frequency that nobody else is using. Such special frequencies must be agreed upon by international arrangement. Indeed, radio astronomers have been allotted a few special frequencies, and it is at these special frequencies that all observations in radio astronomy must be made.

Using one of the frequencies allotted to them, radio astronomers can measure the power received from a source. And they can survey the sky, or a portion of it, to count the number of sources that are more powerful than some assigned amount, say S. That is, they count the number of small patches of the sky that give power readings in their receivers that are greater than S. Denote this number by $N(S)$. How do we expect N to behave as S is varied in this experiment?

This question can readily be answered if we make the following simplifying assumptions:

1. All radio sources are intrinsically alike.
2. The radio sources are distributed uniformly in space.
3. The geometric properties of spacetime over the range of the survey are those of the special theory of relativity.

Making a logarithmic plot of N versus S, we find that the expected behavior of N follows the simple straight line of Figure 13-13. The number N increases as S becomes smaller because smaller S implies increasing distance, and there are many more sources at large distances than there are close by.

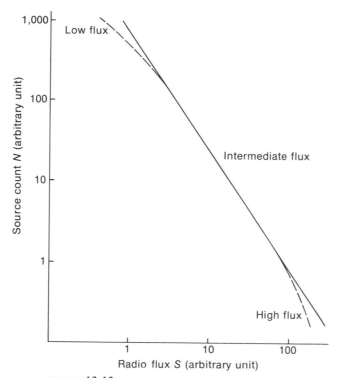

FIGURE 13-13
The straight-line behavior of the number N of radio sources, more powerful than some assigned S, expected on the assumptions of intrinsic similarity and of uniformity of distribution and on the hypothesis that spacetime has large-scale properties similar to its small-scale local properties. Evidence for this kind of behavior is shown in Figure 13-14.

ACTUAL COUNTS OF RADIO SOURCES REVEAL AN INDECISIVE BUT MYSTERIOUS SITUATION

The results of the radio source count carried out at the National Radio Astronomy Observatory at Greenbank, West Virginia, are shown in Figure 13-14. The observations agree very well with the expected straight line of Figure 13-13 except at high and low values of S. Throughout a considerable intermediate range of S, the agreement is good. The departure of the observed counts from the strict straight line of Figure 13-13 is significant and important at low values of S, but it is not important at high values of S. The total number of sources affecting the high S end of Figure 13-14 is about 400. The deviation from the expected straight line of Figure 13-13 would be removed if about 40 more sources were added to this 400. Now, the square root of 400, namely, 20, represents what is called a *standard deviation*, so about two standard devia-

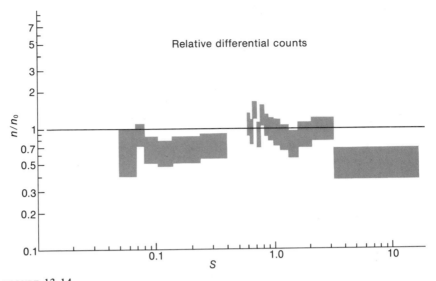

FIGURE 13-14

Radio-source counts in which the range of S has been divided into bins in such a way that equal numbers of sources would be found in each bin *if* the distribution were of the form of the straight line of Figure 13-13. The left-hand scale gives the ratio of the actual count n for each bin to a constant number n_0 that is arbitrarily chosen. The bin at the highest range of S has a deficit of sources. Otherwise, there is no appreciable deviation from the straight-line situation of Figure 13-13 until the left-hand row of bins is reached. The results for these left-hand bins show that $N(S)$ falls off at low S, as indicated by a dotted line in Figure 13-13. (Reprinted courtesy of K. Kellerman, M. Davis, and I. Pauliny Toth and *The Astrophysical Journal*, published by the University of Chicago Press; © 1971 The American Astronomical Society.)

tions are involved in the deviation of the observations at high S from the expected straight line of Figure 13–13. This deviation is well within the range of a normal statistical fluctuation and hence of no far-reaching significance.

The tendency of the observed points to fall below the straight line of Figure 13-13 at low values of S is not a statistical fluctuation, however. The number $N(S)$ of sources at low S is so large that fluctuations have little effect. The falloff of the observed points must be ascribed at low S to a failure of one or more of the preceding three assumptions. It is usually supposed that, at the large distances implied by low values of S, it is assumption 3 that fails. That is, the geometric structure of spacetime, when taken on a large enough scale, is different from the geometry of which we have local experience. This situation is precisely what we expect from the discussion in Section 13-2.

We end this section by discussing a strange and mysterious situation, one that springs on us in the very place where the observations of Figure 13-14 seem to fit the expected line of Figure 13-13, at the intermediate value of S. To come to grips with this issue, we notice first that most of the sources contributing to $N(S)$ are radio galaxies. The QSOs make about a 15 percent contribution to the source counts, and this contribution is not sufficient to have a critical effect on the behavior of N with respect to S.

Now, for galaxies that are not too remote, we can be assured that, because of Hubble's law, the measurements of z values lead to their distances D given by using the line of Figure 13-15. (Notice that this line is established from pairs of z and D values derived from observations of galaxies. Then, with the line established, we no longer trouble ourselves to measure D values any more. Since measuring z is relatively easier than measuring D, what is done—with z known—is simply to read off D in the manner of Figure 13-15.) If z values were available for all the radio galaxies counted at intermediate S values, those for which the behavior of $N(S)$ agrees with the line of Figure 13-13, we would thus know the D values for all these radio galaxies. This procedure would yield the general distance scale corresponding to the intermediate S values and would enable us to assess assumption 3 in relation to this distance scale.

What has actually been done so far is to measure z values for a modest fraction of the radio galaxies that contribute to N at the intermediate values of S. These measured z values lead to quite large D values, indeed, to D values so large that it is a surprise that assumption 3 would seem to hold good on so large a scale. Such a situation would involve a major reassessment of all our ideas on cosmology, a radical position that astronomers would be reluctant to accept unless the alternatives were first eliminated. The alternative most favored by

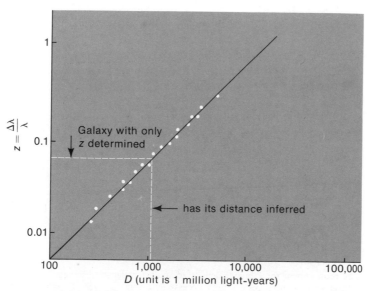

FIGURE 13-15
With the relation between z and D established by observing a number of galaxies, distance values for other galaxies are simply read from the resulting curve once an observed redshift is obtained.

astronomers comes from abandoning assumption 2, the one used in arriving at the straight line of Figure 13-13. Then we could not arrive at this straight line (which actually fits the data at the intermediate S values) *unless assumption 3 were also invalid*. The idea is to have *both* 2 and 3 invalid, but in such a way that the expected behavior of N with respect to S continues, at the intermediate values of S, to follow the straight line of Figure 13-13—a fortuitous compensation for two errors in the previous argument!

This latter alternative requires assumption 2 to break down in the following sense: we must have a higher density of radio sources at large distances than at smaller distances. Does this mean that our own galaxy is situated at the center of some kind of hole in the distribution of radio sources? Yes, but not a spatial hole, a *time hole* of the kind illustrated in Figure 13-16. As we look to greater distances, we look farther back in time. We require sources to have been more frequent in the past than they are today. In the manner of Figure 13-16, we still preserve uniformity in space *at any given moment of time*.

This favored interpretation of the situation is often referred to as an *evolutionary universe* because the distribution of radio sources, although uniform at any given moment, is considered to change with time. As a means of fitting the so-far available z value measurements for a fraction of the radio galaxies to our ideas of the permitted range of local geometry and to the observations of Figure 13-14 at intermediate S values, such an evolutionary universe has one peculiar

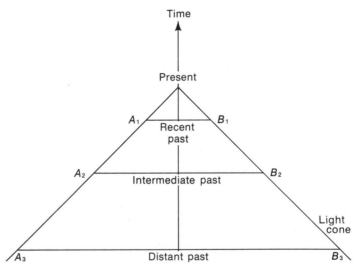

FIGURE 13-16
The requirement for a higher density of sources at larger distances rather than at smaller distances does not necessarily imply any lack of spatial uniformity. We can see more sources at large distances because we are then looking farther back in time. Here the density of radio sources is the same at A_1 as it is at B_1, but the density there is less than at A_2 and B_2, and the density at A_2 and B_2 is in turn less than it is at A_3 and B_3.

defect, namely, the fortuitous compensation just mentioned. At present, there is no plausible theory to explain this compensation. The usual view is that the compensation happens accidentally. A less usual view is that, unlike the minority for which z values are available, most of the sources with intermediate S values are not very far away, and so their optical faintness arises because they happen to be intrinsically faint. Although this further possibility contradicts a mystique that has grown up in radio astronomy, namely, that all radio galaxies are optically very bright and similar to each other, it satisfies the facts without requiring any coincidence. At present, it is the most straightforward way to resolve the problem. It amounts simply to taking the straight-line part of Figure 13-14 at its face value, as showing us that most of the sources at intermediate S values are comparatively close by, within the range of applicability of the local structure of spacetime. If this is so, the observations of Figure 13-14 cannot be considered to have far-reaching consequences for the study of the large-scale properties of the universe.

§13-5. *The Angular Size Test*

In 1958, one of the authors suggested another test that makes use of the techniques of radio astronomy to verify the predictions of the non-Euclidean geometry in a Friedmann universe. This test involves looking at the angular sizes of radio galaxies situated very far away from us.

To understand how this test works, let us take again a simplified example of a two-dimensional space. Imagine a flat creature crawling in Euclidean flat land, say, on a plane floor (see Figure 13-17). Suppose the creature to be looking at a linear object, say, a stick of length l situated at a distance D. What angle will the stick subtend at the observer? Let us simplify the problem still further, and assume that the stick's distance D is very large (more than 100 times) compared to its length l. Then it is a good approximation to say that the angle α subtended by the stick at the observer is given by

$$\alpha = \frac{l}{D} \text{ (radians)}.$$

(To obtain the angle in degrees, multiply by $180/\pi$.)

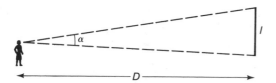

FIGURE 13-17
The object of length l subtends an angle $\alpha \approx l/D$ at the observer if D is very large compared to l.

Thus, as D increases—that is, as the stick recedes from the observer—the angle α decreases in inverse proportion to D. This result is the outcome of the Euclidean geometry that operates in the flatland.

Now place this same flat creature on the surface of a sphere and let the stick (whose length l is small compared to the radius of the sphere) again move away from the observer as in the first case. How will the angle subtended change with distance now?

Assuming that the stick is viewed with light, we note that light travels in straight lines, that is, along the arcs of the great circles on the sphere. Without loss of generality, we can locate the observer at the North Pole and take the stick away from him such that the stick always lies along a latitude circle with one end always on the same meridian, which we can take to be the Greenwich meridian. The other end of the stick does not always lie along the same meridian, however, and the angle α in this case is the angle between the two meridians through the ends of the stick. Notice that as the stick decreases in latitude in its progress toward the equator, the angle α decreases. So just as in Euclidean geometry, the angle subtended by the stick will get smaller and smaller as the stick moves away.

A careful observer will notice, however, that the rate of decrease of α with distance is not as rapid in this case as it is in the Euclidean case! In fact, the decrease will be slower and slower as the stick approaches the equator. It vanishes at the equator and becomes negative thereafter. In other words, α starts *increasing* beyond the equator. This situation is illustrated in Figure 13-18. The broken great circle makes a larger angle with the zero meridian than the continuous one. Figure 13-19 illustrates the two cases we have just discussed.

To return to cosmology, we now replace two-dimensional space by the four-dimensional spacetime that constitutes the Friedmann universe. With what

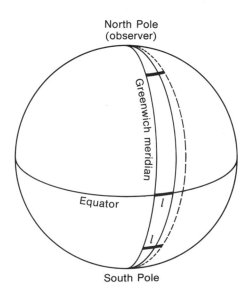

FIGURE 13-18
The angle subtended by the stick at the observer located on the North Pole is a minimum when the stick is at the equator. It is larger if the stick is farther away from the equator. This is the consequence of non-Euclidean geometry.

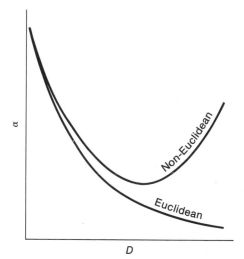

FIGURE 13-19
The variation of α with D in the cases of the Euclidean geometry on the plane and non-Euclidean geometry on the surface of a sphere.

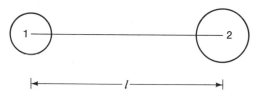

FIGURE 13-20
The separation between the two components of a typical extragalactic double radio source is $l \approx 200$ kpc.

should we replace the stick? A convenient linear structure is a typical extragalactic radio source. As illustrated in Figure 13-20, the separation between the two blobs of radio emission is on the order of 200 kpc. Thus, at a distance of 2000 mpc, the angle subtended by the length l (assumed perpendicular to the line of sight) is on the order of

$$\alpha \approx \frac{200 \text{ kpc}}{2000 \text{ mpc}} \simeq 10^{-4} \text{ radians} \simeq 20^{s},$$

provided, of course, that Euclidean geometry holds. However, in the Friedmann models, this geometry does not hold. Even in case A, where spatial sections are Euclidean, the spacetime geometry of the *four-dimensional* universe is non-Euclidean because these sections expand with time. In Figure 13-21, we show

384

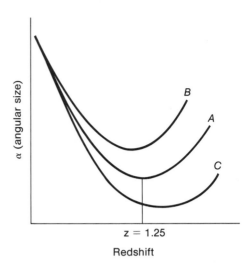

FIGURE 13-21
The angular size plotted against redshift for a family of radio sources of fixed linear size and varying redshifts, drawn separately for the three types A, B, and C of the Friedmann model. The minimum value of α for type A occurs at $z = 1.25$.

how the angular size should vary in the Einstein–deSitter model, as well as in models of type B and C.

Notice that, in all the cases, we have plotted α against z, the redshift of the radio source. In all cases, α has a *minimum* value at a certain value of z. For the Einstein–deSitter model, the minimum occurs at $z = 1.25$. For a B-type model, it occurs at a smaller value of z, whereas for a C-type model, it occurs at $z > 1.25$.

In principle, therefore, this method is a simple one for determining the geometrical type of the spacetime sections, $t = $ constant. In practice, however, several difficulties intervene to make the test far from decisive. Let us look at some of them just to familiarize ourselves with the various uncertainties.

First, we do not always see a radio source with its linear structure perpendicular to the line of sight. Thus, what we see is the projection of the actual source along the transverse direction. Second, not all sources are of the same linear size. The linear size varies considerably between, say, 50 kpc and 500 kpc, and the preceding figure of 200 kpc was only a representative one. Both these effects cause a considerable scatter around the expected curves of Figure 13-21, with the result that it becomes difficult to single out one curve as being the natural (or best) fit to the data.

There is the further difficulty that if the universe is not truly homogeneous but has aggregates of matter in the form of clusters and superclusters, then even the theoretical predictions of Figure 13-21 break down to some degree. The curves shown in the figure become modified, subject to the extent of the inhomgeneity present in the universe, because the gravitational effect of inhomogeneities is to produce a scattering of light rays.

Finally, there is the difficulty of measuring z for radio sources. As mentioned in Section 13-4, the distance of a radio source cannot be directly inferred from its flux density S unless special assumptions are made about its intrinsic power.

A more reliable procedure is that of Figure 13-15: to identify the radio source with an optical galaxy and infer the redshift from the apparent magnitude, or, better still, to measure z for the galaxy directly. This procedure is slow and cannot be applied straightaway to all the radio sources for which α is available.

On the positive side, we can add that the improvement in the sensitivity of radiotelescopes has made the actual measurement of angular size α an easy and accurate procedure. Since 1958, considerable data now exist on α from various radiotelescopes around the world. The first extensive results were presented in 1975 by R. Ekers, and a further large quantity of data has come from the Radio Astronomy Observatory at Ootacamond in Southern India, where the large collecting area at low latitude makes the telescope (see Figure 13-22) well suited to measuring angular sizes by a technique using lunar occultations.

THE ANGULAR SIZE TEST IS ALSO INDECISIVE

The conclusion drawn by both Ekers and the Ooty radio astronomers Govind Swarup and Vijay Kapahi from their first assessment of the data is that the angular size α goes on decreasing without showing a minimum as in Figure 13-21. How should we interpret this result? Should we abandon the Friedmann models and opt for a Euclidean geometry that gives such a continually decreasing result? As in the case of the radio-source counts, this view might be the reasonable one to take. However, as in the case of the radio-source counts, the interpretation favored by astronomers is the one that introduces the concept of an evolutionary universe and argues that the radio sources were of smaller

FIGURE 13-22
The linear array of parabolic antennas that constitute the radiotelescope at Ootacamond in South India. (Photo by Bharat Upadhyaya.)

intrinsic size in the past than they are now. Thus, we again have a fortuitous compensation working, with the smaller linear sizes in the past just happening to compensate for the high redshift turn up of α predicted by the Friedmann model curves of Figure 13-21. In view of the scatter in the data, it is perhaps wisest to look at this test again when more reliable distance indicators for radio sources are likely to become available.

§13-6. The Early Universe

The three tests developed so far in this chapter are hardly effective in telling us what model of the universe to adopt. These tests probe the past history of the universe out to distances corresponding to redshifts of $z \sim 1$ or higher, up to $z \sim 3$ if we consider the redshifts of QSOs to be due to the expansion of the universe.

For example, to see how the redshift is related to past epochs, consider the following situation. Suppose a light wave emitted by a distant galaxy at time t_e reaches us at the present epoch t_p. How much will the redshift be for this light wave? A simple calculation involving light propagation in the non-Euclidean geometry of the Friedmann models gives the formula:

$$1 + z = \frac{Q(t_p)}{Q(t_e)}.$$

In an expanding universe, the function $Q(t)$ increases with t, and hence $Q(t_p) > Q(t_e)$. The redshift is therefore always positive.

Now, consider the Einstein–de Sitter model, for which the preceding formula becomes

$$1 + z = \frac{t_p^{2/3}}{t_e^{2/3}}.$$

If we take $z = 1.25$, say, then we get $t_e = \frac{8}{27}t_p$, which means that when we observe a galaxy of redshift 1.25, we are looking at it at an epoch when the universe was only $\frac{8}{27}$ of its present age. From here on, we often identify past epochs by their redshifts.

The tests described so far attempt to find changes in the universe over fractions of its present age. In an evolutionary universe, the physical nature of its constituents (for example, luminosity of galaxies, density of radio sources, sizes of radio sources, and so on) is supposed to have changed significantly over these time spans.

If we push the time span further back into the past, do we expect to see drastic differences in the state of the universe? Naturally, the Friedmann models lead us to answer this question in the affirmative. If we let t decrease toward zero, $Q(t) \longrightarrow 0$, we arrive at the moment of the *big bang*—the primeval explosion

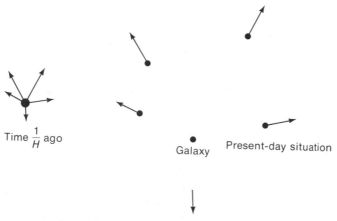

Time $\frac{1}{H}$ ago

Galaxy   Present-day situation

FIGURE 13-23
The quantity $1/H$ is the length of time, going backward, that would be required to bring all the nearby galaxies together if they have always been moving apart at their present speeds.

when the universe is believed to have come into existence. All the matter that we now see in the form of galaxies, QSOs, radio sources, and so forth must have been tightly packed in the early era. The big bang must have happened at a time $\sim 1/H$ ago (see Figure 13-23).

IT IS IMPORTANT TO OBTAIN OBSERVATIONAL INFORMATION CONCERNING THE EARLIEST PHASES OF THE UNIVERSE

The tests described so far do not take us far back in the past. We may therefore ask: Are there other observational indications that could tell us about the universe in the very early stages? To answer this question, we first extrapolate the Friedmann models backward in time.

Let us look first at dust models and ask: What was the density of matter like in the past? Suppose the present density of matter is ρ_p. If the present density is ρ_p, the amount of matter contained in a box of volume V is $V\rho_p$. Now, in the past the box was of the same shape but of smaller size because the universe has expanded. In fact, as space stretches, the linear scale of the box changes in proportion to the expansion factor $Q(t)$. Thus, the volume of the box at an earlier epoch t_1 was

$$V \left[\frac{Q(t_1)}{Q(t_p)} \right]^3 .$$

If, during the period from t_1 to t_p, the box simply expanded and no matter entered or left through its faces, then we see that at epoch t_1, the density ρ_1 was given by the equation

$$\rho_1 \times V\left[\frac{Q(t_1)}{Q(t_p)}\right]^3 = V\rho_p.$$

This relation simply means that the density in the past was higher by a factor equal to the cube of the ratio of the present expansion factor to the past one. Or, to put it in another way, if the redshift of a galaxy at epoch t_1 is now seen to be z_1, then

$$\rho_1 = \rho_p(1 + z_1)^3.$$

What would happen to the density of electromagnetic radiation in the course of expansion of the universe? We have thus far ignored the effect of radiation on the dynamics of the model because at present the energy density u of radiation produces a negligible effect on the rate of expansion of the universe. To express the situation mathematically,

$$u_p \ll \rho_p c^2,$$

c being the speed of light. However, we can still ask: How much larger was the radiation density in the past than it is now? The answer comes out to be*

$$u_1 = u_p(1 + z_1)^4,$$

that is, at an epoch of redshift z_1, the density of radiation u_1 was higher than the present density by a factor $(1 + z_1)^4$.

We have now encountered a curious situation that is best illustrated by Figure 13-24. In Figure 13-24, we have plotted the behavior of the matter and radiation densities on a logarithmic scale. The slope of the *matter line* is 3, while that of the *radiation line* is 4. Thus, even though the former is now at a higher position ($\rho_p c^2 > u_p$), as we consider the epochs further back in the past, the latter overtakes it at the point of intersection of the two lines that we have shown to occur at a redshift of z_c. This critical redshift corresponds to a critical past epoch t_c given by

$$Q(t_c)(1 + z_c) = Q(t_p).$$

As an example, in the Einstein–de Sitter model we have

$$u_1 = u_p\left(\frac{t_p}{t_1}\right)^{8/3}, \qquad c^2\rho_1 = c^2\rho_p\left(\frac{t_p}{t_1}\right)^2.$$

---

*The difference between the $(1 + z)^4$ dependence of radiation and the $(1 + z)^3$ dependence of matter is due to the fact that radiation experiences the redshift effect, causing an extra factor $1 + z$ to appear in the energy density of the radiation.

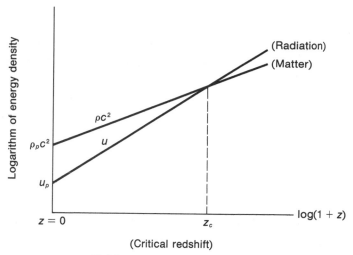

FIGURE 13-24
For epochs earlier than the one denoted by the redshift z_c, the radiation density is higher than the matter energy density.

At $t_1 = t_c$, $u_1 = \rho_1 c^2$. This gives us

$$t_c = t_p \left(\frac{u_p}{c^2 \rho_p} \right)^{3/2}.$$

Thus, we find that even though we were right in ignoring the effect of radiation on the basis of the present-day data, we are not right in ignoring the effect of radiation in the very early stages of a Friedmann universe. Rather, we should invert our priorities and attach more importance to radiation than to matter in the very early stages. The epoch t_c is the transition epoch separating the radiation-dominated early universe from the matter-dominated later universe. Of course, the transition would not have been a crisp and instantaneous one. We expect there to be a grey area around $t = t_c$ when the two energy contributions were comparable. According to the observational determinations of u_p and ρ_p, the redshift around this time would have been about 1,300.

As we saw in Chapter 3, a black-body radiation distribution is assigned a temperature. In the following section, we use temperature more often than redshift to indicate the epochs in the early stages. But is there good reason to believe that the radiation should be in a black-body form? We return to this question later, assuming at present the answer to be affirmative. Making this assumption, we recall that the energy of radiation in each unit volume goes as the fourth power of temperature (see Chapter 3). Hence, we get the simple relation

$$T_1 = T_p(1 + z_1).$$

That is, the radiation temperature was higher in the past compared to its present value by the factor $1 + z_1$. As we go toward the big bang, $Q(t_1) \longrightarrow 0$ and $z_1 \longrightarrow \infty$. Thus, the temperature gets higher and higher and becomes infinite as we near the big-bang epoch, which leads us to the concept of the *hot big bang*.

§13-7. *The Hot Big Bang*

We have so far worked backward in time toward the big bang. We now reverse the procedure and start from the big bang—the hot big bang—and explore the consequences of the early high temperature for the subsequent behavior of the universe.

We begin with the deferred question of whether the radiation in the early stages of the expansion was of a black-body character. As we saw in Chapter 3, black-body radiation arises when we have radiating and absorbing systems operating in an enclosure in such a way that the radiation is not allowed to leave the enclosure. That is, the photons, the carriers of radiation, are not allowed to go far before being scattered or absorbed by matter. Now, in the early universe, just after the big bang, we do not expect matter to have been concentrated into galaxies or other separated aggregates. Rather, we expect the matter to have been distributed in its elementary form as a system of free particles like electrons, protons, neutrons, and so forth. All these particles served to interact with the primordial radiation—to scatter, absorb, and reemit it—so that the system soon took on the black-body form.

At this stage, we make a diversion in our main discussion to consider the so-called random motions of particles of matter. According to the Weyl postulate that we have assumed, there are negligible random motions of the matter in the form of galaxies. A small variable motion would modify the picture of Figure 13-22 to the one illustrated in Figure 13-25. Here the wiggles in the world lines of galaxies correspond to small motions. Calculations show that random motions tend to die out as the universe expands. Conversely, as we look back in time, the random motions of matter in the universe must increase. Since pressure arises from random motions, it follows that although at the present epoch the pressure of matter is negligible, it must have been considerable in the past. It therefore seems unlikely that big structures such as galaxies could have existed in the face of such high pressures. Indeed, it seems more likely that matter was in the form of fast-moving particles with large random motions. Elementary particles with very large random speeds (close to the speed of light) would resemble radiation more than dust.

Thus, we have in the early stages of the universe a hot brew of radiation (photons)—electrons, protons, neutrons, and so forth—all in thermal equilibrium. The brew also contained pairs of particles and antiparticles like electrons and positrons, neutrinos and antineutrinos, and so forth. The overall tempera-

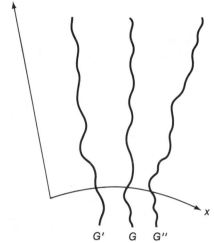

FIGURE 13-25
The wiggles in the world lines of galaxies correspond
to small random motions.

ture of this mixture at about 1 second after the big bang would be as high as
10^{10} K. Let us follow the development of the universe from an *earlier* phase, say,
when it was only 0.01 second old.

At that stage, the universe had a temperature of $\sim 10^{11}$ K. The hot brew
contained all the particles just mentioned, with even neutrinos trapped in
thermal equilibrium. Normally, neutrinos have very little interaction with
matter and can travel undisturbed through several light-years of lead. However,
in the early universe, the density of matter was high enough to hold neutrinos
within the characteristic size of the universe.* The light particles predominate in
this mixture. There is about one proton or neutron for every 1,000 million
photons, electrons, or neutrinos. The predominance of electrons and neutrinos
produce rapid β decay and inverse β decay reactions, for example.

$$\bar{\nu} + p \Longleftrightarrow e^+ + n$$

and

$$\nu + n \Longleftrightarrow e + p.$$

Thus, neutrons constantly change into protons and vice versa, and these
transitions occur so rapidly that the number of neutrons and protons is about the
same.

---

*A typical measure of the characteristic size of the universe is c/H, where H is the value of the
Hubble constant at that epoch. For the early universe, this size is $\sim \frac{1}{2}ct$ where t = age of the
universe.

As the temperature drops with time, the small mass difference between the neutron and the proton becomes significant, and the lower mass of the proton favors its abundance. Consequently, when the temperature has dropped to $\sim 3 \times 10^{10}$ K (the universe is now ~ 0.1 second old), the neutron-to-proton ratio has changed to 0.61 from being near 1.0 at 10^{11} K.

When ~ 1 second has elapsed and the temperature has dropped to 10^{10} K, the density of the universe has also fallen, and neutrinos can no longer be confined in thermal equilibrium. The electron-positron pairs also begin to disappear by mutual annihilation. (At higher temperatures, the annihilation rate was more or less balanced by the creation rate of these pairs.) The neutron-proton ratio is shifted lower to 0.32. However, the neutrons and protons are still too hot and energetic to become bound together by nuclear forces.

When the temperature drops still further to 3×10^{9} K (~ 13.8 seconds after the big bang), it is then cool enough for protons and neutrons to form stable nuclei like helium (He^4). The formation of stable nuclei happens through an intermediate stage of the formation of deuterium (heavy hydrogen), part of the grand scheme conceived first by George Gamow in the late 1940s. Gamow had suggested that the hot big bang and the subsequent conditions in the early universe would be suitable to synthesize the various elements we see today. Starting with hydrogen (p), deuterium (p, n), and helium ($2p$, $2n$), the idea was to go on building heavier nuclei. However, this process could not be carried very far because there are no stable nuclei with five or eight nuclear particles. Later, as stellar astrophysics made progress, it became clear that stars provide much more likely sites for the synthesis of higher elements. Nevertheless, the formation of helium and deuterium in Gamow's cosmological scenario would still be of interest, as we see in Section 13-9.

By the time the temperature of the universe drops to $\sim 3 \times 10^{8}$ K (about 35 minutes after the big bang), the nuclear processes stop, and the proportion of helium to free protons is more or less frozen in the range of from 22% to 28% by mass. There is now one electron for each proton, whether the proton is free, in the form of hydrogen, or bound, in the form of helium. However, the universe is still too hot for electrically neutral atoms to hold together.

The free electrons provide the main blocking agent for radiation. Thus, the universe remains opaque until the radiation temperature drops to ~ 3000 K. At this stage, the chemical binding in atoms is strong enough to hold the electrons predominantly in neutral atoms. With the disappearance of free electrons from the primordial brew, the radiation can travel long distances; that is, the universe becomes optically transparent. By a curious coincidence, this temperature occurs close to the epoch t_c we mentioned earlier—the epoch at which the universe changes over from the radiation-dominated phase to the matter-dominated phase.

Table 13-1 summarizes the essential features of the early universe that we have discussed so far. We now consider possible observational checks on this picture of the early universe.

TABLE 13-1
The important developments of the early universe

Age	Temperature	State of matter	Comments
10^{-2} second	10^{11} K	$n, p, e^-, e^+, \nu, \bar{\nu}$ in thermal equilibrium	$[n]:[p] = 50:50$
10^{-1} second	3×10^{10} K	same as above	$[n]:[p] = 38:62$
1 second	10^{10} K	n, p, e^-, e^+ and photons in thermal equilibrium	neutrinos escape $e^- \text{-} e^+$ annihilations begin to dominate creations $[n]:[p] = 24:76$
13.8 seconds	3×10^9 K	deuterium and helium nuclei begin to be formed	pairs e^-/e^+ disappear
35 minutes	3×10^8 K	e, p and He^4, D	the proportion of He^4 and D frozen $He^4/H \sim$ 22-28% by mass
7×10^5 years	3×10^3 K	neutral atoms begin to be formed	transition from radiation-dominated to matter-dominated universe universe is now transparent to radiation

§13-8. The Microwave Background

In Sections 13-6 and 13-7, we saw how important the role of radiation was in the early history of the universe. While following the progress of the early universe, we left it when the temperature had fallen to ~3000 K, the stage when the matter-dominated phase took over. What happened afterward? As stated earlier, the subsequent role of radiation became a passive one. Radiation no longer appreciably influenced the rate at which the universe expanded. Radiation simply cooled down according to the rule obtained in Section 13-6. Therefore, it follows that even today there should be some radiation around, albeit of very low temperature. And since the process of expansion of the universe does not interfere with the black-body nature of the spectrum, this relic radiation should still have a black-body character.

In Figure 13-26, we have a few typical black-body curves for different characteristic temperatures. Notice that the peak intensity occurs at a wavelength that changes with temperature. It is important for the observer to know what temperature present-day cosmic radiation is expected to have. The observer can then search for the peak intensity at the appropriate wavelength.

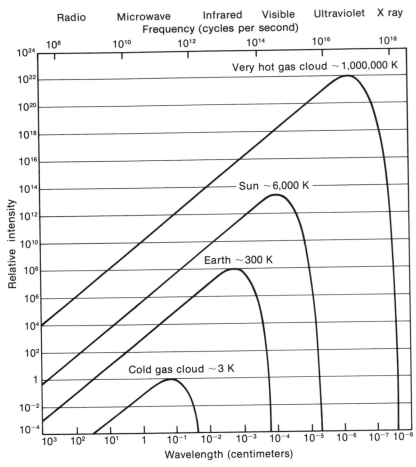

FIGURE 13-26
Examples of black-body curves plotted on a logarithmic scale.

We have already referred to Gamow's pioneering work in the nucleosynthesis of the early universe. Calculations made by Gamow and his colleagues, Ralph A. Alpher and Robert Herman, although not quite correct in all the details available today, led them to predict a cosmic background temperature of ~5 K at the present epoch. From Figure 13-26, we see that the peak intensity in such a case would lie in the microwave range.

Although this tentative prediction was made in 1948, no immediate search was instituted for such a radiation background, partly because at that time the idea of a radiation background was not taken very seriously, and particularly because it failed to synthesize all the elements. The shift from cosmological to stellar nucleosynthesis in the 1950s, culminating in the work of Geoffrey and Margaret Burbidge, William Fowler, and Fred Hoyle, kept the original Gamow picture very much in the background. It was only in 1964 that calculations of cosmological nucleosynthesis were taken up again, by Hoyle and Roger Tayler in England, by P. J. E. Peebles in the United States, and by Ya. B. Zeldovich in the U.S.S.R. However, these theoretical efforts, while clearing up many of the points in Gamow's theory, did not establish the existence of the cosmic radiation background. The discovery of the background came quite accidentally. The discoverers, Arno A. Penzias and Robert W. Wilson, were looking for something else when they happened upon it.

Penzias and Wilson at the Bell Telephone Laboratory had begun in 1964 a series of measurements of radio-wave intensities from the plane of the Milky Way. For this work, they had the use of an antenna having a 20-ft horn reflector of low noise that had been built for communication via the *Echo* satellite. In the course of their measurements, Penzias and Wilson used the microwave wavelength of 7.35 cm because, at this wavelength, the noise from the galaxy was expected to be negligible. They were surprised to find, however, that there was a residual noise that was isotropic (that is, it did not vary with direction). Because of its isotropy, it was not possible to link the noise with a specific nearby source like the galactic center or our neighboring galaxy, M31 (the Andromeda nebula).

After months of careful scrutiny, the noise still persisted, and by early 1965, Penzias and Wilson were able to ascribe to it a notional temperature of 3.5 K.* They were, however, unable to interpret the noise as they were not then aware of the cosmological theory we have been discussing. The news of the discovery reached Princeton, where Peebles immediately grasped its significance. His own theoretical work had already led him to the idea of cosmic relic radiation. He and his senior colleague, Princeton physicist Robert H. Dicke, who had himself set up an experiment to measure this radiation along with his experimentalist colleagues P. G. Roll and D. T. Wilkinson, were greatly excited by the discovery of Penzias and Wilson.

Penzias and Wilson themselves were very cautious in their paper in the *Astrophysical Journal* entitled "A measurement of excess antenna temperature at 4,080 Mc/s." They simply reported their experimental setup and the precautions taken in arriving at the unaccounted for noise. Their paper was followed by a companion paper by Peebles, Dicke, Roll, and Wilkinson giving a cosmological interpretation to the excess microwave radiation of Penzias and Wilson.

Soon afterward, Roll and Wilkinson announced their own results at a wavelength of 3.2 cm. They also found an excess radiation temperature in the range from 2.5 K to 3.5 K. Since that time, the relic radiation has been measured by radio astronomers at a dozen or so wavelengths, ranging from 73.5 cm to 0.33 cm. The experimental points and the theoretical black-body curve through them are shown in Figure 13-27. The best black-body temperature for these points is 2.7 K. The recent work by D. P. Woody and P. L. Richards from Berkeley extends the curve to the short wavelength end also. Although there is

---

*The error bars to this result were ± 1 K.

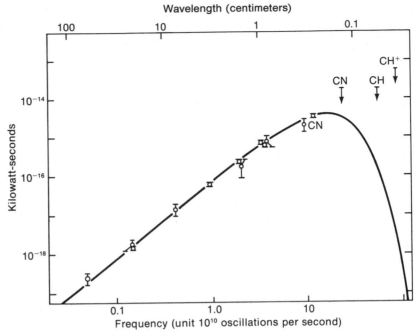

Wavelength (centimeters)

Frequency (unit 10^{10} oscillations per second)

FIGURE 13-27

The black-body curve for a temperature of 2.7 K gives a close fit to the observed points. (With the left-hand scale interpreted in units of kilowatt-seconds, the curve represents the power in kilowatts falling on an area of $1/\pi$ square kilometers.) The points and arrows marked CN, CH, CH^+ provide indirect measurements based on molecular transitions in these systems.

an overall agreement between the Berkeley data and the black-body curve, fine-scale analysis shows discrepancies, whose significance has still to be assessed.

The curve of Figure 13-27 is perhaps the strongest evidence for the type of cosmological picture—the hot big-bang universe—that we are discussing in this chapter. If we believe in this picture, the black-body curve tells us about the state of the universe at least as far back as when the universe first became optically transparent, that is, to redshifts of $z \cong 1,000$. In this sense, the relic radiation takes us much further back in the history of the universe than the other tests mentioned in Sections 13-3, 13-4, and 13-5.

Is the hot big bang the only possible explanation of the cosmic microwave background? So far, no viable alternative theory has been given, but we cannot at this stage entirely shut the door on alternatives. Microwave astronomy is still in its infancy, and we cannot rule out other astrophysical processes giving rise to microwave radiation. Also, the turn over to the right of the maximum intensity point of Figure 13-27, so essential for establishing the black-body character of the curve, still remains to be fully confirmed.

We are now in a position to relate our development of cosmology to an important problem raised earlier, namely, the origin of the element helium. Except for a certain class of star in which helium appears to be largely absent from the surface layers, helium is found in all objects in our galaxy. (The exceptional stars are ones in which helium is very likely segregated by gravity, and hence they are probably not relevant to the problem of the ubiquitousness of helium in the universe.) Helium is also found in neighboring galaxies. Everywhere the abundance seems to be about the same; some 25 percent of the mass of cosmic material consists of helium. If we were entirely sure that all the abundances were closely the same, or even if we could be certain that the abundance never falls below some particular value, say, 20 percent, there would be a strong case for supposing that most of the observed helium was not generated in stars by the processes discussed in Chapter 8. However, there is inevitably some ambiguity in the observations because helium is an awkward element to deal with, essentially because its strongest quantum transitions emit radiation with frequencies in the far ultraviolet, and these frequencies do not penetrate the Earth's atmosphere. Consequently, all measurements of the helium abundance are subject to some uncertainty. When we say that the mass fraction of helium is everywhere about 25 percent, in some places the value could be 30 to 35 percent; in other places, it could be from 15 to 20 percent without doing violence to the data. Thus, we cannot be certain that there is a standard cosmic value for the helium abundance as is sometimes asserted in the astronomical literature. Nor do we know much about the helium abundance in galaxies other than quite nearby ones.

The main argument for supposing that most of the helium originated very early in the history of the universe comes from the quantitative difficulty of explaining an abundance of helium as high as 25 percent in terms of processes within stars. If as much as 25 percent of hydrogen had been converted to helium, and if the resulting energy production escaped from the galaxies as visible light, galaxies would be much brighter than they are observed to be. This argument has to be considered a strong one, strong enough to make it attractive to seek a primeval origin for most of the observed helium content of the galaxies. Consequently, let us discuss the primeval origin of helium, assuming that the particles present at an early time are of the familiar kind.

The relevant particles for the formation of helium are neutrons and protons. Neutrons are unstable through $n \longrightarrow p + e + \bar{\nu}_e$, decaying under laboratory conditions in a characteristic time of about 10 minutes. However, in the early stages of the universe, when the temperature and matter density were high, the probability of neutrons and protons becoming associated together, $n + p \longrightarrow D$, was fairly high. The deuterons so formed become largely converted, by $D + n \longrightarrow T$ and by $D + p \longrightarrow {}^3He$, into tritium (T) and ${}^3He$. Then, ${}^4He$ was formed by the addition of neutrons to ${}^3He$ and protons to T. In the Einstein–de Sitter model, the outcome can be shown to be a helium production amounting to about 28 percent of the mass of the primeval material.

TABLE 13-2
Some light nuclei that cannot be formed by synthesis within stars

Nucleus	Approximate mass fraction
D	2×10^{-4}
^{3}He	6×10^{-5}
^{6}Li	10^{-9}
^{7}Li	10^{-8}

Several other light nuclei are found on the Earth, in the Sun, and in meteorites, with concentrations that are far too high to be explained by synthesis within stars. They are given, together with their mass fractions, in Table 13-2. Processes outside stars, or at the surfaces of stars, involving high-speed particles can explain the origin in kind of these nuclei. Whether it is also possible to explain the mass fractions of these nuclei by such processes remains a matter of controversy, especially for deuterium. Because of its comparatively high mass fraction, D is harder to account for in this way than ^{3}He, ^{6}Li, and ^{7}Li. Hence, D may also have originated from a primeval synthesis.

§13-10. The Age of the Universe

Let us return to Hubble's law. Can we relate the presently observed value of Hubble's constant to the present age of the universe? There is a simple formula that does this very thing. We can denote by $\dot{Q}(t)$ the rate of change of the scale factor Q with respect to time. Then we have the present value of Hubble's constant given by

$$H = \frac{\dot{Q}}{Q}\Bigg|_{t=t_p}.$$

That is, we have to evaluate the ratio on the right-hand side at $t = t_p$. If we know $Q(t)$, we can therefore find H in terms of t_p or, conversely, t_p in terms of H.

For the Einstein–de Sitter model, $Q(t) \propto t^{2/3}$. This gives

$$\frac{\dot{Q}(t)}{Q(t)} = \frac{2}{3t},$$

that is,

$$H = \frac{2}{3t_p}, \qquad t_p = \frac{2}{3}H^{-1}.$$

If we take $H = 75$ km/mpc, we get $t_p \approx 9 \times 10^9$ years.

If we use a type B model, the age comes out to be shorter than this value, whereas if we use a type C model, the age is longer than this value but not greater than $H^{-1} \approx 1.33 \times 10^{10}$ years. We would therefore have a serious problem on our hands should astronomers find objects in the universe older than H^{-1}. At present, estimates of the astrophysical age of stars and galaxies are not precise enough to test this prediction in a rigorous way. For example, the age of our galaxy is between 10^{10} and 1.5×10^{10} years, a range that begins to be difficult to accommodate in the Einstein-de Sitter model or in a type B model but manageable in a type C model. If galaxies older than our own exist, the problem would be even more difficult!

§13-11. Olbers' Paradox Again

Let us see how the expanding universe modifies the calculation made by Olbers. The essential piece of information that Olbers lacked is contained in the phenomenon of the redshift.

Consider, for example, the amount of light L sent out by a remote galaxy each second. The light quanta, by the time they get to us, are redshifted. Thus, a quantum of frequency ν and energy $h\nu$ at the source has the energy

$$\frac{h\nu}{1 + z}$$

at the receiver. Moreover, the time scales are also affected (see Chapter 11). Thus, an interval of time Δ over which this quantum is emitted at the source stretches out to an interval of time

$$\Delta \cdot (1 + z)$$

at the receiver. The result is that the amount of light received per second per unit area at the receiver is not

$$f = \frac{L}{4\pi D^2}$$

as calculated by Olbers, but

$$f = \frac{L}{4\pi D^2 (1 + z)^2},$$

a result quoted at the end of Chapter 12.

Thus, remote shells contribute much less light than was estimated by Olbers owing to their large redshift. When we add such contributions from all shells out to infinity, we get only a small answer, making the sky effectively dark. We can say that the sky is dark at night because the universe expands!

1. Discuss why we expect gravitation to be the most dominant basic interaction in controlling the large-scale structure of the universe.

2. Comment on the use of the other basic interactions (apart from gravitation) to the cosmologist.

3. What are the basic principles and postulates of modern cosmology? To what extent is there an observational justification for them? How do they simplify the task of the theoretician?

4. Describe a procedure whereby observers in different galaxies can set up a system of universal time.

5. Join the spatial positions of three galaxies at a particular moment of universal time. What property does the resulting triangle have in common with a second triangle obtained by joining the same three galaxies at a different moment of universal time? In what respect may two such triangles differ?

6. Discuss the information concerning the triangles of Problem 5 given by Einstein's theory of gravitation.

7. In what important respect do astronomers seek to use observational procedures to overcome the limitations still present in cosmology even after the use of Einstein's theory?

8. In a particular type-*B* Friedmann model, the relation between the flux received from a galaxy of redshift z and its luminosity is given by

$$f = \frac{L}{4\pi(c/H)^2 z^2},$$

whereas for a particular type-*C* Friedmann model, it is given by

$$f = \frac{L}{4\pi(c/H)^2 z^2 (1 + \frac{1}{2}z)^2}.$$

Taking the definition of apparent magnitude given in the text (see Figure 12-15), estimate the differences between the predicted magnitudes by the type-*B* and type-*C* models at $z = 0.1, 0.3, 1.0, 3.0$, and 10.

9. Comment on the effectiveness of the magnitude-redshift test using galaxies to decide the type of universe we live in.

10. Discuss the role of QSOs in the magnitude-redshift test.

11. What is the $N(S)$ relationship obtained from the counting of radio sources?

12. If a logarithmic plot of the observed $N(S)$ relationship had turned out to be a certain straight line, what inferences would you have drawn?

13. What inferences can in fact be drawn from the observed $N(S)$ relationship?

14. What is an evolutionary universe?

15. Show that the plot of log N versus log S under the three assumptions given in Section 13-4 is a straight line of slope -1.5 in a Euclidean universe.

16. Describe how the angular size test can, in principle, measure the effects of non-Euclidean geometry.

17. In the type-A model, α depends on the redshift by the formula

$$\alpha = (\text{constant}) \cdot \frac{(1 + z)^{3/2}}{(1 + z)^{1/2} - 1}.$$

Plot α against z and show that the minimum of α occurs at $z = 1.25$.

18. Outline the way in which observational difficulties prevent a clear-cut conclusion from the angular size test.

19. Give an account of the first hour in the existence of the big-bang universe.

20. Describe the events leading to the discovery of the microwave background.

21. The discovery of the microwave background is often described as the most significant cosmological discovery since Hubble's law. Give reasons.

22. Describe how the abundance of certain elements may carry the signature of the early history of the universe.

23. Discuss the age of the universe. Why would an increase in the value of Hubble's constant present a problem for the big-bang models?

24. If the microwave background curve turns out not to have a black-body character, what would this imply for the hot big bang?

25. Show how the expansion of the universe affects Olbers' paradox.

Chapter 14
Inertia and Cosmology

§14-1. Introduction

The position we reached at the end of Chapter 13 represents the general view held by most astronomers about the origin and the structure of the universe. This picture has many commendable features. The simplest models based on Einstein's law of gravitation exhibit the observed expansion of the universe and account for Hubble's law. Moreover, this cosmological theory is not mere speculation; it has inspired many interesting observational tests in optical and radio astronomy. And the implicit faith placed by cosmologists in the extrapolation of known physical laws to the extreme conditions operating in the early stages after the big bang has paid rich dividends in the form of an understanding of the microwave background radiation and the origin of the observed abundances of He and D.

These successes have led to something of a feeling of complacency, however, a feeling we find dangerous to the future growth of cosmology. An impression has been created that the picture of a hot big bang is more or less right, and now all we have to do is fill in details. This conditioning of mind shows itself in many ways. Any nonstandard approach to cosmology is considered outlandish, and any observational data that do not conform to the standard picture are considered suspect, a situation that has led to ignoring genuine difficulties with the standard picture.

It seems undeniable to us that a difficulty is associated with the instant $t = 0$ of the big bang. Why was the universe created at $t = 0$, all at one time? Why can we not extend the history of the universe prior to the big bang? With so many other possibilities available, why did the primeval explosion make a homogeneous and isotropic universe? In the early epochs, the range of communication between different parts of the universe was very short. How then did the microwave background become as homogeneous as the data indicate?

Notice that according to relativity, the geometry of spacetime at the big bang ($t = 0$) was singular in much the same way that a singularity of spacetime arose in the problem of gravitational collapse discussed in Chapter 11. Since the laws of physics break down at such a singularity, the cosmologist is helpless to answer most of the questions just raised. We have to be content with the remark: it is all due to the initial conditions. This is another way of saying that "the universe is what it is, because it was what it was," a comment due to T. Gold.

In Appendix D, we describe the steady-state theory that attempts to do away with the singularity and also brings the creation of matter within the realm of physics. However, this theory is at present under a cloud because of its seeming inability to account for the observed microwave background.

In this chapter, we outline another approach to these problems; this approach views the standard big-bang picture from a different angle and leads to a theory wider in perspective than the general theory of relativity. It has the additional advantage of incorporating Mach's principle, a principle that seeks to relate the inertia of matter to the large-scale structure of the universe. We begin with a discussion of this principle.

§14-2. Mach's Principle

Let us go back to Chapter 10 and take a critical look at Newton's laws of motion. The quantities occurring in the statements of these laws are (1) velocity, (2) acceleration, and (3) force. How do we measure these quantities to verify the laws of motion?

First, we notice that any measurement of velocity or acceleration must be relative. Take the following statement: a car is moving in a northerly direction with a speed of 55 miles per hour. This velocity measurement is clearly intended to be relative to the surface of the Earth. If we take into account the fact that the Earth rotates around its axis, that it goes around the Sun which in turn goes around the galaxy, we come up with a considerably larger speed for the car in an altogether different galactic direction. For, in that case, we would be using a nonrotating coordinate frame of reference located at rest in the center of our galaxy—nonrotating with respect to the most distant observable galaxies.

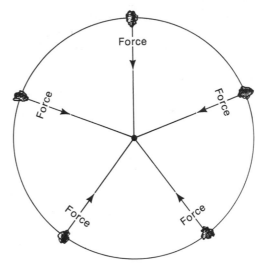

FIGURE 14-1
To keep a stone whirling around in a circle, a force must be applied to the stone. This force must always be *toward the center of the circle*.

The same situation applies to acceleration. Take the example of the stone being whirled around in a circle by a string tied to it. Figure 14-1 is a reproduction of Figure 10-2, which looks at the stone from the frame of reference in which the center of the circle (the other end of the string) is at rest. In Figure 14-2, on the other hand, the stone is at rest, and the center of the string moves in a circle about it.

Clearly, before we start applying Newton's laws, we have to decide what reference frame we are to use for measuring velocities and accelerations. In the example of the stone in the first frame, we argue that the tension in the string supplies the force that accelerates the stone. If m = mass of the stone, v = uniform speed of the stone in the circle, r = radius of the circle, and T = tension in the string (force), we write the second law of motion in the following form:

$$\text{mass} \times \text{acceleration} = \text{force},$$

which gives

$$m \times \frac{v^2}{r} = T.$$

So far so good. What happens in the second reference frame? The stone is at rest and not accelerated. Thus, the left-hand side of the preceding equation is zero.

What we find somewhat disconcerting is that the right-hand side is apparently not zero. The force on the stone is still toward the center and is given by T. In the absence of any other force, how can we make the right-hand side equal to the left-hand side?

It seems, therefore, that the second law of motion does not apply to all reference frames. Newton was aware of this problem, and after considerable thought he formulated the postulate of *absolute space*.

In absolute space, Newton singled out a unique reference frame in which his laws of motion were assumed to hold good. A frame that is accelerated relative to absolute space is one in which we encounter the type of difficulty found in the second reference frame of the stone example. To resolve the difficulty for such a nonabsolute situation, we use the prescription given by Newton: invent *fictitious forces* to balance the equation of motion. In the example of the stone, we have to invent a force of $-T$, that is, a force of magnitude T in the opposite, outward direction. This fictitious force is usually referred to as *centrifugal force*. It is fictitious because there is no real source for it; we have to invent it to balance our books in the nonabsolute frame. According to Newton, the absolute frame for the stone example is that of Figure 14-1.

Such fictitious forces are called *inertial forces* because they are proportional to the inertia of the system under consideration. In the case of the stone, the magnitude of the centrifugal force is mv^2/r, which is proportional to the mass m of the stone. Notice that the fictitious forces arise only for frames accelerated relative to absolute space. For a reference frame in uniform motion relative to absolute space, no inertial forces are necessary. Reference systems in uniform motion are called *inertial frames*. We encountered such reference systems in Chapter 10. Thus, we can extend the applicability of Newton's laws beyond absolute space to the entire class of inertial frames. An accelerated frame (relative to absolute space) is called a *noninertial frame*.

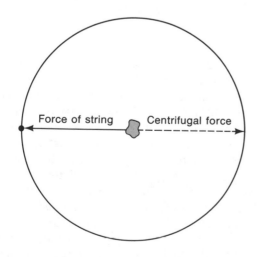

Force of string   Centrifugal force

FIGURE 14-2
In the frame of reference of the stone, the stone has zero acceleration. Consequently, we have to invent a force equal and opposite to the force of tension along the string. This *centrifugal force* is shown by the dotted line.

FIGURE 14-3
Ernst Mach (1838–1916). (From *Ernst Mach, Physicist and Philosopher*, ed. R. S. Cohen and R. J. Seeger, published by D. Reidel, Holland.)

While Newton's absolute space remained an abstract concept, in the last century the Austrian philosopher-scientist Ernst Mach (see Figure 14-3) noticed a significant astronomical coincidence that seemed to give a concrete status to Newton's concept. Suppose we want to measure the Earth's rotation relative to absolute space. To make such a measurement, we first have to calculate what fictitious inertial forces arise from the circumstance that, in a laboratory experiment, we invariably use a frame of reference fixed on the Earth's surface. Although the centrifugal force arises from the spin of the Earth, it is rather small because the angular speed of rotation of the Earth (one rotation of 360° in 24 hours) is small. A somewhat larger fictitious force, known as the *coriolis force*, arises in the manner illustrated in Figure 14-4, where we have a simple pendulum (a bob suspended by a string from a point) moving to and fro in a vertical plane. The effect of the coriolis force, due to the Earth's rotation, is to turn the

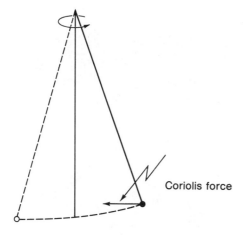

Coriolis force

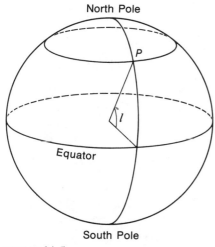

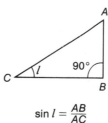

$$\sin l = \frac{AB}{AC}$$

FIGURE 14-5
At the point P with latitude l, the Foucault pendulum will complete one revolution in $1/\sin l$ days. The value of $\sin l$ is equal to the ratio AB/AC for the right-angled triangle shown at the right.

plane of oscillation of the pendulum around its vertical axis. If the pendulum is designed so that it is free to oscillate in any vertical plane, we find that, at a geographical latitude l, the plane of oscillation will complete one revolution about the vertical direction in

$$\frac{1}{\sin l} \text{ days}$$

as shown in Figure 14-5. Such a pendulum is called the Foucault pendulum (see

FIGURE 14-6
A working model of the Foucault pendulum. (Photograph
courtesy of the California Academy of Sciences.)

Figure 14-6). With a Foucault pendulum, an experimenter can measure the
angular velocity of the Earth simply by multiplying the angular rate of the
turning of the pendulum by sin l. What is remarkable is that the answer obtained
in this way agrees very well with the answer that would be obtained had we
looked at distant stars going around us. In other words, *the Earth's rotation
relative to distant stars is closely the same as its rotation relative to Newton's
absolute space.*

IS THE PROPERTY OF INERTIA RELATED TO THE DISTANT PARTS OF THE UNIVERSE?

409

§14-3. Units and Dimensions

Noting this coincidence, Mach argued that Newton's absolute space is in fact determined by the frame of distant stars. And since the concept of inertial forces is linked with this special frame, Mach went on to conjecture that the property of inertia is itself somehow linked with the background of distant stars. Remove the background, he argued, and we would have no concrete way of fixing the absolute space on which the laws of motion are based. Since inertia is proportional to mass, we have to argue that the mass of a body is not an intrinsic property of the body itself, as postulated by Newton, but related to the distant parts of the universe. This concept is known as *Mach's principle*. The interpretation of "distant parts" has changed since the nineteenth century. Extragalactic astronomy has shown that the distant galaxies provide an even better approximation to the absolute space of Newton than the distant stars in our own galaxy. Physicists differ in their evaluation of this coincidence and of the importance to be attached to Mach's principle. In this chapter we explore some of the cosmological consequences of taking Machian ideas of mass and inertia seriously. We feel a coincidence that has stood the test of time for nearly a century deserves to be investigated further.

§14-3. Units and Dimensions

Science deals with many different physical quantities: mass, velocity, force, angular momentum, electric charge, magnetic field, and so forth. Each quantity is expressed in convenient units so that the resulting number representing the quantity is not too large. For example, we can express the mass of a human being in pounds or kilograms. However, these units are not convenient to describe the mass of a star. For a star, a convenient unit is the mass of the Sun ($M_\odot$), which equals nearly 2×10^{30} kg.

THERE IS ONLY ONE BASIC DIMENSION

A little thinking shows us that all the physical quantities can be described in units that are constructed from powers of length (L), mass (M), and time (T). For example, velocity has a unit of (L/T), electric charge has a unit of ($L^{3/2}M^{1/2}/T$), the gravitational constant has a unit of (L^3/T^2M), and so on. The reason that scientists use so many different units—for example, the dyne, joule, volt, gauss, and so on—is again for the sake of convenience. It would be very cumbersome to keep track of powers of L, M, and T at each stage. This practice should not, however, mask the basic dependence of all physical units on L, M, and T.

Carrying the reduction process further, is it necessary to have three basic units, one for L, one for M, and one for T? Two important developments of twentieth-century physics have obviated the necessity to have three independent basic units. One of these, the special theory of relativity (see Chapter 10), has shown the existence of a fundamental velocity in nature. This fundamental velocity is the velocity of light, whose magnitude is

$$c = 2.997929(\pm 0.000004) \times 10^{10} \text{ cm/second.}$$

If we take into account the existence of this important result, is it not natural to use a system of units in which $c = 1$? With such a system, we can define the unit of length in terms of the unit of time or vice versa:

$$1 \text{ second} = \{2.997929(\pm 0.000004) \times 10^{10}\} \text{ cm.}$$

With $c = 1$, we can dispense with the second as an independent unit of time. All velocities are now dimensionless quantities; that is, they are pure numbers. Special relativity puts an upper limit of 1 on the magnitudes of all velocities (see Figure 14-7).

The second important discovery was that of quantum theory, which brought in another constant of nature, Planck's constant, h, or the related constant, $\hbar = h/2\pi$. The quantum uncertainty principle

$$\Delta x \, \Delta p \gtrsim \hbar$$

tells us that no measurement of position (x coordinate in Figure 14-8) and

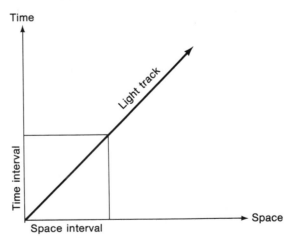

FIGURE 14-7
With the help of a light ray, we can translate any space interval into a time interval and vice versa. Taking the speed of light as unity ($c = 1$) amounts to establishing such an equivalence between the units of space and time. The light track makes an angle of 45° with the space (and also the time) axis.

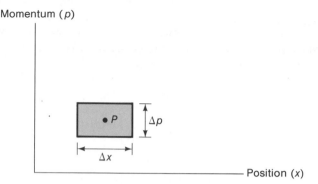

Momentum (p)

Position (x)

FIGURE 14-8

Classically, we can specify the position (x) and the momentum (p) of a particle exactly, and in the x-p diagram we can describe the particle by the point P. Quantum theory tells us that this accuracy is impossible to achieve. At best, we can locate P in a rectangle (shaded in the figure) of area $\hbar$.

momentum (denoted by p in Figure 14-8) of a system can be infinitely precise. The preceding relation places a *lower limit* on the product of the uncertainty Δx in the measurement of position and the uncertainty Δp in the measurement of momentum. If we try to improve the precision of our measurement of x—that is, if we try to reduce Δx—then we have to pay the price of increasing Δp.

Now the momentum has the dimensions of mass × velocity, and we have already seen that, by taking $c = 1$, we make velocity dimensionless. Thus, the *uncertainty rectangle* of Figure 14-8 has one side with the dimension of length ($\Delta x \sim L$) and the other side with the dimension of mass ($\Delta p \sim M$). If we now set $\hbar = 1$ and make the area of the shaded rectangle of Figure 14-8 equal to unity, we get a relation between the mass unit M and the length unit L:

$$ML \sim 1,$$

that is,

$$L \sim M^{-1}.$$

In these units, the uncertainty principle can be written in the form

$$\Delta x \, \Delta p \gtrsim 1.$$

It is therefore possible to express length in a unit of gram^{-1}. The same applies to time, which we have already expressed in length units by setting $c = 1$. By stating the usual value for $\hbar$,

$$\hbar = 1.05443(\pm 0.00003) \times 10^{-27} \, \text{g/cm}^2/\text{second} = 1,$$

we are simply giving an equation, between gram, centimeter, and second that, together with the earlier relation between centimeter and second, leaves only one independent unit. We choose this unit to be the *mass*. We then write the unit for all other quantities in terms of mass only:

$$\text{Length} \sim M^{-1},$$
$$\text{Time} \sim M^{-1},$$
$$\text{Energy} \sim M,$$
$$\text{Electric charge} \sim M^{\circ},$$
$$\text{Magnetic field} \sim M^{2},$$
$$\text{Gravitational constant} \sim M^{-2}.$$

We can say that a unit expressible as M^n has the dimension n in this system.

Thus, it all comes down to choosing a unit for M. We can choose the gram, or we can choose the mass of an elementary particle like the electron or a proton, *provided* we are sure that the mass of an elementary particle is constant. If we follow the Newtonian precept that inertia is an intrinsic property of matter, then we have an intrinsic value for our mass unit. We may convince ourselves, or simply postulate in the absence of any contrary evidence, that the mass of the electron m_e is the same everywhere at all times. This gives us a definite framework in which all units are then fixed.

BUT A FRAMEWORK USING VARIABLE PARTICLE MASSES CAN LEAD TO SIMPLIFIED GEOMETRY

When we discuss length measurements or geometric relationships in the distant parts of the universe, the preceding framework assures us that the units of measurement used there are the same as they are in our laboratory. When we consider spectral lines emitted by atoms in distant galaxies, we can assume that the same numbers are relevant there as those that would be found in similar processes here and now.

Nevertheless, we must not forget that our apparently secure framework rests only on an assumption of Newtonian physics, an assumption that becomes distinctly shaky if we attach significance to Mach's principle. If m_e is the mass of the electron here and now, m_e has its currently observed value in relation to the present-day background of distant galaxies. We *do not* have any guarantee that, in a general situation, the distant background will always conspire to give the same value for m_e everywhere and at all times. Thus, we must build into our physical description of the universe the possibility that m_e is variable, that is, our measurement framework is not rigid but allows m_e to be scaled by an environmental factor.

Such a development need not always complicate the situation. By having the freedom to adopt different length scales at different spacetime points, this alternative picture enables mathematicians to choose a simpler form of geometry than would otherwise be possible. Thus, it can be shown that in most cases the following two pictures are mathematically equivalent:

Particle masses constant + complex universal geometry
$$\Longleftrightarrow \text{particle masses variable + simpler universal geometry} \cdots \text{(E)}$$

The cases where the two pictures differ crucially are in special instances where, at particular points of spacetime, all particle masses happen to be zero in the new picture. The corresponding version of the vanishing of the particle masses in the first picture is a singularity of spacetime geometry.

In the previous chapter, we gave a description of the universe using the first picture. In the present chapter, we use the second description since changing to this alternative way of describing the universe makes many aspects of cosmology easier to understand, especially those aspects concerned with the origin of the universe. The origin of the universe is just such a singularity of spacetime geometry. The origin of the universe thus appears as a great mystery in the first picture, but it is no mystery at all in the second.

§14-4. The Meaning of the Expansion of the System of Galaxies

Starting from the first picture, consider the positions $G_1, G_2, \ldots, G_n$ at a certain moment of universal time, say, t, of a set of n galaxies, and suppose we join G_1 to G_2, G_2 to $G_3, \ldots, G_{n-1}$ to G_n, and G_n to G_1 by straight lines to form a polygon. We can do the same thing for the same galaxies at a different moment of time, say, t', thereby obtaining a second polygon. What the cosmological principle permits us to deduce (by a fairly sophisticated piece of mathematics) is that the second polygon must have the *same shape* as the first polygon. What we *cannot* deduce, however, is that the two polygons have the same scale; the scales can be different, as in Figure 14-9. Our procedure for setting up a system of universal time permits us to construct a polygon from any set of galaxies, and

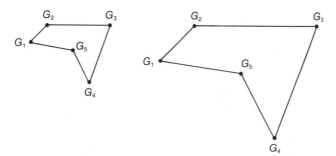

FIGURE 14-9

Under the usual postulates of cosmology, the shapes of two polygons, obtained by joining the spatial positions of a set of galaxies at two different moments of universal time, must be the same. But the usual postulates of cosmology do not require the two polygons to have the same scale.

this polygon has a definite shape whatever the moment of universal time at which we choose to construct it.

In Chapter 13, we used the scale function $Q(t)$ to define the scale of a polygon at time t. Thus, in Figure 14-9, $Q(t)$ for the polygon on the left was smaller than that for the polygon on the right. In the expanding universe picture, $Q(t)$ increases with time with the figure on the left of Figure 14-9 preceding the figure on the right. Einstein's theory of gravitation enables us to calculate certain aspects of the behavior of $Q(t)$ as a function of cosmic time.

Given this brief recapitulation of the situation in the first picture, let us consider three galaxies. The polygon is then a triangle, and the triangle increases its scale (Q increases) as time goes on (see Figure 14-10). In the past, the triangle was smaller than it is now. In the future, the triangle will be larger than it is now. This situation prompts the question: If we go far enough back into the past, was the triangle ever shrunk to nothing at all as in Figure 14-11? The answer is yes because Q was once zero. This answer follows in Chapter 13 from Einstein's theory of gravitation, as we saw from Figure 13-3 for the Einstein–de Sitter model and from Figure 13-4 for the three geometrical cases A, B, and C.

Now exactly what do we mean by $Q(t)$ varying with time? How do we measure a change in the scale of our galaxy polygon? The answer is that our basic standard, whether of an interval of time or of a spatial length, is set by the wavelength of the radiation from a specified quantum transition of some chosen atom. In turn, the wavelength of the radiation is determined by the masses of the

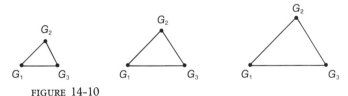

FIGURE 14-10
In an expanding universe, the triangle formed by joining the positions of three galaxies was smaller in the past than it is now; in the future, it will be larger than it is now.

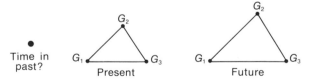

FIGURE 14-11
Figure 14-10 raises the question: Was there a time in the past when the triangle was shrunk to a point?

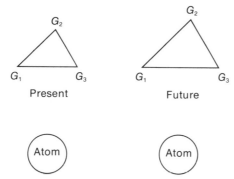

FIGURE 14-12
When we say that the size of the triangle of Figure 14-10 increases with time, we mean that the *ratio* of the length of any side of the triangle to a length scale determined by atomic sizes (which are grossly exaggerated!) is increasing, whereas the scale itself stays the same.

particles that constitute the atom, especially by the mass of the electron. When we calculate the frequency, say, ν, of the radiation emitted in the transitions of a certain atom, say, the Hα transition of hydrogen, ν is proportional to the electron mass. Thus, writing m for the electron mass, we have ν proportional to m. Now the wavelength of the radiation is c/ν, and hence it is proportional to the reciprocal of the electron mass, $1/m$. It is $1/m$ that determines our physical scale, expressed by the sizes of atoms and by the unit of time given by radiation from atoms, that is, the unit of an atomic clock. When we say that distances between galaxies increase with time, we mean distances measured with respect to $1/m$. We mean that our galaxy polygons increase with respect to $1/m$ as the unit as in Figure 14-12.

§14-5. An Alternative Description of the Expansion of the Universe

The significance of the equivalence given in (E) of Section 14-3 will now be clear. Its left half requires $1/m$ to stay fixed, in which case the galaxy polygons increase according to the scale factor $Q(t)$. But if we go by its right half, we can assume the galaxy polygons stay fixed so that $Q = 1$ at all times, in which case $1/m$ must decrease as in Figure 14-13. Can we devise physical experiments that distinguish between the situations depicted in Figures 14-12 and 14-13? Local experiments cannot do so because they occupy intervals of time so short that any change of $1/m$ during them would be quite negligible. But suppose a terrestrial experimenter were able to live for a very long time, and suppose he could take radiation from a certain atom—for example, the H and K oscillations of

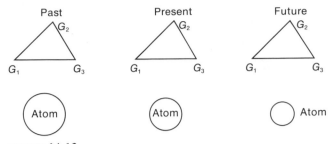

FIGURE 14-13

An alternative to Figure 14-10 would be to keep the triangle fixed in size and to suppose that the atomic dimensions (again grossly exaggerated) decrease with time.

calcium—and store the radiation for future reference. Then, after a long time had elapsed, the stored radiation could be compared with new radiation from the same kind of atom. Would the oscillation frequency of the old radiation be the same as that of the new radiation?

THE HUBBLE REDSHIFT CAN BE INTERPRETED AS ARISING FROM PARTICLE MASSES VARYING WITH THE EPOCH IN A NONEXPANDING UNIVERSE

There is a sense in which this experiment *can* be carried out, and this situation is illustrated in Figure 14-14. Light from a distant galaxy takes a very long time to reach us, and we can think of passage through space as a form of storage. When we receive light from such a galaxy, we can examine this old light, light that for some galaxies was generated billions of years ago. What do we find? We find that the old light does *not* have the same oscillation frequencies as the new light that is generated currently in the terrestrial laboratory. Should we regard this as a confirmation that the particle masses were different billions of years ago than they are today? Why bother with the Doppler shift and the idea that galaxies are rushing apart from each other? The redshift observation we have just described can be interpreted in this very different way, in terms of a change of $1/m$.

Physicists become uneasy at the thought of two *different but indistinguishable* interpretations of a phenomenon. They react to such a situation by arguing that if there are two indistinguishable ways of describing an observation, then, however apparently different they may seem, the two ways must really be the same. So, instead of asking which of Figures 14-12 and 14-13 is the correct picture, we should regard them as the same picture, *and we should adjust our physical ideas to make them the same.*

What would the foregoing imply? It would imply that the mass of a particle must be determined by its relationship to other particles according to certain rules that are stated with full mathematical precision. These rules must be chosen so that our physical theories (for example, of gravitation) lead to exactly

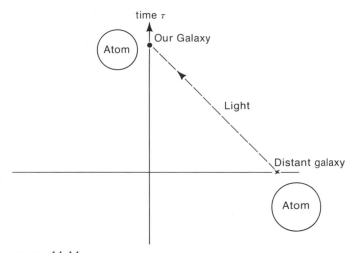

FIGURE 14-14
We can say that light from a distant galaxy was generated at a time when the masses of particles were smaller than they are at present. For this reason, radiation from a specified atomic transition is red-shifted with respect to present-day radiation from the same transition.

the same observable results for Figure 14-13 as for Figure 14-12. This demand for precise equivalence has necessitated considerable technical development, the details of which need not concern us here. For now, it is sufficient to know that Figures 14-12 and 14-13 *can* be made rigorously equivalent to each other. Then, we can proceed using Figure 14-13 instead of 14-12 as our description of the universe.

A remarkable simplification emerges immediately. When we adopt the picture of Figure 14-13, the universal geometry becomes the same as local spacetime geometry, *and it remains the same no matter whether the spacelike behavior of our galaxy polygons in Figure 14-12 is of type A, B, or C.* Whatever the former situation, we now have the convenient situation that the geometry of the whole universe is the same as the local geometry of special relativity.

The Einstein–de Sitter model has another important simplification. In the new picture, the galaxies are uniformly spaced as in Figure 14-15. Here it is useful to introduce the idea of a smoothed-out universe. The matter in the galaxies of Figure 14-15 can be imagined to be smoothed out so that the density of particles is the same everywhere. When this is done, we can draw a picture like Figure 14-15, but the world lines now represent individual particles instead of galaxies as in Figure 14-16. Clearly, the individual particles of Figure 14-16 will be much closer together than the galaxies of Figure 14-15. (The decrease of separation would be much more marked than can be indicated in the figures.)

Notice that the average separation of particles in Figure 14-16 provides us with a new scale of length, say, L, that we can use for measuring spatial lengths instead of using the scale determined by atoms. The scale L has the advantage of

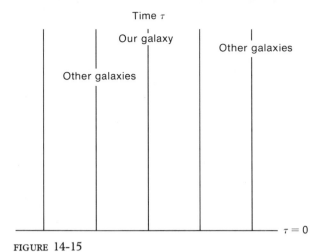

FIGURE 14-15
In this alternative picture to Figure 14-13 and in the Einstein–de Sitter model, the universe is static.

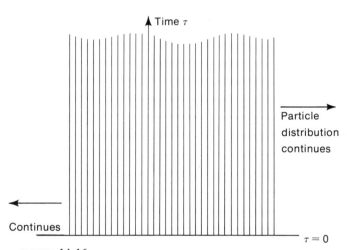

FIGURE 14-16
We can think of the material of the galaxies as being smoothed out into a uniform background of particles.

being the same at all times, whereas the scale set by radiation from atoms would now have the disadvantage of changing with time because the masses of the particles making up the atoms are variable. Indeed, this variation is just the redshift effect appearing again in a new guise, as we already noted. Notice too that the scale L is physical. We can also establish a physical scale for our time unit simply by requiring light and other forms of radiation to propagate at $45°$

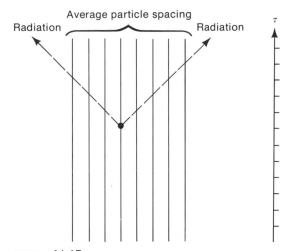

FIGURE 14-17
The average distance between particles in the smoothed-out model of Figure 14-16 gives a spatial scale. The scale in which time is measured is chosen so that radiation propagates at an angle of 45° to the time axis as in this figure.

in Figure 14-16, as in Figure 14-17. When measured in this way, we denote the time by τ. Another result that can be proved exactly is that, for time defined in this way, the particle masses in the Einstein–de Sitter model vary proportionately to τ^2. Let us see if we can gain an insight into the way this result arises.*

MACH'S PRINCIPLE CAN BE GIVEN A MATHEMATICAL FRAMEWORK

We recall at this stage the ideas of Mach described in Section 14-2. Without giving a mathematical theory, Mach argued that the observed property of inertia of a particle is not an intrinsic property of the particle but a property acquired by it because of its interaction with the background. In the picture we are now attempting to develop, we arrive at the type of conclusion that Mach's principle would have warranted: the mass of the typical particle depends on the large-scale structure of the universe. We now go deeper into this concept and seek to formulate the interaction that could describe *quantitatively* how the inertia of a particle arises from the rest of the particles in the universe.

The mass of particle a at point A in Figure 14-18 is to be regarded as made up of contributions that travel at an inclination of 45° from other particles, of which particle b in Figure 14-18 is an example. We also regard the interactions,

---

*Instead of worrying about the following details, the reader may prefer to turn immediately to Section 14-6.

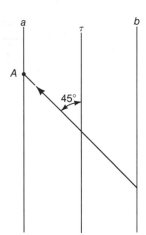

FIGURE 14-18
The mass of particle *a* at point *A* is made up of contributions from other particles *b*. These contributions travel at an angle of 45° to the time axis.

at any rate for the moment, as coming from the past as light does. First we ask: How will the contribution of such an interaction depend on the distance r between particles a and b? Clearly, we expect the contribution to be less when r is large and the two particles are widely separated than it will be when r is small. This expectation suggests some kind of inverse relationship, but will the contribution be proportional to the simple reciprocal, $1/r$, or to the inverse square, $1/r^2$, or to some other form? At first, we might suppose that the inverse square law, $1/r^2$, would be correct as it would be for the intensity of radiation from a source at distance r. But the intensity of radiation itself depends on the square of a more basic quantity, often referred to as the *amplitude* of the radiation, that behaves as $1/r$. The interaction we are seeking behaves like amplitude, not like intensity, and so it varies like $1/r$. Therefore, the situation is that, to determine the mass of particle a at point A in Figure 14-18, we have to add $1/r$ contributions for all other particles, of which particle b in Figure 14-18 is an example. Let us see how to go about making this addition, remembering that the particles are uniformly spaced with a separation L.

With center at point A in Figure 14-18, draw a set of spherical surfaces with radii L, $2L$, $3L$, and so on. Because we show only one of the three dimensions of space in Figure 14-18, we cannot draw these spheres in the usual way. With two of the three spatial dimensions suppressed, the spheres appear as simple points as in Figure 14-19. The spheres eventually reach particles from which there is no contribution because, for interactions propagated at 45° in our diagram, there is a limit to how far back we can go into the past *since none of the particles exists before time zero*. Writing τ for the time at point A, we reach such particles for a spatial distance $r = \tau$. Evidently, then, the contributions to the mass of particle a at point A are limited to particles reached by the finite series of spheres L, $2L$, and so on, the last sphere in this series having a radius that is less than τ by no more than the small distance L. The situation is illustrated in Figure 14-19.

Next, we consider the contribution to the mass from all those particles that are

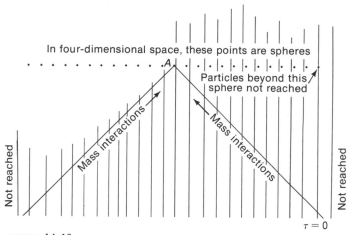

In four-dimensional space, these points are spheres

Particles beyond this
sphere not reached

Mass interactions

Mass interactions

Not reached

Not reached

$\tau = 0$

FIGURE 14-19

A set of uniformly spaced spheres with center at A. Because two of the three spatial dimensions are suppressed here, the spheres appear as simple points. There is a last sphere reaching particles that contribute to the mass of particle a at A.

reached by the sphere of radius $(n + 1) L$ but are not reached by the sphere of radius nL. Here, n is to be thought of as a large integer but not so large that $(n + 1) L$ exceeds τ. All the particles in question are essentially at a spatial distance nL from point A, and so each will make a contribution to the mass at A that is proportional to $1/nL$. The number of such particles will be proportional to the volume between the two spheres, which is $4\pi n^2 L^3$ to sufficient accuracy. Hence, the total contribution from all the particles reached by the sphere $(n + 1) L$ but not by the sphere of radius nL must be proportional to the product $(1/nL) \times 4\pi n^2 L^3$, that is, to $4\pi n L^2$.

The last step in the argument is to sum the contributions for all such pairs of spheres. The contributions vary from one pair to another like the integer n, and so the required summation is proportional to the series.

$$S = 1 + 2 + \cdots + k.$$

Here, kL is the radius of the last sphere in Figure 14-19.

There are many ways of obtaining this sum in a compact form. Notice that the first and the last member of the series add up to $k + 1$. The second and the last-but-one member add up to $2 + k - 1 = k + 1$. In fact, if we form pairs of members equidistant from both ends, we will always get the same answer: $k + 1$. How many such pairs are there? If k is even, there are $\frac{1}{2} k$ such pairs so that $S = \frac{1}{2} k(k + 1)$. If k is odd, there are $\frac{1}{2}(k - 1)$ such pairs together with the central member of the series, which is $\frac{1}{2}(k + 1)$. Hence, $S = \frac{1}{2}(k + 1) + \frac{1}{2}(k - 1)(k + 1) = \frac{1}{2} k(k + 1)$. In either case, we get the same answer. For large

k, we can approximate S as simply $\frac{1}{2}k^2$. In addition, we can write $k = \tau/L$ since L is small compared to τ. Substituting $k = \tau/L$, we get $S = \frac{1}{2}(\tau/L)^2$, which is proportional to τ^2. This answer is the one we are seeking. We now understand how it comes about that the mass of a particle can vary with time, and we have seen in an elegant way why a redshift effect is observed in the radiation from distant galaxies.

§14-6. The Redshift-Magnitude Relation of Hubble and Humason

We now aim to prove a still more ambitious result, namely, the relation between the magnitude and the redshift that was plotted in Figure 13-6. Consider light received at time τ from a galaxy at distance r as in Figure 14-20. Since the scale for measuring τ was defined by the requirement that light propagate at an angle of 45° in this figure, the light from the galaxy must start its journey at time $\tau - r$ in order to be received at time τ. Now, from what we have just seen, particle masses at the time of emission of the radiation must be proportional to $(\tau - r)^2$, whereas the particle masses at the time of reception of the radiation are proportional to τ^2. We can express this result for electrons by writing $m(\tau - r)$ and $m(\tau)$ for the masses at times $\tau - r$ and τ, respectively, with $m(\tau - r)$, $m(\tau)$, satisfying the equation

$$\frac{m(\tau)}{m(\tau - r)} = \frac{\tau^2}{(\tau - r)^2}.$$

Using a similar notation, we write $\nu(\tau - r)$, $\nu(\tau)$, for the frequencies of the radiation emitted by a certain transition of a certain atom; for example, $\nu(\tau - r)$

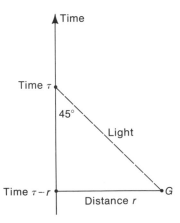

FIGURE 14-20
Light received at time τ from a galaxy at spatial distance r must start its journey at time $\tau - r$.

could be the frequency of the Hα transition of hydrogen emitted at time $\tau - r$, and $\nu(\tau)$ would be the frequency of Hα emitted at time τ. From what was said earlier about the relation of the emitted frequency of the radiation to the electron mass, namely, ν proportional to m, we see that

$$\frac{\nu(\tau)}{\nu(\tau - r)} = \frac{m(\tau)}{m(\tau - r)}.$$

Now, the left-hand side of this equation is just the quantity $1 + z$, so

$$1 + z = \frac{m(\tau)}{m(\tau - r)} = \frac{\tau^2}{(\tau - r)^2}.$$

To obtain the different values of the redshift z for different galaxies, as determined by an observer living at time τ, we simply keep τ fixed in this formula and vary the distance r.

The next problem is to find how the magnitudes of such galaxies depend on r. The apparent brightness of a galaxy of intrinsic luminosity $\mathcal{L}$ at distance r is just $\mathcal{L}/4\pi r^2$ because our universal geometry is now the same as our familiar local geometry. However, in using this simple result, we must remember to choose $\mathcal{L}$ for time $\tau - r$, the moment of emission of the light received at time τ. Now $\mathcal{L}$ can be shown to behave like the *square* of the particle masses. Thus, $\mathcal{L}$ is proportional to $m^2(\tau - r)$ and so to $(\tau - r)^4$. [Luminosity means energy emitted per unit time. Energy behaves like the particle masses and so is proportional to $m(\tau - r)$. Unit time is proportional to $1/m(\tau - r)$. Hence, energy divided by unit time is proportional to $m^2(\tau - r)$].

THE REDSHIFT-MAGNITUDE RELATION FOLLOWS IN A VERY SIMPLE WAY IN THE MINKOWSKI UNIVERSE WITH CHANGING PARTICLE MASSES

It follows from the preceding paragraph that the energy flux f is proportional to $(\tau - r)^4/r^2$. In Chapter 12 (see Figure 12-15), we saw that the magnitude, say, M, is defined by*

$$M = -2.5 \log f + \text{a certain constant,}$$

which can be written in the form

$$M = -2.5 \log \left[\frac{(\tau - r)^4}{r^2 \tau^2} \right] - 5 \log \tau + \text{a certain constant.}$$

---

*The usual symbol for apparent magnitude is m. In order not to confuse with the symbol for mass, we are using M instead of m for apparent magnitude.

For τ constant, the term $-5 \log \tau$ can be taken with the last term on the right-hand side. Thus, we have

$$M = -2.5 \log \left[\frac{(\tau - r)^4}{r^2 \tau^2} \right] + \text{a constant,}$$

the constant being the same for every galaxy observed at the time τ.

The last step in the argument comes from combining this equation for M with the preceding equation for z. From $1 + z = \tau^2 / (\tau - r)^2$, it is easy to see that

$$\frac{r}{\tau} = 1 - \frac{1}{\sqrt{1 + z}}.$$

Now start by specifying z and calculate r. Then use r to calculate M. The result for various starting values of z leads to the curve given in Figure 13-6 and reproduced here as Figure 14-21.

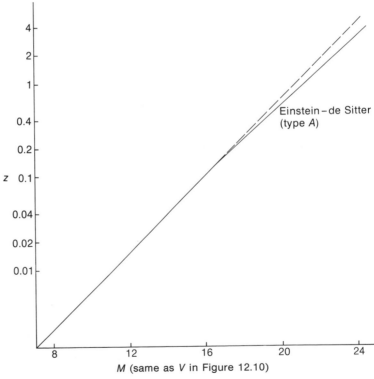

FIGURE 14-21
The z-M relation for the Einstein–de Sitter model.

This result is a considerable achievement since an extended technical course on relativity and cosmology is usually considered necessary in order to obtain it. Here we have used only very straightforward considerations.

§14-7. The Early Universe

We now have a clear picture of the structure of the universe in terms of the Einstein–de Sitter model. We have a simple understanding of the reason galaxies exhibit the redshift phenomenon: because particle masses increase with time. Our galaxy polygons no longer shrink to zero at the initial moment $\tau = 0$. Geometry is simple at all times. In particular, there is no geometrical problem at $\tau = 0$; the geometry of special relativity applies at $\tau = 0$ just as it does at any other moment of time. Notice too the important point that all our particles are at rest. We have no need for the Doppler effect here. There is no conceptual difficulty of the kind we encountered in Chapter 12 where we considered the erroneous idea, which often arises out of the Doppler interpretation of the redshifts, that because all galaxies appear to recede from our own galaxy, we must therefore be at a center of the universe. No center appears in Figure 14-16. We also see what makes the moment $\tau = 0$ so peculiar. Particle masses are zero at $\tau = 0$. They are zero because there were no interactions preceding $\tau = 0$.

Consider the situation when τ is small but not strictly zero. Because the masses of particles are then small, radiation emitted by atoms has low frequencies. Thus, an atom that now emits visible light in the terrestrial laboratory would emit even radio waves if τ were sufficiently small. Since the frequencies of oscillation of any radiation left over from the early history of the universe will not have changed with the passage of time, the geometry being now of simple Euclidean form, we expect that such radiation, if it exists, will be observed only at low frequencies. We can calculate that the frequency distribution would have a form of the kind shown in Figure 13-26, but we cannot, as yet, calculate which of the various possibilities it would follow. Indeed, we saw in Chapter 13 that this radiation is none other than the microwave background first observed by Penzias and Wilson in 1965. Of the various curves in Figure 13-26, the lowest one for 3 K comes closest to the observations. An extension of the picture we have so far presented leads to a further understanding of the origin of this radiation, as will be seen in Section 14-9.

In many modern cosmological discussions, it is assumed that radiation was already present at $\tau = 0$. Although this idea may seem somewhat peculiar since nothing that could have generated the radiation is supposed to have existed before $\tau = 0$, it is an idea with interesting consequences. Even though the radiation was of only low frequency, it must have dominated the behavior of matter just because particle masses were so close to zero at small τ. The radiation could have produced all manner of esoteric particles that now can be generated only by powerful accelerators like the one at Batavia, Illinois (Plates VII and VIII). The nature of the particles existing very near $\tau = 0$ would be exceedingly complex, and the particles would belong to what has been called the *hadronic era* of the universe.

Some physicists have wondered whether such complexities could have had an effect that is still observable in the present-day world. Could a bunching of particles near $\tau = 0$ have led to galaxies being formed? This question may seem ambitious, but we must bear in mind that other apparently more straightforward attempts to explain the origin of galaxies have not had very much success.

§14-8. The Present Cosmological Dilemma

The picture of the Einstein–de Sitter model as we have developed it so far is summarized in Figure 14-22. Although the general scenario is satisfactory in terms of observable features, such as Hubble's law or the microwave background, it nevertheless has unsatisfactory features. The one that strikes us immediately when we look at Figure 14-22 is that all particle world lines end abruptly at the epoch $\tau = 0$. Why should the lines suddenly end in this way? As far as we can see, the geometry at $\tau = 0$ is the same as at any later epoch, $\tau > 0$.

Suppose that, instead of letting there be nothing before $\tau = 0$ in Figure 14-22, we seek to avoid the artificial breaks in the paths of the particles by extending them to earlier times, going backward indefinitely as in Figure 14-23. We are immediately in trouble because our method of calculating the mass of a particle leads to an infinite result. Thus, in the previous section we found the mass of a particle to be proportional to a finite series, $S = 1 + 2 + 3 +$

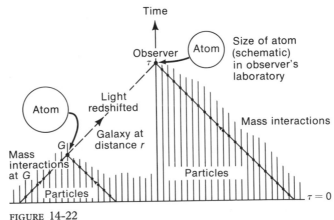

FIGURE 14-22
The summary of results for the Einstein–de Sitter model.

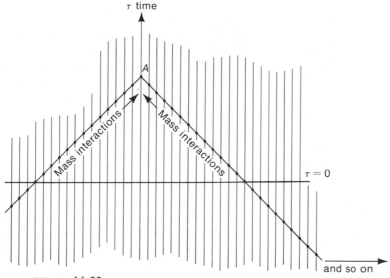

FIGURE 14-23
We can seek to avoid the breaks in the paths of particles by extending them backward indefinitely.

$\cdots + k$, in which the last member was defined in the manner shown in Figure 14-19: the product of the integer k and the average interparticle spacing L is close to τ so that, to sufficient accuracy, $k = \tau/L$. But in the situation of Figure 14-23, we have no such means of terminating the contributions to the mass from distant shells; that is, the series of spheres introduced in Section 14-5 with radii $L, 2L, 3L$, and so on can go on indefinitely, with the result that the calculated mass becomes proportional to a series $S = 1 + 2 + 3 + \cdots$, which goes on indefinitely.

We see from this abortive attempt to avoid the broken ends of the paths in Figure 14-22 that the result that the particle mass, m, is proportional to τ^2—a result necessary for the Einstein–de Sitter model—arose precisely because all the particles were taken to begin abruptly at $\tau = 0$.

THE PROBLEMS OF SPACETIME SINGULARITY AND THE ORIGIN OF THE UNIVERSE CAN BE AVOIDED

Let us accept the lesson we learned from Figure 14-23, return to Figure 14-22, and attempt to find a physical explanation of the abrupt beginning of the particle paths of Figure 14-22. Doing these things leads us to the problem we consider in Appendix D and illustrated there in Figure D-3. We shall see that a mathematical law (determined by an action-principle method) can be set up to describe the paths of particles with broken ends. Once we permit this mathematical law to operate physically, paths with broken ends at other values of τ also arise as in Figure D-3, and the general solution of the resulting physical problem leads completely away from the big-bang cosmologies. Once the paths with broken ends are described by a sensible physical principle, we arrive inevitably at the steady-state model of Appendix D. Yet, the existence of the microwave background appears to vitiate the steady-state model. The dilemma is, then, that normal physical and mathematical reasoning seem to lead us to a contradiction between theory and observation.

Many people are happy to accept this position. They accept Figure 14-22 without looking for any physical explanation of the abrupt beginning of the particles. The abrupt beginning is deliberately regarded as *meta*physical—that is, *outside* physics. The physical laws are therefore considered to break down at $\tau = 0$ *and to do so inherently*. To many people, this thought process seems highly satisfactory because a "something" outside physics can then be introduced at $\tau = 0$. By a semantic maneuver, the word "something" is replaced by "god," except that the first letter became a capital, God, in order to warn us that we must not carry the enquiry any further.

Attempts to explain phenomena by means of metaphysical intrusions into the world have always failed in the past. At the beginning of the nineteenth century, it was thought impossible to synthesize organic molecules by normal chemical processes. Now a whole industry is based on doing so. The origin of life was another supposed breakdown of the physical laws, and this point of view also appears to have collapsed. It is true that phenomena that have forced a widening of the physical laws have been discovered in the past. The discovery of radioactivity is one example. But widening the physical laws does not change their basic logic. Of course, one can argue that the origin of the universe is by its very nature a special case. Although to many this last contention appears respectable, we prefer to rely on past experience. We do not believe that an appeal to metaphysics is needed to solve *any problem of which we can conceive.*

Is the time ripe for the solution of the dilemma just described? This question is perhaps the most crucial of any in astronomy, perhaps of any in physics. We explain why we believe that inevitably the time must be ripe.

It is a curious aspect of scientific research that no matter what stage may have been reached, whether the sophistication is that of 1800, 1900, 1950, or 1980, perfect understanding always seems to be just around the corner, although to the workers at any one time it is clear that the similar confidence of earlier generations was wholly misplaced. Why do we have this illusion of perfect truth always waiting around the next corner in what is obviously a long and tortuous road? Because we cannot conceive of a problem until we are close to its solution. The solution to the problems of which we can conceive do indeed lie around the next corner of the road. We take no account of the problems that will torment future scientists for the good reason that we cannot yet conceive of them.

We take it, then, that since the problem associated with $\tau = 0$ in Figure 14-22 is one of which we can conceive and one that can be formulated, the time must be ripe for its solution. What line should we take? Should we perhaps return to the steady-state model and simply refuse to bow to the criticism based on the microwave background? This was the attitude we took ten years ago toward criticisms based on the counting of radio sources, and it has turned out that those criticisms were not so strong as they first seemed. Perhaps the criticisms based on the microwave background will also turn out to have been overstated and weaken with the passage of time? Possibly so, but at this stage, after ten years or more of attempting to "tough it out," we prefer to try to break out of the restraining cycle of the argument just described. Let us attempt to do so at what may at first seem an unprofitable point, the extension of Figure 14-22 to Figure 14-23, the extension that led to infinite mass values. Let us see if the infinites can be avoided in some way.

We know that there can be both positive and negative contributions to an electrical field, depending on the signs of the electric charges of the contributing particles. Aggregates of matter contain particles that have both positive and negative charge (protons and electrons) and hence make both positive and negative contributions to the field.

Let us try a similar idea for a field giving rise to particle masses, with positive and negative contributions coming from distinct, very large-scale aggregates, large even compared to the distances of remote galaxies. There is to be an important difference from the electrical case, however. An individual aggregate generates either a systematically positive contribution or a systematically negative contribution, not a mixture of the two. For example, suppose we consider positive and negative contributions to be separated on a cosmological scale at the time $\tau = 0$, as in Figure 14-24. Then, provided we make one important change of method, we arrive at calculated particle masses proportional to τ^2 just as we require for the Einstein–de Sitter model.

In Figure 14-25, we have a mass interaction of the form that was used in Section 14-5. The portions of the paths of particles a and b shown in the figure are considered to be on the same side of $\tau = 0$ (the positive side, say) and with τ at point A on the path of particle a at a later time than at point B on the path of particle b. In our previous method of calculation, the mass of particle a at point A was considered to receive a contribution from point B, *but the mass of particle b at point B was not considered to receive a contribution from A;* that is, the method of calculation had an undesirable asymmetry. Let us make the interaction of Figure 14-25 symmetric. In Newtonian terminology, action and reaction are made equal and opposite. This change is indeed a major improvement of method, and it leads to the required result that the particle masses are proportioned to τ^2.

The dilemma just set out is therefore resolved by the symmetric system of Figure 14-25. The need for the particle paths to have broken ends is then avoided. But why is there a switch in the sign of the interaction at $\tau = 0$? Why is this particular moment so special? These are legitimate questions, and it is encouraging to find that they can be answered by a quite powerful generalization of the specialized arrangement of Figure 14-24.

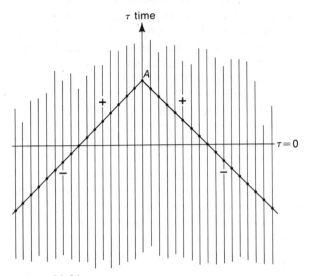

FIGURE 14-24
Illustrating the hypothesis that positive and negative mass interactions are separated at time $\tau = 0$.

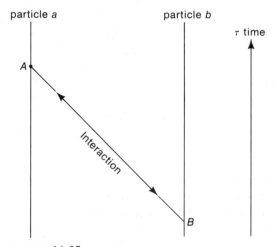

FIGURE 14-25
The situation of Section 14-5, where the mass of particle a at point A received a contribution from the (earlier) point B of particle b, but the mass of b at B did *not* receive a contribution from A, is made symmetric so that the interaction goes equally from A to B.

Consider the situation shown in Figure 14-26 where there are more than two + and − aggregates. (As always, we note the illustrative limitations imposed by the need to represent what is really a four-dimensional situation in terms of only two dimensions.) To determine the sign of the mass interaction between points A and B on the paths of particles a and b, shown schematically in Figure 14-27, we multiply the signs appropriate to the regions in which A and B happen to fall. If A and B are both in + aggregates, then the sign of the interaction is $(+1) \times (+1)$ and so is positive. If A and B are both in − aggregates, the sign is determined by the product $(-1) \times (-1)$ and so is again positive. But if one point is in a + aggregate and the other in a − aggregate, the sign is given by $(-1) \times (+1)$, which is negative. This meaning is the one to be attached to the + and − signs of Figure 14-26.

Figure 14-26 is a schematic representation of the universe on a scale much larger than the portion of the universe accessible to practical observation. Indeed, we are to think of our observations of all the galaxies, even the most remote ones visible in the largest telescopes, as occupying only a comparatively *small element* of just one of the aggregates of Figure 14-26; to be more specific, let us say a + aggregate.

The interactions on a particle at an arbitrary point anywhere in the universe will in this picture be a complicated addition of contributions from all the various aggregates. If we make the sensible assumption that *on the average* − aggregates are as important as + aggregates, the combined effect of all interactions at an arbitrary place is as likely to be negative as positive. Regions where the contributions add to a positive total will be separated from regions with a negative total by surfaces on which the + and − contributions just cancel each other. The mass of a particle at the points on such surfaces will be zero.

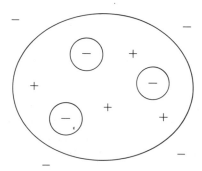

FIGURE 14-26
Schematic representation of large scale + and − aggregates.

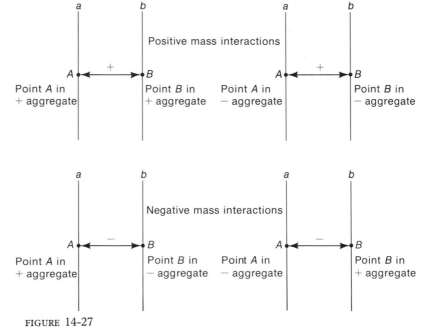

FIGURE 14-27
The rule for determining the sign of the mass interaction between point A of a and point B of b is that the sign is $+$ if both the points A and B are in aggregates of the same type; otherwise the sign is $-$.

In an extremely general way, we have now arrived at a crucially important result: the essential lesson we learn from the study of cosmology, especially from the study of the redshifts of the galaxies, is that *we happen in our element of the universe to lie near such a surface of zero mass*. Notice that the surface in question need not be a single unique surface as we took it to be in Figure 14-24. To emphasize the point that our observations range over only a small element of the universe, let us show in Figure 14-28 that Figure 14-24 is indeed only a small element of Figure 14-26, noticing that our studies of cosmology have been concerned with only this tiny region of a much vaster universe.

On the scale of this vaster universe, we can no longer use the symmetries enunciated in Chapter 12. Instead, we must grapple with the full complexity of Riemannian geometry. It is only on the scale of the small triangle of Figure 14-26 that this so-called principle of cosmology holds good. Indeed, it is just because we happen to be close to a zero-mass surface that it does so. Discovering the full-scale world geometry will therefore be a very difficult task; it will be far from easy to obtain the detailed forms of the zero-mass surfaces shown schematically in Figure 14-26.

Even so, it is surprising how much insight we can gain into the effects of the local zero-mass surface shown in Figure 14-24. Suppose light travels in the same

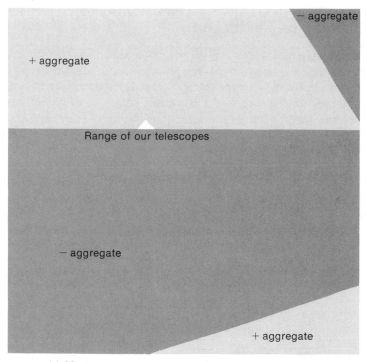

FIGURE 14-28
Our studies of cosmology may be concerned with only a small element of a much vaster universe.

sense on *both sides* of this surface, and suppose that galaxies and stars also exist on the other side. Would we then expect to be able to observe these other galaxies? Unfortunately, we could not expect to observe them directly because all the light from the stars of such galaxies must be strongly scattered, absorbed, and blurred at times close to $\tau = 0$ by particles lying between the stars. This strong blurring effect is caused by the smallness of the particle masses near $\tau = 0$. However, the blurred radiation would continue into our half of the zero-mass surface and would indeed be observable. It would have just the black-body form we studied in Section 14-7. It would, moreover, have an energy content determined by the conversion of hydrogen into helium within the stars on the other side. This energy can be calculated and shown to lead to a temperature on the order of $3\,\mathrm{K}$ for the blurred radiation, just what the observed microwave background is found to have. Therefore, the existence of the microwave background (so apparently damaging to the steady-state model) favors the ideas presented here quite strongly. We are not required to postulate its existence ad hoc as one must do in the usual big-bang cosmologies. With some justice, we might argue that the microwave background demonstrates the existence of the other side of the local zero-mass surface.

Inertia and Cosmology

The concept of galaxies existing on the other side may also be related to problems of the clustering of galaxies on our side. Astronomers have long suspected that galaxies are more clumped together than might reasonably be expected if only random factors are at work. Figures 14-29 to 14-33 relate to other problems that we discussed in preceding chapters. They may receive full solution only when conditions on the other side are incorporated into our astronomical theories.

Much of what we have learned about astronomy in general is well brought out by Figures 14-30 to 14-32, which are all of NGC 1097, a galaxy best seen in the southern hemisphere. These photographs show how, by taking exposures in various ways, we discover the different structural properties of the galaxy. Figure 14-30, in the line Hα of hydrogen, shows a bright central nucleus and disc surrounded by remarkably thin spiral arms. The bright condensations,

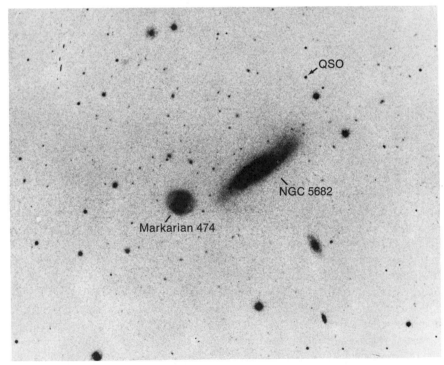

FIGURE 14-29
The galaxy NGC 5682 has a redshift $z = 0.0073$, Markarian 474 has $z = 0.041$, and the QSO has $z = 1.94$. This association could be due to chance, but then it is strange that these objects should be so peculiar in their individual forms. (Courtesy of H. Arp, Hale Observatories.)

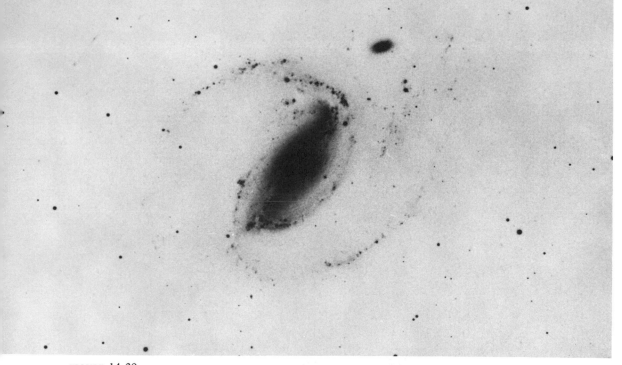

FIGURE 14-30
The galaxy NGC 1097 in the line Hα of hydrogen. Notice the remarkably thin outer spiral arms with many HII regions strung out along them. (Courtesy of H. Arp, Hale Observatories.

FIGURE 14-31
An intermediate exposure of NGC 1097 suggesting a swirling rotary motion. (Courtesy of H. Arp, Hale Observatories.)

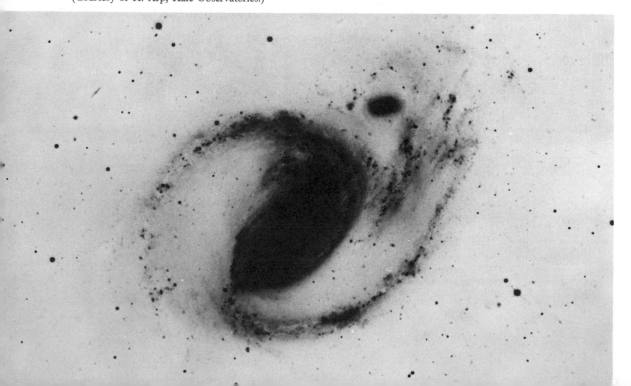

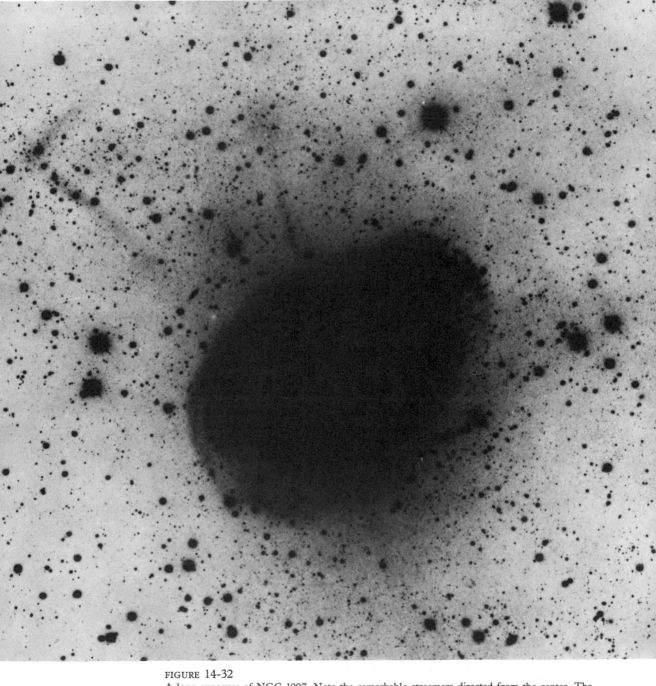

FIGURE 14-32

A long exposure of NGC 1097. Note the remarkable streamers directed from the center. The streamer at upper left has a chain of five soft images, possibly small subsidiary galaxies, seemingly associated with it. (Courtesy of H. Arp, Hale Observatories.)

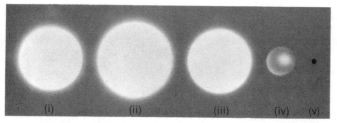

FIGURE 14-33
The expansion and subsequent contraction of a local object without
internal pressure follows rules that are strikingly similar to those of a
world geometry of type *B*. The local object begins by expanding (i)
to maximum size (ii), then falls back (iii), and continues to fall (iv)
until it eventually becomes a black hole (v).

strung like beads along the arms, are gaseous nebulae (Section 9-4), glowing
clouds of ionized hydrogen made visible by associations of bright young stars
not more than a few million years old. In the intermediate exposure of Figure
14-31, the arms broaden and strengthen in appearance but still have myriads of
gaseous nebulae. The pattern now suggests a dramatic swirling motion for the
whole galaxy. In the long exposure of Figure 14-32, the central detail has been
burned out of the photographic plate, but now a profusion of objects—star
clusters and what are perhaps faint small galaxies—can be seen surrounding the
whole system. Two streamers from the center to the upper left are easily detected, and a counterstreamer in the opposite direction from one of these is
probably also present. The leftmost upper streamer has a chain of five soft
images, apparently small subsidiary galaxies, strung along its outer part just
before the streamer makes a sudden curve to the right. This chain appears to be
associated with the streamer and hence with the main galaxy.

The streamers suggest the presence of an explosive object at the center of the
galaxy, an object perhaps similar in its physical properties to a QSO. The
general halo of luminosity surrounding the galaxies could be an overlap of many
streamers, older than the ones we see, that are almost certainly very young
compared to the age of the galaxy itself—and probably young even compared to
the rotation period of the galaxy. We are probably observing the effect of
explosions that occurred during the last 10 to 100 million years.

§14-10. Black Holes and White Holes

We end this chapter with a discussion of black holes and white holes within the
wider framework of Mach's principle. In Chapter 11, we saw that a black hole is
formed by the gravitational collapse of a massive object and that a white hole

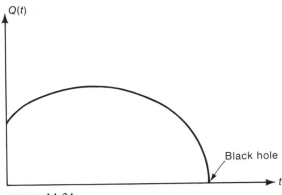

FIGURE 14-34
The $Q(t)$ curve for such a local object reaches a maximum and then turns over and decreases to zero. This phenomenon produces a black hole.

may be regarded as a time reversal of gravitational collapse—an implosion changed to an explosion.

Suppose we start, as in Chapter 11, with an extended homogeneous ball of dust, initially with a modest *outward* motion. The ball will expand for a while and then contract as gravitation asserts itself. Gravitational collapse to a black hole and then to a spacetime singularity occurs as is illustrated in Figure 14-33. An overall scale factor Q that changes with time can be assigned to the object. The time t of Figure 14-33 is measured by an observer at the surface of the object. The scale factor $Q(t)$ has the same behavior as the scale factor in a cosmology of type B (Chapter 13), except that Q does not start from zero as it does in Figure 13-4. Instead, we have the behavior of Figure 14-34. The critical aspect of Figure 14-34 is that Q declines to zero, which means that the object collapses into a point where, according to usual ideas, it ceases to exist.

BLACK HOLES AND WHITE HOLES ARE LINKED TOGETHER

The question now arises whether we can proceed further as we did in the cosmological case. Can we avoid the object's ceasing to exist in somewhat the same way that we resolved the dilemma of the origin of the universe? We resolved the latter dilemma by changing from the picture of Figure 14-12 to that of Figure 14-13 and by replacing Figure 14-22 by Figure 14-24—that is, by adding a second half to the universe. The answer for a local object is that we can indeed proceed in a similar way. The second half that is thus introduced can be referred to as a *white hole*. The structure developed by a local collapsing object consists, then, not just of a black hole but of a black hole together with a white hole.

The spatial scale of a black hole/white hole (BH/WH) ranges from a few kilometers for a body with mass of stellar order to about 0.01 light-year for a body with mass of galactic order. The nature of the zero-mass surface requires the following sequence of events. The object collapses into a black hole as the zero surface is first approached. Immediately on crossing the surface, the body becomes a white hole. This first white hole does not persist, however, but changes for a second time to a black hole as a second condition of zero mass is reached by the particles of the body. Thereafter, a second white hole emerges, and it could well be observable as an outburst of the kind that is found in QSOs and radio galaxies.

This double sequencing of a white hole following a black hole is illustrated in Figure 14-35, where the closed surface describes a zero-mass surface, one of many that might exist in the wider picture of the universe described in Section 14-9. The world lines of matter enter this closed region where a switch in the sign of the mass interactions takes place. *It is this switch of sign that brings about a change from general relativity.* The change shows itself in the connection between the + outer region and the − inner region. *General relativity holds good separately in either region,* but the fit of what mathematicians call boundary conditions is different across the closed surface.

Before the world lines enter the closed zero-mass surface of Figure 14-35, the situation is that of a black hole. As the world lines enter the bubble, the particles behave like a white hole. But after an initial white-hole phase, there is a further black-hole situation as the world lines cross the surface for a second time. After this second crossing, the world lines have the appearance of a white hole to an external observer O in the + region. For O, who may not be able to look into the closed surface, the second white hole appears unconnected with the first black hole, whereas from the point of view of an observer on the object, they are

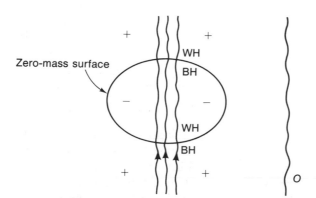

FIGURE 14-35
A set of particles entering and leaving a zero-mass closed surface represents two pairs of black hole/white hole (BH/WH). To a remote observer O, the second WH may seem unconnected to the first BH. The zero-mass surface arises from a switch in sign of the mass interaction within the surface.

connected in the manner of Figure 14-35. Thus, a black hole may be connected with a white hole.

The second topic we want to consider here is that of spacetime singularity. We saw in the case of the big-bang universe that Einstein's theory requires the universe to start from a spacetime singularity. The singularity occurs as the end point of gravitational collapse or as the beginning of a white hole. Is the appearance of singularities in these cases an inevitable aspect of Einstein's theory?

In cosmology, we made simplifying assumptions—the cosmological principle and the Weyl postulate—about the symmetries of the universe. In our discussion of the gravitational collapse of a homogeneous dust ball, there were also many symmetries. Did the singularities of spacetime arise because of these assumptions of symmetry?

In the late 1950s and the early 1960s, many theoreticians believed that, by going to more complicated matter distributions, they might avert the appearance of a singularity. For example, rotation introduces centrifugal forces perpendicular to the axis of rotation. These forces could withstand the force of gravity. But, of course, they are ineffective *along* the rotation axis. Would other anisotropies help? If the universe expands at different rates in different directions, the medium is sheared in much the same way that an iron bar develops shear when twisted. Shear in the cosmological medium, however, makes matters worse and helps the appearance of singularities. Recourse to inhomogeneities also did not help. In fact, powerful theorems by Roger Penrose, Stephen Hawking, and Robert Geroch in the second half of the 1960s showed that none of these devices would help—that the appearance of singularities is an inevitable feature of general relativity except under highly specialized or esoteric cases.

In going from Figure 14-12 to Figure 14-13, we left the usual description of general relativity and found that, as far as the Einstein–de Sitter model is concerned, the spacetime singularity at $t = 0$ in the relativistic picture becomes replaced by the zero-mass surface at $\tau = 0$ in the new picture. Although the two are mathematically equivalent for describing our part of the universe, the new picture has the advantage of leading us to a much more general view of the universe that not only extends to the other side of $\tau = 0$ but also shows that our observable universe may be a small section of a larger universe consisting of many zero-mass surfaces.

The starting point of this generalization was the Einstein–de Sitter model, which has many symmetries. This model again raises the question: Could we remove the singularities of the Einstein–de Sitter model because of its high degree of symmetry? It is certainly true that the geometry we obtained in our new picture—the geometry of special relativity—owed its simplicity to the symmetries of the Einstein–de Sitter model. If we abandon this model and go to other more general models of relativity, such as those with shear and rotation and with inhomogeneity, we will not be able to arrive at the simple geometry of special relativity in our new picture *but in all these other cases our new geometry can be shown to be nonsingular.* Where there was singularity in the usual relativistic models, we now have a zero-mass surface. Thus, we are back at our general picture of a universe containing zero-mass surfaces described in Figure 14-26, and we therefore have the following equivalence:

Particle masses constant + spacetime singularity(ies)
$$\Longleftrightarrow \text{Particle masses variable} + \text{zero-mass surface(s).}$$

If we proceed in this equivalence from right to left, it becomes immediately clear why singularity is such an inevitable feature of general relativity. Going back to our system of units, and recalling that the length unit goes as

$$L \sim M^{-1},$$

we see that, as M becomes zero, L must go to infinity. Any attempt to keep the length unit rigid as is done in general relativity by changing the rules of geometry is therefore going to be futile as we approach a zero-mass surface. The price we pay for our insistence on carrying our rigid rod all the way to $m = 0$ shows up in the impossible constraints put on the geometry, so that a breakdown finally occurs at $m = 0$. This situation is the spacetime singularity of general relativity.

Rather than read any metaphysical significance into spacetime singularities, we should recognize them for what they are: the result of using rigid units where they are not permitted mathematically.

General Problems and Questions

1. A stone is whirled around in a circle of radius 100 cm so that it completes 50 revolutions per minute. Calculate the acceleration of the stone in the rest frame of the center of the circle.

2. In Problem 1, calculate the centrifugal force. Explain why it arises. (The mass of the stone may be taken as 100 g).

3. Write an essay on absolute space.

4. Calculate how long the plane of oscillation of the Foucault pendulum would take to complete one revolution at (1) the North Pole, (2) New York City, (3) Los Angeles, and (4) Colombo.

5. Explain the underlying observations that led to Mach's principle.

6. In terms of L, M, and T, write the units of (1) the electric field, (2) Planck's constant, (3) pressure, (4) magnetic potential, and (5) impulse.

7. Express the quantities in Problem 6 as powers of M in the case $c = 1$, $\hbar = 1$.

8. With $c = 1, \hbar = 1$, and the electron mass as a unit for M, give numbers for the following: (1) the mass of the Sun, (2) the speed of sound at normal temperature and pressure, (3) 1 Volt, (4) Hubble's constant ($H^{-1} \cong 1.3 \times 10^{10}$ years), (5) one horse power, and (6) the "radius" of the hydrogen atom.

9. Write an essay on the use of units in science.

10. Discuss how the redshift can be explained in the picture of the *non*expanding universe with variable particle masses.

11. Illustrate how geometry can be simplified by the use of variable particle masses.

12. In the reinterpretation of the Einstein–de Sitter model as a nonexpanding universe with variable particle masses, calculate the mass of the hydrogen atom in a galaxy with redshift 3.

13. Discuss the microwave background in terms of the state of the universe close to $\tau = 0$.

14. Describe the role of black holes and white holes in relation to the surfaces of zero mass.

15. Comment on the status of spacetime singularities in general relativity. What reinterpretation of such singularities can be given in terms of the ideas of this chapter?

Suppose we restrict our attention to measurements of distances and choose, as is usually done, a rectangular system of coordinates $0X$, $0Y$, $0Z$ as in Figure A-1, which shows typical examples of three points P, Q, and R lying on a curve. The z coordinate, measuring distance from the plane $X0Y$, is positive for P, zero for Q, and negative for R. Three coordinates x, y, z are needed to specify completely the location of P, a typical point in space.

Suppose next that we are given the three coordinates of P to be (x_P, y_P, z_P) and of Q to be (x_Q, y_Q, z_Q). What can we say about the distance PQ? If we accept the statement that Euclid's geometry holds in space, we have the Pythagorean theorem to give us the answer:

$$PQ^2 = (x_Q - x_P)^2 + (y_Q - y_P)^2 + (z_Q - z_P)^2.$$

When P and Q lie in the plane $X0Y$ (that is, their z coordinates are zero), the answer is the familiar one of the right-angled triangle in a plane shown in Figure A-2. Also, PQ^2 is an *invariant*, that is, we would have gotten the same answer for PQ^2 if we had used another rectangular coordinate frame.

Now let us consider the four-dimensional picture of space and time. To specify an event completely, we must state *when* as well as *where* it occurred. For a typical event P, we now need four coordinates: x, y, z, and t. In Newtonian

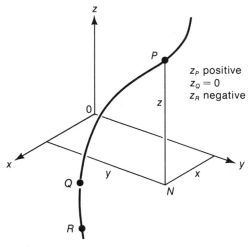

FIGURE A-1
A rectangular coordinate system. Three coordinates, x, y, z, as needed to specify a typical point P in space.

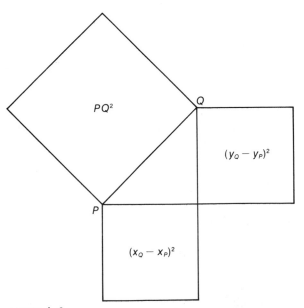

FIGURE A-2
In Euclidean geometry, the square of the long side of a right-angled triangle is equal to the sum of the squares of the other two sides:

$$PQ^2 = (x_Q - x_P)^2 + (y_Q - y_P)^2.$$

This is the famous theorem of Pythagoras.

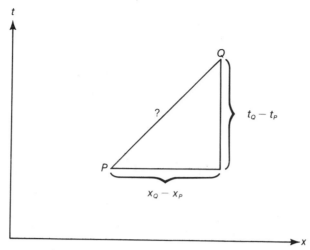

FIGURE A-3
In the spacetime diagram, the rule for measuring PQ is:

$$S^2_{PQ} = (t_Q - t_P)^2 - x_Q - x_P)^2.$$

dynamics, t was simply the universal absolute time common to all particles. It was the same whether measured by one observer or by another in relative motion. In special relativity, it becomes important to specify not only the t coordinate but also the observer who is using the clock to measure it. As mentioned in the text, it is customary to choose an inertial observer.

Suppose we now have two events P and Q with their y and z coordinates the same (that is, $y_P = y_Q$ and $z_P = z_Q$) but with different x and t coordinates. In Figure A-3, we again have apparently a right-angle triangle with sides $x_Q - x_P$ and $t_Q - t_P$. What is the length PQ? Is it given by

$$PQ^2 = (t_Q - t_P)^2 + (x_Q - x_P)^2,$$

as suggested by the Pythagorean theorem? If we were to proceed in this way, we would be treating time in the same way as space. All four dimensions of *spacetime* would be on the same footing. From all aspects of experience, we know this to be wrong. Subjectively, we are aware that time is somehow different from space. Yet, if we do not accept the Pythagorean theorem as giving a physically significant quantity when time is involved, how do we proceed?

Einstein's first great contribution to physics, which he made in 1905 while working as a clerk in the Swiss Patent Office, was equivalent to requiring that there be a *sign difference* between time quantities and spatial quantities. Thus, writing $S^2_{PQ} = (t_Q - t_P)^2 - (x_Q - x_P)^2$, the physical relationship between points P and Q, is to be determined by S_{PQ}, *not* by $PQ^2 = (t_Q - t_P)^2 +$

$(x_Q - x_P)^2$. This may seem a simple enough step in itself, but it turns out to have profound consequences as we shall see.

In a case where the y and z values of the points P and Q are also different, we keep the same sign for all the spatial dimensions but give an opposite sign to the time dimension. The physical relationship between P and Q is now determined by S_{PQ}, where

$$S^2{}_{PQ} = (t_Q - t_P)^2 - [(x_Q - x_P)^2 + (y_Q - y_P)^2 + (z_Q - z_P)^2].$$

The spatial part of $S^2{}_{PQ}$, within the square brackets, has the same structure as it has in ordinary three-dimensional Euclidean geometry.

In this equation for $S^2{}_{PQ}$, it is implied that both t and x, y, z are measured on the same standard, say, by using a light wave of a standard type. However, it turns out that our characteristic time scales and space scales are such that a light wave that is suitable for the one is not suitable for the other. Thus, the basic time unit—the second—is obtained from a light wave resulting from certain transitions of the caesium atom, whereas the basic space unit—the centimeter—is obtained from transitions of the krypton atom. Since different units are used for t and for x, y, z, we need to introduce a *conversion factor* that we denote by c. Thus, we rewrite $S^2{}_{PQ}$ as

$$S^2{}_{PQ} (t_P - t_Q)^2 c^2 - [(x_Q - x_P)^2 + (y_Q - y_P)^2 + (z_Q - z_P)^2].$$

The conversion factor, $c = 2.997929 \times 10^{10}$ cm/second, is none other than the speed of light. This expression for $S^2{}_{PQ}$ must be applied by practical physicists using seconds for measuring time and centimeters for measuring spatial distances. More fundamentally, however, we can use the same unit for time and space, which amounts to putting $c = 1$ in accordance with the form of $S^2{}_{PQ}$ given first.

Next, we notice that it is possible for $S^2{}_{PQ}$ to be zero even though P and Q are distinct points. This is a quite unfamiliar circumstance, that the distance can be zero between separated points. Taking points P and Q to have the same y and z values, the distance will be zero whenever the square of the ordinary spatial distance $(x_Q - x_P)^2$ happens to be equal to the square of the time difference $(t_Q - t_P)^2$. Let us think of P as a fixed point and ask for all the points Q such that the distance P to Q is zero. The result is shown in Figure A-4. The point Q must always lie on one or the other of two lines drawn at an angle of $45°$ to the direction of increasing time.

Now let P be a source of radiation. In what directions does the radiation travel? The answer is along the lines of Figure A-4, but only along the segments in the forward sense with respect to time.

Nothing is changed in principle if we also add the two remaining dimensions y and z. We have $S^2{}_{PQ}$ zero when the square of the ordinary spatial distance, $(x_Q - x_P)^2 + (y_Q - y_P)^2 + (z_Q - z_P)^2$, is equal to the square of the time difference, $(t_Q - t_P)^2$. Once again, radiation emitted from an isotropic source at

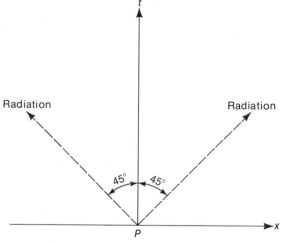

FIGURE A-4
When the same radiation wave is used to measure both
time intervals and spatial distances, radiation from a source
at P inevitably travels at 45° in the x, t diagram.

P travels to all points Q such that $S^2{}_{PQ}$ is zero. The points Q form a cone in the forward time sense, but here we cannot easily draw it because the full four dimensions of spacetime are not readily represented on a two-dimensional piece of paper. This light cone is shown schematically in Figure 10-10.

We have now attached physical meaning to what was previously only a statement. When we wrote

$$S^2{}_{PQ} = (t_Q - t_P)^2 - [(x_Q - x_P)^2 + (y_Q - y_P)^2 + (z_Q - z_P)^2],$$

one could have responded with a laconic "So what? Why not some other expression for the four-dimensional distance?" For mathematics, this question is an entirely valid one, but not for physics since we have now used our statement to make an assertion about the world that may or may not be true. Our assertion is in fact true, provided we do not extend the light cone from the source P to points Q such that $\sqrt{(x_Q - x_P)^2 + (y_Q - y_P)^2 + (z_Q - z_P)^2}$ is too large—that is, that $t_Q - t_P$ is too large.

However, the rule for computing $S^2{}_{PQ}$ changes in the presence of gravitation, and this change leads to the non-Euclidean geometries that form the basis for Einstein's law of gravitation. Thus, near a massive object or over very large distances (of billions of light-years), we may have to abandon the geometry of special relativity and look for more complex geometries.

We end this appendix with one application of the preceding expression for $S^2{}_{PQ}$. Suppose P and Q are two events on the world line of a particle moving with uniform velocity v relative to the inertial frame of measurement of $t, x, y,$ z. Then, since velocity denotes the rate of change of distance with time, we have

$$(x_Q - x_P)^2 + (y_Q - y_P)^2 + (z_Q - z_P)^2 = v^2(t_Q - t_P)^2,$$

and hence,

$$S^2{}_{PQ} = (t_Q - t_P)^2(1 - v^2).$$

Now the geometry of special relativity tells us that $S^2{}_{PQ}$ is invariant with respect to all inertial frames, just as in the three-dimensional Euclidean space PQ^2 was the same in all rectangular frames of measurement. So let us choose the reference frame in which the particle is at rest. Let t'_P and t'_Q be the time coordinates of P and Q in this new frame. Since the particle is at rest now, $x'_P = x'_Q, y'_P = y'_Q, z'_P = z'_Q$. Hence,

$$S^2{}_{PQ} = (t'_Q - t'_P)^2.$$

Equating the two values of $S^2{}_{PQ}$, we get

$$t'_Q - t'_P = (t_Q - t_P)\sqrt{1 - v^2}.$$

Note that when we go to practical units of measurement (centimeters and seconds) and replace v by v/c, the time dilatation is the same one we had arrived at earlier in Section 10-6. We mention a practical example of time dilatation next.

Fast-moving particles are constantly hitting the Earth's atmosphere from outside. These particles, known as *cosmic rays*, hit the atoms of the outer atmosphere at speeds very close to that of light, generating other fast-moving particles in their collisions. One of these other particles is the muon, written as μ. The muon can also be studied in the laboratory. It is found to decay spontaneously into several particles (of which the electron is one). *When a muon is traveling at comparatively slow speeds,* the characteristic time for this decay is about 2 millionths of a second, 2×10^{-6} second. Muons generated by cosmic rays are mostly produced at a height above the Earth's surface of about 30 km. Now it takes light 10^{-4} second to travel 30 km, and the muons cannot move downward toward ground level faster than this. A naive view argues that since the travel time of 10^{-4} second is much longer than the measured characteristic decay time of 2×10^{-6} second, the cosmic-ray muons must decay before reaching the ground, and therefore they cannot be observed in an ordinary laboratory. Experiment shows this naive deduction to be wrong. Many muons do in fact reach ground level. Why?

When muons are studied at low speeds in the laboratory, their proper times are not much different from our own proper time. A clock measurement made by us of 2×10^{-6} second is then a fair estimate of the muon's own proper decay time. But our clock measurement of 10^{-4} second for the travel time downward from a height of 30 km has no relevance at all to the decay of fast-moving muons, *that decay with respect to their own proper times*. To obtain proper time for a fast-moving muon, we must multiply the clock time of 10^{-4} second by a square-root factor, $\sqrt{1 - v^2}$, where v is the speed of travel. Evidently, for v close to the speed of light, that is, close to 1, this square-root factor is small, and the result of multiplying 10^{-4} second by $\sqrt{1 - v^2}$ becomes less than 2×10^{-6} second. Thus, in the muon's own proper system, there is insufficient time for decay (for most of them, at any rate). Consequently, fast-moving muons should reach ground level, as in fact they are found to do. This experiment shows decisively that our present way of thinking is correct.

Appendix B
Gravitational Radiation

It may seem surprising that, in our two chapters on gravitation, we have made no reference to the phenomenon of gravitational radiation. There are two reasons for this omission. Conceptually, gravitational waves are much more complicated to visualize than their electromagnetic counterparts. Secondly, there is as yet no agreement among experts that gravitational radiation has been detected in the laboratory. We briefly elaborate these points here.

The history of electromagnetism tells us that it was out of Maxwell's basic equations that the concept of an electromagnetic wave emerged. The propagation of an electromagnetic wave represents periodic changes of the electric and magnetic fields in space and time. These periodic changes appear to travel across space with a finite speed, the speed of light. Soon after this theoretical picture became clear, the waves were actually detected in the laboratory by H. Hertz. The time interval separating the theoretical concept (in the 1860s) and its experimental verification (in ~ 1888) was less than three decades.

Difficulties in the case of gravitational waves begin as soon as we ask the question: What disturbances should propagate in the form of a wave? We have already seen that gravitation according to Einstein is intimately connected with spacetime geometry. Accordingly, a gravitational wave should reflect changes of geometry through space and time. The difficulty comes in identifying precisely the physical parameters that are supposed to change periodically in a wave motion.

To give one example, when we switch on a wireless transmitter, we begin to generate electromagnetic waves. When we switch the transmitter off, the waves cease. If we switch on a source of gravitational waves, keep it running for a while, and then switch it off, not all geometric disturbances disappear. Some parameters describing the geometry when the source was operating do not die out. We presently clarify the characteristics of such a source.

The behavior of gravitational waves is somewhat clearer when the gravitational effects themselves are small. In the language of general relativity, this means when the geometry does not differ much from special relativity. There is considerable parallelism between such weak field waves and electromagnetic waves. Like electromagnetic waves, weak gravity waves propagate with the speed of light. The primary source of electromagnetic waves is a changing electric dipole moment. The primary source of gravitational waves is a changing *quadrupole* moment of the matter distribution (see Figure B-1). Just as the photon is the carrier of electromagnetic radiation, the *graviton* is the carrier of gravitational radiation.

The transition from weak fields to strong fields evidently makes for great complications. As yet, there is no exact solution describing gravitational waves radiated by a compact source in the strong-field case. The *nonlinearity* of Einstein's equation poses severe mathematical problems.

Another difficulty from the point of view of the detection of gravitational waves is the smallness of the energy of gravitational radiation from any laboratory source. To see this difficulty, consider a typical example. Suppose we have a cylindrical steel beam of radius 1 m, length 20 m, density 7.8 g/cm³, and mass ~490 tons. Let the beam rotate about its middle with an angular velocity of ~28 radians/second. (At this revolution rate, the centrifugal force and the tensile strength of the beam are in balance.) Such an arrangement radiates

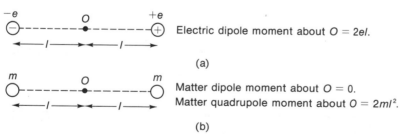

Electric dipole moment about $O = 2el$.

(a)

Matter dipole moment about $O = 0$.
Matter quadrupole moment about $O = 2ml^2$.

(b)

FIGURE B-1

In (a), the system of electric charges has a dipole moment equal to $2el$ along the line joining them. In case (b), the dipole moment of the matter distribution about the midpoint O is zero, but the quadrupole moment is $2ml^2$. (The quadrupole moment is a tensor and has components with respect to a pair of directions. In this case, the only nonzero component is along the identical pair of directions joining the two masses.)

gravitationally with a power of only $\sim 10^{-22}$ erg/second. To generate a power of 1 W, we would need about 100 billion billion billion such sources.

Astrophysics provides larger sources. Typical examples are massive objects undergoing gravitational collapse, exploding stars like supernovae, and, on an even larger scale, the exploding nuclei of galaxies. Binary stars are probably the most numerous sources of gravitational radiation. Typically, the output rate is from $\sim 10^{30}$ to 10^{32} ergs/second, and the energy of flux at the Earth is from $\sim 10^{-10}$ to 10^{-12} erg/cm^2/second. Although this flux is very much larger than the flux in the preceding laboratory example, it is still several orders of magnitude below the sensitivity of laboratory detectors.

It therefore came as a surprise to most theoreticians when, in 1972, Joseph Weber of the University of Maryland, a pioneer in gravitational wave-detector experiments, reported positive results from several observations. Weber's apparatus consisted of bars of aluminum 153 cm in length, 66 cm in diameter, and 1400 tons in weight. Each bar was suspended by a wire in a vacuum and was mechanically decoupled from its surroundings. Whenever gravitational waves hit such a bar on the broadside, the bar oscillates. The mechanical vibrations of the bar are converted into electric oscillations by coupling it to electric circuits through so-called piezoelectric strain transducers. To avoid recording local (nongravitational) disturbances, Weber set up another similar apparatus at the Argonne National Laboratory near Chicago. Only those events that were simultaneously recorded in both places were included in the analysis. According to the early reports, gravitational waves seemed to come from the general direction of the galactic center or from the opposite direction. The cause of these waves was not known, but their magnitude (if they were gravitational waves) implied very large disturbances in the galactic center (if that were the source). Later experiments by others have failed to reproduce Weber's results. Weber, on the other hand, still continues to get similar coincidence events as before. Perhaps more elaborate and more sensitive future detectors will resolve the controversy that exists in this area today.

Recently there has been indirect evidence to show that gravitational radiation exists. This comes, not from the measurements using gravitational wave detectors, but from radio astronomy. In late 1978, J. H. Taylor, L. A. Fowler, and P. M. McCulloch announced a result (see *Nature*, 277, 437, for details) from their radio measurements of binary pulsar PSR 1913 + 16 that imply the existence of gravitational radiation. To understand this work, we go back to our discussion of gravitational radiation from binary stars.

If we take this picture seriously and apply it to binary pulsar PSR 1913 + 16, we arrive at the following conclusion. As a result of energy loss by gravitational radiation, the orbits (around each other) of the two compact stars shrink and the orbital period *decreases*. As we mentioned earlier, the effect is expected to be small and would have remained undetected but for the remarkable accuracy now possible with present radio measurements. Using the 305-m radio telescope at Arecibo, Puerto Rico, Taylor and his colleagues were able to measure the rate of decrease of the orbital period, and this decrease does turn out to be on the order predicted by the preceding picture of gravitational radiation. Although such a measurement still leaves room for alternative explanations *not* requiring gravitational radiation, the general consensus seems to be in favor of this picture.

The measurements of this pulsar have also confirmed what had been noticed earlier in 1975, the precession of the perihelion of the binary orbit. We have discussed this general relativistic effect for the orbit of the planet Mercury around the Sun (see Figure 10-15). The perihelion of Mercury's orbit precesses at the rate of 43^s per century. By contrast, the precession rate observed in the case of PSR 1913 + 16 was as much as $4.226°$ per year. This large effect indicates how important general relativity is when gravitational effects are strong. In the binary pulsar, the two component masses are highly dense neutron stars in close proximity. Their gravitational effects are therefore much stronger than in the Sun-Mercury system.

Appendix C
The Doppler Effect

Two observers A and B are in relative motion with speed v. For simplicity, we take the motion to be of recession in the direction AB. Observer B sends a light signal to A, judging the light to have frequency ν, with a wavelength λ given by $\lambda\nu = c$. What will be the frequency measured for the light by A?

Suppose B is at a distance of 1 light-second from A at the moment of emission of a particular wavecrest (see Figure C-1). If A sets a clock to zero, $t = 0$, on reception of the wavecrest, then according to A's point of view, the wavecrest started from B at $t = -1$. From the point of view of B, the next wavecrest will be emitted at a time $1/\nu$ later than the previous wavecrest. If this were also A's point of view (it is not quite so, as we see in a moment), then A would argue that B moved a distance v/ν along AB during the interval between the successive wavecrests, and hence the second wavecrest would reach A at the time

$$\frac{1}{\nu} + \frac{v}{\nu c} = \frac{1}{\nu}\left(1 + \frac{v}{c}\right).$$

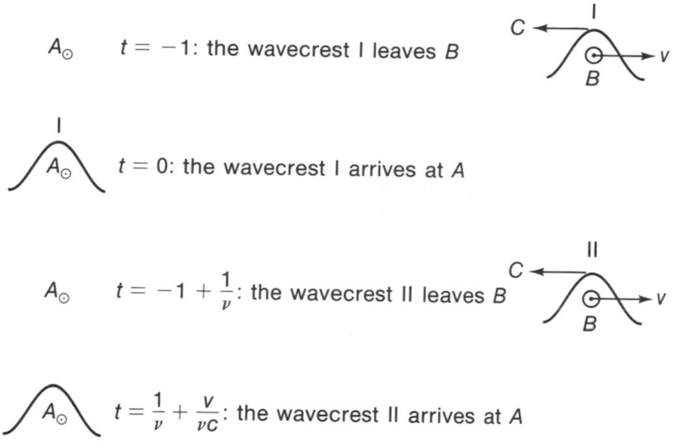

$A_\odot$ $t = -1$: the wavecrest I leaves B

$A_\odot$ $t = 0$: the wavecrest I arrives at A

$A_\odot$ $t = -1 + \dfrac{1}{\nu}$: the wavecrest II leaves B

$A_\odot$ $t = \dfrac{1}{\nu} + \dfrac{v}{\nu c}$: the wavecrest II arrives at A

FIGURE C-1
Schematic description of the Doppler effect. (The relativistic correction has not yet been incorporated.)

The frequency seen by A, which is simply the rate at which wavecrests pass by A, would be given by

$$\tilde{\nu} = \frac{\nu}{1 + \dfrac{v}{c}}.$$

This situation is illustrated in Figure C-1.

This result would be correct if we were using Newton's concept of time. However, as we saw in Chapter 10, this concept is not quite correct and has to be replaced by the time concept of the special theory of relativity. This theory tells us that, because B is moving relative to A, his clock will not run at the same rate as A's clock. In the preceding derivation, we argued that B would send the second wavecrest $1/\nu$ times later than the first wavecrest. However, this is the time by B's clock, not by A's clock. In Appendix A, we saw that B's clock in fact runs slower as seen by A according to the factor

$$\sqrt{1 - \frac{v^2}{c^2}}.$$

Consequently, the time between successive wavecrests, as measured by A, needs to be corrected by multiplying the frequency $\tilde{\nu}$, just obtained in the Newtonian calculation, by this factor. (Compare this case with the muon decay example in Appendix A.)

Hence, the correct answer for the frequency measured by A is given by

$$\nu_A = \frac{\nu}{1 + \dfrac{v}{c}} \sqrt{1 - \frac{v^2}{c^2}} = \nu \sqrt{\frac{1 - \dfrac{v}{c}}{1 + \dfrac{v}{c}}} = \nu \sqrt{\frac{c - v}{c + v}},$$

and the Doppler redshift is therefore given by

$$1 + z = \sqrt{\frac{c + v}{c - v}}.$$

If B were approaching A with velocity v, the preceding frequency relation would be changed to

$$\nu_A = \nu \sqrt{\frac{1 + \dfrac{v}{c}}{1 - \dfrac{v}{c}}}.$$

These formulas were given earlier, without proof, in Chapter 10.

Appendix D
The Steady-State Model
of the Universe

We saw in Section 13-10 that the age of the universe—the big-bang universe—cannot exceed H^{-1}. When Hubble first measured H^{-1}, the value, around 1.8×10^9 years, was embarrassingly low. Even the geological age of the Earth ($\approx 4.5 \times 10^9$ years) is two and one-half times larger. The ages of the oldest stars and galaxies were not well known at the time, but they were expected to be greater than 1.8×10^9 years.

This apparent discrepancy was one of the reasons that prompted some theoreticians to search for cosmological models other than those given by Einstein's theory of relativity. One outcome of such efforts was the so-called steady-state model proposed in 1948, independently by Herman Bondi and Thomas Gold and by Fred Hoyle. In resolving the age difficulty, this model went to the other extreme; the universe was without a beginning and hence infinitely old.

Bondi and Gold approached the cosmological problem from a deductive point of view, arguing as follows. Suppose we make a cosmological observation (like the redshift, say) that involves looking at a very remote galaxy. Since the observation is via light, we are observing the galaxy as it was in the remote past. In order to make a valid comparison between this remote part of the universe and our local neighborhood, we have to make a further assumption. We have to assume that the same laws of physics operated there as they do here. In the case of the redshift, for example, we identify the spectral lines from the galaxy on the basis of the atomic physics that operates here in a terrestrial laboratory. If we do not make this assumption, the entire basis for a comparison collapses. Of course, this does not preclude the possibility that a different set of laws might in fact operate elsewhere in the universe. If they do, the cosmologist's job is made the more difficult. The cosmologist's predicament can be compared with a person searching at night for a missing coin in an ill-lit street. The only spots he can search meaningfully are those under the few street lamps. If he is lucky, his coin may be in such an illuminated spot, but it may be in the dark somewhere else.

Turning now to the standard big-bang picture, Bondi and Gold argued that the state of the universe in the early stages was so much different from its present state that the assumption of the same physical laws was severely strained. After all, the universe contains everything, including its basic laws. What guarantee do we have that the laws were the same then as they are now if the universe was so different? To guarantee that the laws of physics were the same, the universe would have to stay the same. To give credence to this idea, Bondi and Gold formulated the concept of the *perfect cosmological principle*.

We saw in Section 12-4 that the ordinary cosmological principle requires the three-dimensional spaces $t = $ constant to be homogeneous. The assumption of the ordinary cosmological principle is sufficient to insure that, at any given cosmic epoch, the same laws of physics operate over all the three-dimensional space of galaxies. However, we need to compare (observationally) two galaxies at different t times. Hence, it is also necessary that the laws of physics are the same at different epochs. To insure this situation, the perfect cosmological principle states that the spaces $t = $ constant are physically similar. The concept of homogeneity is thus extended to the time dimension.

The resulting picture of the universe is therefore an unchanging one, as the name *steady state* implies. This does not exclude motion, however. A steadily flowing river presents the same picture at all times, but it is not static. In the same way, the system of galaxies need not remain static; it can be of the expanding type. Indeed, by using Olbers' paradox (Sections 12-5 and 13-11), Bondi and Gold deduced that the darkness of the night sky admits only the second of the following three possibilities for resolving the paradox in a steady-state situation.

1. The universe is static.

2. *The universe is always expanding.*

3. The universe is always contracting.

Moreover, since all observable parameters must be the same at all epochs in the steady-state universe, Hubble's constant must also be the same at all values of cosmic time. This result implies an expansion function $Q(t)$ of the form

$$Q(t) = \exp(Ht),$$

where H is the Hubble constant. Notice that, unlike the situation for the big-bang models, *no* gravitational theory has been used to arrive at the form of $Q(t)$. The form $Q(t)$ has been deduced from the perfect cosmological principle.

This principle also requires that, in the steady-state model, the average separation of all galaxies must always be the same. If either this or any other observable property were dependent on the epoch, the steady-state condition would be contradicted. How then, since the distance between every pair of galaxies increases like $Q(t)$, are we to meet this requirement? It can be met if new galaxies are steadily being formed in the manner of Figure D-1. In this picture, new galaxies are born at the rate necessary to compensate for the increases of the distances between already existing galaxies. As the already existing galaxies spread out, new galaxies are born in the spaces between them.

How is this new matter created? This question was not answered by Bondi and Gold beyond the statement that it was a deduction of the perfect cosmological principle. In fact, it is somewhat unfair to throw this question at only the

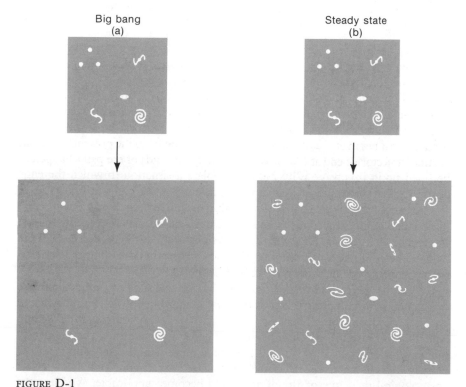

FIGURE D-1
In (a), the already existing galaxies expand away from one another, this being the situation for big-bang cosmologies. In (b), new galaxies are born at the rate necessary to compensate for the steadily increasing distances between already existing galaxies. The density of galaxies is greatly exaggerated here.

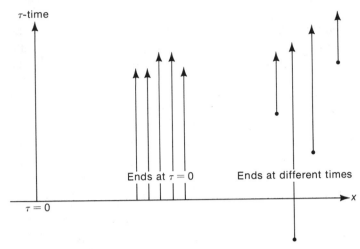

FIGURE D-2
Instead of thinking that the ends of the particle paths all fall at one moment of time (as in the big-bang model), why not contemplate the ends falling at different moments of time, even moments *before* $t = 0$?

steady-state theory. In the big-bang models, all matter is created at $t = 0$, and no explanation is given as to how and why this happens.

It was the concept of the creation of matter that led Hoyle to the idea of a steady state. The phenomenon of creation implies the existence of broken world lines. In Figure D-2, the left-hand set of arrows describes the world lines of particles that begin at $t = 0$. Going backward in time from the present epoch, all particle trajectories end at $t = 0$. Why should all the ends of the paths happen to be lined up in this way? Why not contemplate a situation in which the ends occur at different values of t, as in the right-hand set of paths in Figure D-2?

The ends of the particle paths occur within the universe and must therefore be understood in physical terms. The kind of explanation that is needed is shown schematically in Figure D-3, where interactions occur at the broken ends of the paths. These interactions cannot be electrical or gravitational but must constitute a new kind of field, as physicists describe it.

Given the general idea of Figure D-3, it is possible to build a precise mathematical theory using methods that are standard in physics, methods based on an action principle. There are standard ways of relating such an action principle to Einstein's gravitational theory and of then working out the consequences for cosmology. When this is done, a quite remarkable result emerges, namely, that the universe settles itself into a steady-state condition. The perfect cosmological principle of Bondi and Gold becomes applicable. We have the logical equivalence:

Ends of particle paths at different values of t

⟺ Perfect cosmological principle.

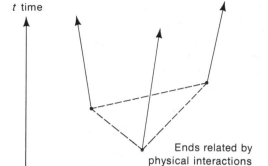

FIGURE D-3

In a theory that assumes the ends of particle paths
falling at different moments of time, we consider
physical interactions to relate the end of one particle to the ends of other particles.

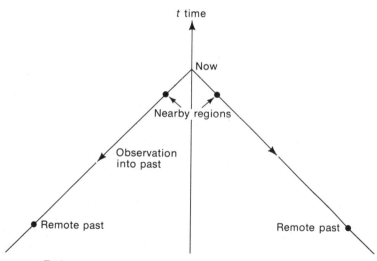

FIGURE D-4

By being compared with the state of affairs at a great distance—that is, at a significantly earlier epoch—the steady-state model can be tested for any astronomical property.

The Exposure of the Steady-State Model to Disproof

The steady-state model is more readily subject to test than any of the other
cosmological models, the reason being illustrated in Figure D-4. When we make
observations using any form of radiation, we look back into the past along the

light cone. Accordingly, our observations sample the conditions that existed at earlier epochs, and if we go back along the light cone far enough, it is possible to examine the state of affairs at times significantly earlier than the present. If any astronomical property is found to have been significantly different in the remote past, as shown in Figure D-4, from what it is in nearby regions, we will have disproved the contention that the universe is steady in all respects.

Two difficulties stand in the way of using this logically straightforward method as a test of the steady-state model. If the universe is changing with time and if the steady-state model is wrong, then it will be better in principle to disprove it by looking back a long way along the light cone than by looking back only a small way. Changes during a comparatively short time are likely to be much less than changes during a long time, and inherent sources of inaccuracy in our observations will always obscure small changes. Inevitably, then, we must expect to be concerned with observations at great distances. Unfortunately, however, objects at great distances are usually very faint and therefore difficult to observe with precision. Spurious effects that arise from working at the limits of our instruments must not be misinterpreted as evolutionary changes.

A second difficulty comes from an uncertainty in the steady-state model itself. On what scale is the universe supposed to be steady? Clearly not when taken on the scale of the solar system, or on the scale of our galaxy, or even on the scale of clusters of galaxies. The protagonists of the steady-state model have never had an unequivocal way of answering this question. Bondi and Gold supposed the universe to be steady on a scale not much greater than the distances of nearby galaxies. Yet, it would be possible, without sacrificing the most important concepts of the theory, to consider that the "scale of steadiness" should be set much larger than this. Since in theory there is no limit to either space or time, the scale on which we elect to work is rather arbitrary. Steadiness, in the sense of maintained properties, might only set in when regions containing many millions of galaxies are considered. By extending the scale sufficiently, one could thus frustrate attempts to disprove the theory along the lines just suggested.

Let us consider these issues in terms of an explicit example, the counting of radio sources discussed in Section 13-4. The latest observational data were shown in Figure 13-14. If the observations are taken to require an evolutionary model for the universe, then the steady-state theory in the restricted form of Bondi and Gold is disproved. There are ambiguities in the data, however, that were explained in Section 13-4, so that the issue remains open, particularly if the scale of steadiness is taken to be large.

If it were not for the existence of the microwave background discussed in Section 13-8, the steady-state model might reasonably be judged to stand well at the present time. The origin of the microwave background is associated in the big-bang cosmologies with the origin of the universe. No such explanation can be given in the steady-state model, however, because *there is no origin of the universe in the steady-state model*. For the large-scale features of the universe, one epoch cannot be different from another epoch. This steady condition would be violated if the universe had an origin.

How, then, in the steady-state model, can we explain the existence of the microwave background? The observed background consists of radio waves with frequencies ranging from about 10^9 oscillations per second up to about 3×10^{11} oscillations per second. Many radio sources that generate such waves are known. So why should not the sources—that is, discrete objects like radio galaxies and QSOs—be responsible for the background? The difficulty with this apparently straightforward approach to the problem is that the known sources, counts of which are plotted in Figure 13-14, fail to give an adequate intensity, particularly at the high-frequency end of the background. A conceivable recourse would be to postulate the existence of a very large number of *undetected sources,* sources of very low intrinsic emission, say, with only a millionth of the intrinsic emission of the sources we do observe. There would need to be about 10^{14} such weak sources, which is about 10,000 times the total number of visible galaxies. Most astronomers find it unpalatable to assume the existence, not just of a new class of source, but of a class with a very great number of members. The criticism is a fair one. Even so, we must be on our guard against the ever-present tendency in astronomy to imagine that nothing exists in the world except the things that happen to be observable with present-day instruments. This point of view has been repeatedly proved wrong, and doubtless it will turn out to be wrong again. But will it turn out to be wrong about the existence of a profusion of weak radio sources that would save the steady-state model? Although an affirmative answer here seems doubtful, it is nevertheless relevant to notice that the steady-state model has not been rigorously disproved in the sense of a strict mathematical result. Other possibilities for resolving the difficulty, besides that just mentioned, also exist. More observational information is needed before their plausibility can be assessed.

Glossary

Absolute magnitude: The absolute magnitude of a star is the magnitude the star would have if, instead of being at its actual distance, it were located at a distance from us of 10 pc, which equals 32.6 light-years. A similar definition of absolute magnitude is used for galaxies and other objects, such as globular clusters, although in fact another galaxy could not fit inside our galaxy at only 10 pc away from us.

Albedo: The sunlight that falls on a planet, satellite, or smaller body in the solar system is partly reflected and partly absorbed. The reflected fraction is the albedo. Since the light is not reflected uniformly in all directions, a practical calculation of the brightness of a planet or satellite must allow for the fact that the reflection depends on the angle between the direction from the Sun to the object and the direction from the Earth to the object. This dependence is known as a *phase effect*.

Amino acid: A molecule with H, NH_2, and COOH, and a side chain R bonded to a carbon atom. The structure of R differs from one type of amino acid to another. There are many amino acids, of which 20 play a critical role in the biochemistry of life.

Ångstrom (Å): A unit of length equal to 10^{-8} cm.

Astronomical unit (AU): The average distance of the Earth from the Sun, close to 149.6 million kilometers.

Atomic number: The total number of protons in an atomic nucleus is called the atomic number.

Atoms: Bulk materials can be broken down into atoms, of which 103 different kinds are known. Of these, 20 undergo comparatively rapid radioactive decay and are thus not found in appreciable quantities in the materials of our everyday world. Atoms consist of a nucleus that is very small in size (10^{-13} to 10^{-12} cm) and contributes most of the mass, surrounded by a cloud of electrons that extends farther out (to about 10^{-8} cm). The nucleus is made up of protons and neutrons. In a neutral atom, the number of protons is equal to the number of electrons. This number determines the chemical properties that define the type of the atom; for each value of this number, the atom is that of a distinct element.

Baryons: There are many forms of baryon, of which the proton and neutron are examples. Together with six others, the proton and neutron form an important family of eight baryons. The proton is the stable form for an isolated baryon, the neutron decaying into a proton in about 1,000 seconds, the other six decaying in a very much shorter time

(10^{-10} second). When several baryons are associated together, the stable form consists of a mixture of protons and neutrons. Such mixtures form the nuclei of atoms. These properties arise from the curious circumstance that the masses of the neutron and proton are nearly equal, whereas the masses of the other baryons are all much larger than that of the proton or the neutron.

Beta processes: The nucleus of an atom with any given total number of baryons adjusts the numbers of the protons and neutrons until the nucleus takes its most stable form. This happens by means of beta processes that can either change a proton into a neutron or change a neutron into a proton.

Binary stars: A binary consists of two stars that move around a common center known as the barycenter. The line joining the stars always passes through the barycenter, and the orbit of each star is an ellipse with the barycenter as one of its two foci. Binary stars probably originate in the condensation process that gives rise to the stars themselves.

Black hole: Since gravitation tends to pull matter together, large-scale associations of material tend to evolve in such a way that the matter becomes more and more condensed, until in the ultimate limit a black hole is formed. Conventional ideas of geometry do not apply in the vicinity of a black hole, and it is this breakdown of conventional ideas that leads to the limit of the condensation process. The size of a black hole depends on the mass of material involved. A black hole with a mass equal to that of the Sun would be about 6 km across, and one with a mass equal to that of our galaxy would be about 0.1 light-year across. No light can escape from the surface of a black hole.

Black hole types: The most general type of black hole within the framework of Maxwell's electromagnetic theory and Einstein's general theory of relativity is the so-called Kerr-Newmann black hole, which is characterized by its mass, electric charge, and angular momentum. If the black hole has no electric charge and no angular momentum, it is the simplest type of black hole and is called the Schwarzschild black hole. A black hole with mass and angular momentum but no charge is called the Kerr black hole, whereas a black hole with mass and charge but no angular momentum is called the Reissner-Nordström black hole. All these types of black holes are named after the scientists who obtained their mathematical description by solving Einstein's equations of general relativity.

Blueshift: Radiation from a moving object is said to be blueshifted when lines in the visible spectrum are shifted (*see* Doppler shift) toward the blue end of the spectrum. This happens when the object in question is approaching the observer.

Bolometric magnitude: When all the radiation from an object is included in the measurement of its magnitude, the result is called the bolometric magnitude. Thus, the bolometric magnitude includes contributions from the radio band, the infrared, visual and ultraviolet light, x rays, and γ rays. Such measurements are difficult to make because the atmosphere of the Earth is opaque, except in the visual and near infrared and in the radio band. Thus, the bolometric magnitudes mentioned in astronomical literature are almost always estimates rather than direct measurements.

Carbon–nitrogen cycle: The carbon–nitrogen cycle is a sequence of nuclear reactions whose net effect is to convert hydrogen into helium. It is one of the two main processes responsible for the production of energy in most of the stars we see in the sky. The carbon and nitrogen atoms promoting the cycle are only rarely changed (to oxygen), and even then there are further nuclear reactions that eventually regenerate nitrogen. Thus, the carbon and nitrogen atoms behave catalytically.

Celestial sphere: The celestial sphere is an imaginary sphere with its center at the center of the Earth. It is usually considered to have a radius larger than the solar system but not as large as the distances of the stars. The precise radius of the celestial sphere is irrelevant to its purpose, however, which is to provide a surface on which all astronomical objects are projected. The celestial sphere moves with the Earth's motion around the Sun, causing the projected positions of stars and all other distant objects to show slight annual wobbles (*see* Parallax).

Cepheid variables: A star that undergoes regular radial pulsations, with its radius oscillating by about 10% and its luminosity by about 100%, the Cepheid variable is detected observationally from its characteristic variations of luminosity, the so-called light curve. The average intrinsic luminosity is related to the period of the variation in a known way. Hence, an observational measurement of the period is equivalent to a measurement of the intrinsic luminosity. Knowledge of the intrinsic luminosity, taken with the observed apparent luminosity, allows the astronomer to calculate the distance of the star. Cepheid variables can thus be used as distance indicators. The typical surface temperature is not much different from that of the Sun. Periods range from about 2 days up to about 100 days, and intrinsic luminosities range from about 100 to about 10,000 times the luminosity of the Sun.

Chemical elements: *See* **Atoms**

Color temperature: The relative intensity distribution of the various colors (frequencies) in the light emitted by a star depends on the temperature of its surface. Observation of the relative intensity distribution can therefore be used to calculate the surface temperatures of stars, and such calculated temperatures are called color temperatures. Because the absorption of light by the atmosphere of the Earth prevents a ground-based observer from making measurements in the ultraviolet and far infrared, this method of calculating temperature becomes subject to error for both very hot stars and very cool ones.

Coma: Coma is the failure of a telescope's objective, whether it is a lens system or a mirror, to bring objects that are slightly off the telescope's axis to a sharp focus. Correcting systems are used to minimize this defect in all large optical telescopes. The corrector in a conventional reflecting telescope consists of a series of glass lenses placed near the focal plane. In Schmidt-Kellner telescopes, the corrector is a sheet of glass placed across the telescope tube. Coma is not the only optical defect of the astronomical telescope, but it is the most serious.

Comet: A comet is thought to consist initially of one or more icy bodies with a total volume that is typically about 1,000 cu km. Ordinary water ice and dry ice are important components, but other normally volatile materials are also present. Embedded in these ices are myriads of small refractory particles that remain as a swarm if the ices become evaporated when a comet approaches the Sun. This approach can happen because the highly elliptic orbits of comets are rather easily perturbed (changed in form), mainly by the gravitational effect of the planet Jupiter and perhaps also by the influence of nearby stars. The remarkable appearance of spectacular comets arises from the evaporation of the ices.

Constellations: From ancient times, the stars of the northern night sky have been divided into groups known as constellations. The groups were usually decided by chance associations that happened to suggest the shapes of objects or people. From these chance associations, and from myths, the constellations received names that are retained in modern astronomy because the names are convenient labels for specific bits of the sky. The constellations of the far south were named in modern times and not always so imaginatively as those in the north.

Cosmology, principles of: Literally, cosmology means the study of the cosmos, that is, of anything outside the Earth. In recent years, however, cosmology has come more and more to mean the study of the largest-scale aspects of the universe. To facilitate this study, it is usual to make two simplifying postulates known as the principles of cosmology. One principle holds that the same physical laws apply in every locality and at all times; the other principle states that the large-scale structure of the universe is the same in all directions (isotropy) and is also the same when seen from different localities (homogeneity).

Declination and right ascension: The declination and right ascension of an astronomical object are its latitude and longitude values measured in the following way. The object is considered to be projected on the celestial sphere. Projected also are the Earth's equator and the annual path along which the Sun moves in the sky. This gives two great circles on the celestial sphere, known as the celestial equator and the ecliptic, respectively. The declination and right ascension values for the projected astronomical object are then given by measuring from *and* along the celestial equator. The measurement along the celestial equator is taken from the point where the sun (and so the ecliptic) crosses the equator at the spring equinox.

Doppler shift: Given an atom of a specified element and a specified quantum transition of it, radiation generated by that transition in an assembly of such atoms has a well-defined frequency and wavelength. If an observer takes such an assembly that is at rest with respect to himself, he can use the frequency and wavelength of the radiation to establish units of time and of length; time measured by means of such a unit is called the observer's proper time. A second observer in motion with respect to the first can proceed in the same way, using similar atoms that are at rest with respect to himself, and that are therefore moving with respect to the first observer. The units of time and of length thus obtained by the second observer are not the same as those of the first observer. The Doppler shift expresses the relation between the two systems. It does so in terms of the relative motion between the two observers.

Eclipse: The light from a distant object is said to be eclipsed when a nearer object crosses the line of sight from the Earth to the first object. A solar eclipse occurs when the Moon comes between the Earth and the Sun. A binary eclipse occurs when one star in a binary comes between the other star and the line of sight to the Earth. Such binaries are said to be eclipsing. The word eclipse is also used when the Earth blocks the light from the Sun to the Moon, giving a lunar eclipse—that is, an eclipse for an observer situated on the Moon. Inconsistently, however, the word eclipse is not used when the Moon crosses the line of sight from the Earth to a star; in this situation, the star is said to be occulted.

Ecliptic: *See* **Declination**

Electron: The electron is an electrically charged particle belonging to the family of leptons, a family different from the baryons. Electrons surround the nuclei of atoms, forming shell structures wherever possible. The electrons left over when all possible shells have been completed play a critical role in determining the chemical properties of the atom. Successive shells contain 2, 8, 8, 18, 18, and 32 electrons. Only in the heaviest atoms are all these shells completed.

Energy: The electrical energy consumed by a device with a power rating of 1 kilowatt, used for a time of 1 second, is the unit of energy known as the kilowatt-second. The kilowatt-second is related to the unit used most frequently in physics, the erg, by the equation

$$1 \text{ kilowatt-second} = 10^{10} \text{ ergs.}$$

Energy can exist in many forms. When energy is said to be consumed, we mean that it has been changed from a useful form to a less useful form, not that energy has been destroyed. Some forms of energy, electrical energy, for example, are readily converted to other forms, but the conversion is usually not reversible. Gasoline and air are converted in a motor to water, carbon dioxide, and heat, but it is not easy to change the water, carbon dioxide, and heat so produced back again into gasoline and air. The concept of the usefulness of various forms of energy is dealt with in the branch of physics known as thermodynamics.

Equinoxes: There are two equinoxes each year, occurring when the position of the Sun on the celestial sphere is at one of the two points where the ecliptic intersects the celestial equator. The spring equinox occurs about March 21, the autumnal equinox about September 23. The lengths of day and night are usually considered to be equal at the equinoxes. This is not strictly true because the Sun is not a mere point of light, and particularly because of the twilight caused by refraction of sunlight in the atmosphere of the Earth.

Escape velocity: The minimum velocity needed to fire a projectile from a massive body for the projectile to escape to infinity, that is, to leave forever the body from which it was fired. Because the body attracts the projectile gravitationally, the escape velocity is essentially the velocity required to overcome this attraction. The escape velocity from the surface of the Earth is about 7 miles/second.

Evolution of stars: As nuclear reactions go on inside stars, their chemical composition changes; atoms change from one kind to another. The change affects both the structure of the star and the availability of nuclear fuel. The changing condition of the star is referred to as its evolution.

Fission: The electrical repulsion between the positive charges of the protons in the nuclei of atoms would cause them to fly immediately apart if it were not for the nuclear force that attracts protons and neutrons to each other. There is, however, a limit to the binding effect of the nuclear force. When the total number of neutrons and protons in the nucleus increases above about 210, the electrical repulsive effect becomes strong enough to cause the nucleus to disgorge bits of itself in the form of helium nuclei, a process known as alpha decay. As the total number of neutrons and protons is further increased, the electrical repulsion becomes still stronger, until, instead of merely losing helium nuclei, the nucleus divides into two pieces of comparable size. The latter process is called fission. Fission becomes more important than alpha decay when the total number of neutrons and protons exceeds 260. Fission can be induced by firing neutrons into a heavy nucleus. Such induced fission is more important than alpha decay for a total neutron–proton number exceeding about 230. Man-made reactors depend on induced fission.

Flare: Electrical energy is stored in the solar atmosphere at levels above those that give rise to normal sunlight. Conditions can arise in which the stored electrical energy, or a portion of it, is suddenly discharged. This process, known as a flare, is more complex than the discharge of a spark or of a lightning stroke, but the cases are analogous.

Flux: When a flat surface is exposed to a distant source of radiation, and when the direction to the source is perpendicular to the surface, the power that falls on a unit area is defined as the flux received from the source. Flux is like the power rating of a man-made device: it must be multiplied by a time interval to give energy. For a spherically symmetric (isotropic) source of radiation, the flux varies according to the inverse square of the distance from the center of the source

Focal length: When a lens or mirror is used to bring a distant object to a focus, the object being on-axis, the distance of the focal point from the lens or mirror is the focal length of the system. Dividing the focal length by the diameter of the lens or mirror gives the focal ratio (f ratio).

Force: The lack of the correct concept of force held up the development of physics for 2,000 years. A force acts on any body that does not move with constant speed in a straight line.

Frequency: The frequency of a continuing sequence of regular oscillations is the number of them that occurs in a unit duration of time. Frequency can be thus defined for many phenomena, such as the regular pulsations of a star, the bobbing of a float on a regular train of water waves, or the radiation emitted by a certain quantum transition of a certain kind of atom. Radiation of this last type can cause an electrified test particle, an electron, say, to undergo a sequence of oscillations. By observing the test particle, we can measure the frequency of the radiation.

Fusion: The term fusion is applied in astronomy to nuclear processes in which two or more nuclei come together in a reaction, or reactions, whose products contain a nucleus with more neutrons and protons than any one of the ingredients. Heavier nuclei are thus produced from lighter ones. Provided the nuclei so produced do not have an atomic number greater than about 60, fusion reactions yield energy. Fusion reactions take place within stars, where they supply energy. The most effective energy source comes from the fusion of hydrogen into helium.

Galactic coordinates: Like the system of right ascension and declination (*see* Declination), a system of galactic longitude and latitude can be set up on the celestial sphere. The galactic equator is the circle in which the central plane of our galaxy intersects the celestial sphere, and the galactic prime meridian is defined by the direction from the solar system to the center of the galaxy.

Galaxies: A galaxy is a large collection of stars occupying a region of space well separated from other large collections of stars. A galaxy may also contain much gas and dust, or it may contain esoteric objects like black holes, but such nonstellar contents are not necessary to its definition. There is no clearly agreed-on lower limit to what constitutes a large collection of stars. Conventionally, a lower limit seems to be set at about 10 million; a collection of a million stars would usually be described as a cluster rather than as a galaxy. Whether or not a galaxy contains gas and dust is relevant to its type classification. Galaxies with more than a few percent of their mass as gas and dust usually have flattened spiral forms, and are thence known as spirals. Galaxies substantially without gas and dust are usually described as ellipticals, although this gas-and-dust criterion for separating spirals and ellipticals is not strict. Galaxies tend to occur in groups or clusters, with numbers ranging from about 5 on the low side to about 1,000 on the upper side.

Gamma rays: Gamma rays consist of radiation of very high frequency, a hundred thousand or more times greater than the frequency of visible light.

Globular cluster: Globular clusters contain so many stars, 100,000 to 1 million in a typical cluster, that their inner parts usually appear on photographs as an amorphous distribution rather than as separated stars. The amorphous distribution usually has a more or less circular shape, and it is from this that the name is derived. In our galaxy, there are on the order of 100 of these clusters. In giant galaxies, there can be as many as 1,000 of them.

Gravitation: Gravitation causes matter to pull together, to condense into a tighter and tighter clump, unless this tendency is resisted by some other effect, for example, by pressure within the material. In Newton's theory, gravitation is described as a force, and an explicit mathematical formula is given whereby the gravitation between particles of matter can be calculated. In Einstein's theory, however, gravitation is not a force like pressure, but it is instead described by means of a non-Euclidean geometry.

Half-life: Nuclei that undergo radioactive decay do so in a way that depends on the half-life (every radioactive species has its own characteristic half-life). After a time interval equal to the half-life, one-half of the nuclei have decayed. After a further time interval equal to the half-life, one-half of the remainder have decayed, leaving one-quarter in the original form. After yet another time interval equal to the half-life, one-half of the remainder have decayed, leaving one-eighth in the original form, and so on. (The number of nuclei of the radioactive species is assumed in these statements to be so large that statistical fluctuations can be ignored.)

Hertzsprung–Russell diagram: The Hertzsprung–Russell diagram is basically a plot in which the ordinate (vertical axis) of a point represents the intrinsic luminosity of a star, and the abscissa (horizontal axis) represents the *reciprocal* of its surface temperature. Instead of the intrinsic luminosity, the absolute magnitude may be plotted, and instead of the reciprocal of the surface temperature, the spectrum type may be used (since the spectrum of a star is related in a known way to its surface temperature).

HI and HII regions: Hydrogen atoms are about 10 times more numerous in the gas clouds of our galaxy, and in those of most other galaxies, than all other kinds of atoms combined. Regions where the hydrogen atoms are mostly neutral (electron and proton associated together) are known as HI regions. Regions where the hydrogen atoms are mostly ionized (electron and proton separated) are known as HII regions. Left to themselves, ionized atoms change into neutral atoms, emitting radiation when they do so. Consequently HII regions tend to disappear unless there is a rejuvenating agent. The rejuvenating agent is usually ultraviolet light from a hot star, which is absorbed by neutral atoms, causing a separation of electrons from protons. HII regions are systematically hotter and more diffuse than HI regions.

Hot big-bang universe: A model of the universe that follows from the mathematical equations of Einstein's general theory of relativity after a few simplifying assumptions. There are in fact many such models and they all have a spacetime singularity that is identified with the origin of the universe. In the hot big-bang model, it is believed that the universe started with a big explosion and with a very high temperature. As it cooled down, the light elements, such as hydrogen, deuterium, and helium, were cooked in the thermonuclear reactions in the first few seconds after the big bang. A relic of the hot era is believed to have been observed as a radiation background in the microwave part of the electromagnetic spectrum.

Ion: In a neutral atom, the number of electrons surrounding the nucleus is equal to the number of protons in the nucleus. Since the electron and proton have electric charges of equal magnitude but of opposite sign, the total charge of the neutral atom is zero. When the numbers of electrons and protons are not the same, the atom is said to be an ion. The more usual ionized condition is for the electron number to be less than the proton number, giving a positive ion. In rare cases, the electron number is greater than the proton, and the ion is said to be of negative type. These designations arise from the convention according to which the proton charge is taken to be positive and the electron charge taken to be negative.

Isotope: The number of protons in the nucleus of an atom determine the chemical element; the number of neutrons in the nucleus can be changed without the chemical nature of the atom being changed, except in certain fine details. Nuclei with the same number of protons, but different numbers of neutrons, are called isotopes of one another.

Laser (light amplification by stimulated emission of radiation): Normally atoms emit radiation independently of each other. It is possible, however, to set up a special condition in which radiation from an initial number of atoms causes other atoms to emit radiation and to do so in the same direction as the initial radiation. The result is a directed beam of radiation that continues as long as the atoms continue to maintain the special condition. When the radiation emitted is visible light, the system is called a laser. If radio waves are emitted, the system is referred to as a maser (m for microwave, replacing l for light).

Light cone: A pulse of light emitted at a particular moment from a point source travels in four-dimensional spacetime along a three-dimensional cone known as the light cone.

Light-year: The distance traveled by light in a year of 3.156×10^7 seconds is the light-year, about 9.46×10^{12} kilometers.

Local group: Galaxies tend to exist in groups or clusters, ranging from about 5 members on the low side to about 1,000 members on the upper side. Our galaxy is a member of a cluster with some 20 members known as the local group.

Logarithm: The equation $x = 10^{\log x}$ expresses the relation between any positive number x and its logarithm, $\log x$. It is by no means simple to use this equation to find $\log x$

when an explicit value of x is specified, or to find x when an explicit value of $\log x$ is specified. Such calculations can be avoided, however, by using an already prepared table of logarithms. Or one can simply use a pocket calculator that solves the required equation in a fraction of a second.

Luminosity: The luminosity of a star, a galaxy, or some other astronomical object is like the rating of a power station, and we could use the same unit of power in both cases, so many megawatts, for example. To obtain the energy supplied in a specified time interval, the power rating must be multiplied by the time. Thus the energy emitted by a star of luminosity L in a time interval t is the product Lt.

Mach's principle: A hypothesis advanced in the late nineteenth century by the scientist-philosopher Ernst Mach. Mach argued that the *inertia* of a body is not an intrinsic property of the body; rather it is an environmental effect that owes its origin to the large-scale structure of the universe. Mach gave persuasive arguments for this hypothesis, but he did not give a mathematical theory for it.

Magnitude: When the word magnitude is used by itself, it usually means apparent magnitude. The apparent magnitude is a somewhat indirect way of stating what the flux (*see* Flux) from an astronomical object is measured to be. The observer first decides on a unit of power in terms of which the flux is then measured, say, F. The next step is to take the logarithm of F, $\log F$ (*see* Logarithm). The resulting value, $\log F$, is next multiplied by -2.5, giving $-2.5 \log F$, and a certain number, c, is finally added to give the apparent magnitude, $-2.5 \log F + c$. The number c to be added is fixed by convention so that a specified standard object will turn out to have a prestated magnitude, say, $+1$. This conventional number to be added depends on the power unit in terms of which the flux was measured. With fluxes in kilowatts per square centimeter, the number is about -36.5. It is clear that an increase by a factor 10 in the measured flux changes the magnitude by -2.5; an increase by 100 in the measured flux changes the magnitude by -5; and so on.

Main sequence: Stars that generate the energy they radiate by fusing hydrogen into helium in their central regions lie in a particular zone of the Hertzsprung–Russell diagram. When many such stars with variable masses are plotted in this diagram, the resulting points fall in a band that ranges from lower right to upper left. This band is the main sequence. If the plot is restricted to stars that have only recently formed so that little of the initial hydrogen has been converted to helium, the points fall on a line instead of covering a band. This line is known as the zero-age main sequence.

Meridian: A meridian is an arc of a great circle extending from pole to pole on the celestial sphere.

Meteorite: The region of the solar system between Mars and Jupiter contains a number of small planetoids or asteroids, as well as a swarm of still smaller bodies. These bodies are constantly subject to gravitational perturbations from Jupiter and Mars; these perturbations change the orbits of the bodies, as do their occasional collisions with each other. From time to time, particularly because of Mars, these changes cause one of these bodies to take on an orbit that crosses the path of the Earth. Collision with the Earth is then possible. Such fragments hitting the Earth can penetrate through the atmosphere to ground level, where they can be retrieved. Retrieved fragments, known as meteorites, are divided into two classes, one class being of stony material, the other of an iron-nickel alloy. The stones can be subdivided into many varieties, the chemical analysis of which has provided much important information concerning the early history of the solar system.

Meteor: With the evaporation of the ices that form the nucleus of a comet, a swarm of tiny sub-pinhead solid particles is left, following and spreading out along the orbit of the comet. During the course of the year, the Earth crosses several orbits of such evaporated comets, and small particles then enter the terrestrial atmosphere in large numbers. Because of their small size, these meteors, as they are called, do not penetrate to ground level; they are vaporized by friction with the atmosphere, and they show at night as momentary streaks of light in the sky.

Microwave background: The microwave background consists of radiation whose main wavelength distribution is from about 0.5 millimeters to about 20 centimeters, and which reaches the Earth smoothly from all over the sky. The radiation is believed by most astronomers to be a relic from a hot, dense condition of the universe, associated with what is often called the big bang. It provides a new and important datum in the study of cosmology. The variation of flux with wavelength is consistent with a Planck distribution at a temperature of about 3 K.

Milky Way: The Milky Way is a colloquial name for the part of our galaxy that can be seen in the night sky. The many distant stars along the plane of the galaxy produce a diffuse band of light crossing the sky in a great circle. Not all the galaxy can thus be seen because of absorption of light by dust clouds.

Molecules: Although the charges of the electrons and protons in a neutral atom cancel each other, the electrical force that the atom exerts on an external charged particle is not exactly zero because the electrons and protons of the atom have different spatial positions. Although the electrical force between well-separated neutral atoms (as in a diffuse gas) is weak, the force is comparatively very strong when the atoms are close beside each other (as in a liquid or solid). Indeed, the electrical force can bind neutral atoms together into composite structures called molecules. The simplest molecules, like CO and H_2, contain only two atoms, whereas complex biological molecules like proteins contain many thousands of atoms. Chemistry and biology are concerned with the permutations and combinations of these composite structures of atoms.

Neutrino: The neutrino is a particle with no electrical charge and with at most a very small mass; it belongs to the same family of particles as the electron, the leptons. There are several kinds of neutrino, two forms having been identified so far by experiment.

Neutron: A particle without electrical charge, and with a mass about a tenth of a percent greater than the mass of the proton. The neutron is the partner of the proton in the nuclei of atoms.

Neutron stars: As their name implies, neutron stars are stars composed predominantly of neutrons. Typically, they have a radius of about 10 kilometers, although in mass they are comparable with the Sun. A chunk of their matter the size of a small sugar cube contains about 10^8 tons of neutrons. Under these extreme conditions, neutrons may take on the structure of a solid, so that starquakes analogous to earthquakes may occur. This possibility is supported by certain observations of pulsars, which are thought to be rotating neutron stars. Neutron stars are believed to be formed as a final stage in the evolution of supernovae. They are not much bigger than a black hole of the same mass. It is possible that supernovae also give rise to black holes.

Nuclear force: The nuclear force acts between baryons. It is the force that binds protons and neutrons together in the nuclei of atoms. Unlike both gravitation and the electrical force, the nuclear force operates only at the short range of about 10^{-13} centimeter. Should a neutron or proton escape from a nucleus, the nuclear force ceases to act on it.

Nuclear reaction: When two nuclei overlap each other, as they may do for a short while in a collision, the nuclear force connects the protons and neutrons in one nucleus with those in the other nucleus. This may lead to a redistribution in which the protons and neutrons have a changed arrangement after the collision. The nuclei that emerge from the collision are then different from those that entered it, and a nuclear reaction is said to have taken place.

Nuclear shells: The neutrons and protons in a nucleus tend to form (separately) into shells, as do the electrons that surround the nucleus. But whereas the electron number in atoms with closed shells are 2, 10, 18, 36, 54, 86, the proton and neutron shell numbers are 2, 8, 20, 28, 50, 82, 126, 184. The latter are often referred to as magic numbers.

Open cluster: Open clusters usually contain a few hundred stars that appear well separated from each other in photographs. The stars of open clusters are believed to have had a common origin.

Parsec: A star at a distance of 1 parsec would appear to undergo an oscillation through an angle of 2 arc seconds in its position on the celestial sphere due to the yearly motion of the Earth around the Sun. A distance of 1 parsec is equal to 3.26 light-years.

Perihelion: The point in its orbit when a planet, asteroid, or comet is closest to the Sun is its perihelion. There is a corresponding definition for a body moving around the Earth, the word perigee then being used.

Perihelion rotation: According to Newton's theory of gravitation, a planet would follow a strictly elliptic orbit around the Sun if the planet and Sun were alone. According to Einstein's theory of gravitation, the planet does nearly the same thing, except that the long axis of the planet's elliptic path turns very slowly, an effect known as perihelion rotation. The effect is largest for the planet Mercury. Observations of the orbit of Mercury have confirmed Einstein's theory. In practice, the problem is complicated by the fact that Mercury and the Sun are not alone. The other planets also cause a perihelion rotation, and this further effect, predicted by both theories, has to be allowed for before the decision between the two theories can be made.

Periodic table: The periodic table is an arrangement of the elements (the different kinds of atom) according to the increasing number of protons in the nucleus. It is used to arrange the table so that elements in each column have the same number of electrons left over after all possible electron shells have been completed. The advantage of this arrangement is that elements in the same column then have similar chemical properties.

Photosphere: The atoms emitting the radiation that escapes from a star are nearly all contained in a thin shell. It is usual to think of this shell as a surface. For a spherically symmetric star, the surface is a sphere, the photosphere, concentric with the star. The term is most often used in relation to the Sun, which is indeed a nearly spherically symmetric star.

Planck curve: Imagine a closed box with walls that are maintained at a fixed temperature. Because the walls emit and absorb radiation, a distribution of radiation with respect to frequency builds up within the box. This distribution, called a Planck distribution, depends on only the temperature. The distribution can be displayed on a graph by plotting the energy content of small unit steps in the frequency. The resulting curve is known as a Planck curve.

Planet: There are nine planets moving around the Sun. A planet is not just any body that moves in a nearly elliptic orbit around the Sun. A minimum size for a planet is implied by the circumstance that the largest of the smaller bodies in the region between Mars and Jupiter, which also move around the Sun, are called minor planets or planetoids, or asteroids. The minimum size for a planet has never been precisely defined, however.

Precession: The rotation of the Earth causes it to bulge slightly at the equator. The gravitational forces that the Moon and Sun exert on the slightly flattened form of the Earth then produce a twist, technically called a couple, that causes the axis of rotation to precess like that of a spinning top. The effect of the precession is to cause the rotation axis to move slowly around the surface of a cone, completing a circuit in about 26,000 years. Thus, the poles (the directions in which the rotation axis meets the celestial sphere) move slowly in circles on the celestial sphere. The direction toward the center of either of these circles is perpendicular to the plane of the Earth's orbital motion around the Sun.

Precession of the seasons: The celestial equator moves with the motion of the celestial poles. Therefore the two points where the celestial equator intersects the ecliptic also move. Since it is the positions of these two points that determine the spring and autumnal equinoxes, the seasons of the year change slowly with respect to the motion of the celestial equator; the positions of the Earth at midsummer and midwinter interchange approximately every 13,000 years.

Prominence: The Sun has an atmosphere, the lower part of which is called the chromosphere, the upper part the corona. The corona extends far out, to several diameters of the Sun; indeed, the corona may not end at all (*see* Solar wind). The gases of the inner part of the corona, and still more of the outer part, are too diffuse and too hot to emit much visible light. Occasionally, however, owing to unusual effects of the gravitational and magnetic forces that control the solar atmosphere, a region of the gas becomes compressed, cooling and emitting visible light as its density increases. The compressed gas can then be seen and is known as a prominence since it belongs to a region located high above the normal surface of the Sun.

Proper time: *See* **Doppler shift**

Proton: A particle with an electric charge equal in magnitude but opposite in sign to the charge of the electron. The proton is 1,836 times more massive than the electron, however, and it experiences the nuclear force, which the electron does not. The proton is the partner of the neutron in the nuclei of atoms.

Proton–proton chain: The proton–proton chain is a sequence of nuclear reactions whose net effect is to convert hydrogen into helium. It is one of the two main processes responsible for the production of energy in most of the stars we see in the sky, and it is the more important of the two in stars of small mass. The proton–proton chain is largely responsible for the generation of energy in the Sun.

Protostar: A star that is to be; a star in the making.

Pulsar: Pulsars are believed to be rotating neutron stars (*see* Neutron star). They emit radiation, it is thought, from a relatively small spot on the star; this situation produces an effect rather like the beam of a lighthouse. The distant observer receives a pulse of radiation as the beam sweeps across his position on each rotation of the star. Pulsars emit radiation at all frequencies, ranging from the radio band to γ rays. It is by means of their radio emission that pulsars are most conveniently discovered. About 200 of them are known.

Quantum transitions: Quantum transitions occur both for the electrons surrounding the nuclei of atoms and for the neutrons and protons of the nuclei, the physical principles governing the two cases being essentially the same. The electrons spend most of their time in one or another of a number of states. Occasionally, however, an electron undergoes a transition (jump) from one state to another. When a transition occurs, a quantum of radiation is either emitted or absorbed. When many atoms undergo the same transition, the radiation involved has a well-defined wave structure. The protons and neutrons of the nuclei also have states, and radiation is similarly emitted or absorbed in transitions between the states. Frequencies for nuclear transitions are usually on the order of 1 million times greater than the frequencies for the transitions of the electron states.

Quasar (QSO): A quasar is a compact object radiating with a power output comparable to, or greater than, that of a whole galaxy. Quasars are thought to have an inner structure that is about 0.01 light-year across and may contain a black hole. The lines observed in the spectra (*see* Spectrum) of quasars come, however, from a comparatively diffuse gas that probably spreads across several hundred light-years. About 1,300 quasars are known.

Radar: Distances to some nearby objects can be measured by radar. The method consists in emitting a regular train of well-separated radio pulses toward the object in question. The object reflects a small fraction of the radio power back toward the observer, who measures the time that elapses between the emission of a pulse and the reception of its reflection. Distance is then given by multiplying the speed of radio waves by the elapsed time. The radio waves used are confined to a narrow band of frequencies. Large antennas are employed to direct the waves toward the object under investigation. Radar gives the most accurate method for measuring the size of the solar system, the planet Venus being the most suitable reflecting object for this purpose.

Radiative interaction: The influence of one electrically charged particle on another is known as a radiative interaction. If particle a is made to oscillate at frequency v, the speed of the motion being small compared to the speed of light, the radiative influence of a can make another electrically charged particle b oscillate with the same frequency v. This property is often described by saying that the radiative interaction has a wave structure (*see* Waves). The radiative interaction goes from past to future, not from future to past, a property that is important in establishing the relation of cause and effect.

Radioactivity: An association of protons and neutrons, bound together by nuclear forces, reaches its most stable condition by radioactivity. This usually happens by the emission of helium nuclei (two protons and two neutrons in association), a process known as alpha

decay, or by a process known as beta decay, in which a neutron changes to a proton with the emission of an electron, or by the inverse of the same process, with the emission of a positron. All these changes go by the name of radioactivity.

Radio galaxies: Radio waves of exceptionally large intensity are received from small patches on the sky. Many of these patches, but not all of them, are associated with galaxies, which are thus known as radio galaxies. Radio galaxies emit jets of highly energetic particles, often two jets in opposite directions, that come, it is believed, from quasarlike objects (*see* Quasar) situated at their centers. The radio waves are thought to be generated when the emitted particles impinge on external clouds of gas, with magnetic fields playing an important role in the emission process. The radio waves involved in the detection of radio galaxies are usually of longer wavelength than the waves that carry the main energy of the microwave background (*see* Microwave background).

Redshift: A light wave is said to be *redshifted* if its wavelength increases between its point of transmission and its point of reception. Thus, if the wavelength at transmission is λ and at reception is $\lambda(1 + z)$ (where z is positive), then the amount of redshift is given by z. The term redshift arose from the circumstance that the visible part of the electromagnetic spectrum is red in color at its longest wavelength end. Hence, if the wavelength of any other color is increased, it would shift toward the red end. There are three important possible causes of redshift: (1) the *Doppler effect* that arises when the source of the wave moves away from the receiver of the wave; (2) the *gravitational* redshift that describes the situation when the wave travels from a stronger to a weaker gravitational field; and (3) the *cosmological* redshift that arises from the expansion of the universe, wherein a light wave traveling across the non-Euclidean geometry of spacetime gets redshifted. Typical examples of the three cases are respectively: (1) the redshift in the spectra of stars in our galaxy, which happen to be moving away from us; (2) the redshift in the light of white dwarf stars; and (3) the redshift in the spectra of other galaxies.

Redshift-magnitude relation: The spectrum lines (*see* Spectrum) found in galaxies show a redshift that may be described in the following way. Let λ be the wavelength of a particular line as measured in the terrestrial laboratory, and let $\lambda + \Delta\lambda$ be the wavelength of the same line (that is, arising from the same transition of the same kind of atoms) measured for the galaxy in question. Then $z = \Delta\lambda/\lambda$ is the redshift of the galaxy. The value of z is the same whatever line of the spectrum is chosen. The redshift-magnitude diagram is a plot of the logarithm of z, $\log z$, against the magnitude of the galaxy. Usually, the magnitude is represented as the abscissa (horizontal axis) and $\log z$ as the ordinate (vertical axis). When observations of many galaxies are thus plotted, an approximately straight-line relation between magnitude and $\log z$ is found. This straight line is the redshift-magnitude relation.

Reflection: In a vacuum, light can be made to propagate one way, in any chosen direction, and the same is true for light propagating in a translucent material, provided the material is homogeneous—that is, of the same composition at all places. However, when light propagating in a homogeneous medium encounters another, different homogeneous medium, the one-way condition is destroyed. Not all the light continues into the second medium. Light is turned back at the boundary of the media, and it is said to be reflected. Light that continues into the second medium is said to be refracted. The directions of the reflected and of the refracted light are related to the direction of the incident light by the laws of reflection and refraction, which can be expressed by simple mathematical formulas. These formulas provide much of the information required in the construction of telescopes, microscopes, cameras, and other optical instruments. Similar considerations apply to forms of radiation other than visible light.

Refraction: *See* **Reflection**

Resonant nuclear reaction: A nuclear reaction occurs when the neutrons and protons in two colliding nuclei redistribute themselves during collision (*see* Nuclear reaction). The reaction is said to be resonant if the sum of the energies of the colliding nuclei plus the energy of their relative motion happens to coincide with the energy of a state of the compound nucleus, which is simply the nucleus that would be formed by taking all the protons and neutrons of both colliding nuclei and putting them together into one nucleus.

Right ascension: *See* **Declination**

Salpeter function: Stars differ one from another in mass—that is, in the number of atoms they contain—at the time of their birth, but not in a random way. The Salpeter function is a mathematical formula that describes the relation of the number of stars to their masses.

Satellite: Bodies that move around planets are known as satellites. Thus, the Moon is a satellite of the Earth. A rocket payload expelled from the Earth that continues to move around the Earth is an artificial satellite, but a rocket payload escaping from the gravitational influence of the Earth is a space vehicle. The term satellite galaxy is sometimes used for small galaxies that are controlled by the gravitational influence of a larger galaxy.

Solar constant: The flux of the Sun, taken at the Earth's mean distance from the Sun, about 149.6 million kilometers, is the solar constant, about 1.39 kilowatts per square meter. The solar constant includes radiation of all frequencies. Since it has so far been measured from the Earth's surface, allowance must be made for ultraviolet light and infrared that have been absorbed by the terrestrial atmosphere. This difficulty is minimized by making measurements at a high-altitude desert station.

Solar cycle: A complex of phenomena that occur in the outer part of the Sun—sunspots, prominences, flares, the shape of the corona, magnetic fields—all show cyclic change with an average period of about 22 years. This solar cycle is believed to be related to the convection of material that takes place below the photosphere, but the precise cause of the cycle is not well understood. Successive cycles are neither of equal duration nor of equal intensity. For a while in the eighteenth century, the cycle largely died away.

Solar system: All the bodies that are moving around the Sun—planets, satellites, comets, asteroids—belong to the solar system.

Solar wind: The outer atmosphere of the Sun has no end. Its particles become less dense with increasing height above the photosphere, at first rapidly, and then more slowly, ultimately becoming a low-density wind of outward-moving particles. The density in the wind as it reaches the Earth varies markedly with time, sometimes being as low as 1 atom per cubic centimeter and sometimes as high as 1,000 atoms per cubic centimeter. The high-density situations are associated with the occurrence of flares and other disturbances in the lower parts of the solar atmosphere.

Solstice: Throughout the year, the axis of rotation of the Earth maintains a nearly constant direction relative to the stars, but the direction of the line from the Earth to the Sun changes because of the Earth's motion around the Sun. The summer solstice occurs when the angle between the direction of the Sun and the axis of rotation is least. The winter solstice occurs when the angle is greatest.

Spectral classification: The most prominent spectrum lines of different atoms correspond to quantum transitions between states of different energy values; certain of the lines show up prominently only within certain temperature sequences. The appearance (or nonappearance) of these lines in the spectrum of a star therefore gives information about the surface temperature of the star. The classification of stars, essentially in a temperature sequence, according to which lines appear (and which do not) is known as spectral classification.

Spectroscopic parallax: Certain spectrum types among stars are associated with certain values of the intrinsic luminosity. This fact is established by working initially with stars whose distances are already known, usually from measurement of their trigonometric parallax (*which see*). Then, it is argued, other stars having similar spectrum types must have similar intrinsic luminosities. Thus, a measurement of their apparent luminosities allows us to calculate the distances of these other stars. Such distance calculations are said to be by the method of spectroscopic parallax.

Spectrum: When the light from an object—star, galaxy, planet, quasar, or whatever—is separated into its constituent frequencies, the resulting distribution of light is said to be the spectrum of the object. The frequencies are separated by a device known as a spectrograph, whose essential component is either a glass prism or, more usually in astronomy, a diffraction grating. Separation of frequencies is essentially a separation into colors, the colors of visible light ranging from violet (the highest frequency) through blue, green, yellow, to red (the lowest frequency). Light emitted by ionized atoms shows a continuous range of color, usually with an excess or deficit of light at certain frequencies that are called lines of emission or absorption. Neutral atoms in a diffuse gas give only emission or absorption lines, although at high density, as in a heated solid substance, neutral atoms can also give a continuous range of color.

States of an atom: The paths that the electrons of an atom may follow can be arranged in bundles that have the property that, if at a given moment the electrons happen to be in the paths of a certain bundle with certain specified probabilities, then the electrons will continue in the same bundle with the same probabilities for a comparatively long time. When arranged in accordance with this comparatively long-term constancy, the bundles are called stationary states, or more simply states. Transitions from one such state to another occur from time to time, however, and radiation is then emitted or absorbed (*see* Quantum transitions).

Steady-state cosmology: Steady-state cosmology is a model for the large-scale features of the universe that is homogeneous for both time and space (*see* Cosmology, principles of). The model requires matter to be created continuously instead of all at once, as is required in the so-called big-bang cosmology.

Supernova: Stars of large mass end their lives with catastrophic explosions in which the power output for about a month is as great as that of a whole galaxy of stars. Thereafter, the emission declines gradually, the luminosity decreasing by about 0.6 magnitude per month. The cause of supernovae appears to be complex. The explosion is believed to start with the collapse of the core of the star, which eventually becomes either a neutron star or a black hole. Then, from the highly collapsed core, a shock wave propagates into the outer regions that have not yet experienced much collapse. The shock wave, which may arise from the absorption of a sudden flood of neutrinos from the core, heats the outer material. The latter material contains much oxygen, which is explosive on sudden heating to temperatures in the range 2 to 3×10^9 K. It is possible that this sequence of events occurs only in a restricted class of massive star, not in all of them.

Synchrotron radiation: Synchrotron radiation is generated by charged particles moving with speeds close to that of light in a magnetic field. The paths of the particles are helices with axes parallel to the direction of the magnetic field. The essential characteristic of synchrotron radiation is that it contains a continuous distribution of frequencies, and that the main energy of the radiation is at frequencies that are much higher than the number of turns per unit time of the particles in their helices. The smaller mass of the electron causes electrons to be much more effective in the emission of radiation than protons, so that electrons are usually considered to be the source of cosmic synchrotron radiation. The name is derived from a man-made device, the synchrotron, in which the effect was first observed.

Temperature: Any closed physical system, one in which energy is neither added nor taken away, tends to share the available energy uniformly between all its components—between particle motions and radiation, for example. In a situation in which sharing has become complete, temperature is a measure of the average energy per component. Temperature is proportional to this average energy, however, not equal to it, the constant of proportionality being determined by the convention that there are to be 100 units of temperature between melting ice and boiling water (taken at a standard pressure of 76 centimeters of mercury).

Trigonometric parallax: The annual motion of the Earth causes the projections of the stars onto the celestial sphere to oscillate over small arcs that are parallel to the plane of the Earth's orbital motion around the Sun. For a star at a distance of 1 parsec (3.26 light-years), the oscillation is through the angle of 2 arc seconds (*see* Parsec). For a star at a distance d parsecs ($3.26d$ light-years), the oscillation is through an angle of $2 \div d$ arc seconds. By measuring the angle of oscillation, the distance d can be inferred. This method of calculating distance, known as the method of trigonometric parallax, gives reliable results for distances up to about 100 parsecs, but it tends to become inaccurate for larger distances, the angles of oscillation becoming then too small for good observational measurements.

Universal time: An immediate simplifying effect of the assumption of homogeneity for the large-scale structure of the universe (*see* Cosmology) is that a system of universal time can then be defined. Observers in different galaxies use similar clocks, based on the same transition of the same kind of atom, and they all set their clocks to read the same time when the Hubble constant H, obtained from galaxies in their individual localities, has an assigned value that is chosen by convention.

Visual magnitude: When the magnitude of an object (*see* Magnitude) is measured in terms of visual light only, the result is said to be the visual magnitude.

Waves: A wave has three basic properties, as follows: (1) At each spatial point, there is an oscillation. The number of oscillations that occur in a unit time is the frequency ν. (2) There is a spatial ordering. If at one moment, at a given place, the wave is up, at a nearby place the wave is down. The spatial separation between adjacent places where the wave is up (or where it is down) is the wavelength λ. (3) A train of waves can propagate. At one moment, the waves have not yet reached a certain point. At a later moment, they have passed the point. These three properties are possessed both by water waves and by the radiation emitted by charged particles. Because of the similarity of these basic properties, it is possible, by thinking about water waves, to obtain insight about waves of radiation. The speed of propagation c of the waves is related to ν and λ by the simple equation $c = \nu\lambda$.

White holes: Exploding objects apparently emerging from a spacetime singularity. These objects may be looked upon as the time-reversed cases of black holes, which are formed by massive objects undergoing a gravitational *implosion*. Unlike black holes, white holes are visible to an external observer, and they are expected to be very bright in the early stages of the explosion.

X rays: X rays consist of radiation of high frequency, between about 100 and 100,000 times the frequency of visible light.

Index